Organische Chemie in Einzeldarstellungen

Band 13

Herausgegeben von

Hellmut Bredereck · Klaus Hafner

Eugen Müller

X28.30Eu/1

DATE DUE

GAYLORD			PRINTED IN U.S.A.

Ch. Grundmann · P. Grünanger

The Nitrile Oxides

Versatile Tools of Theoretical and Preparative Chemistry

Springer-Verlag New York · Heidelberg · Berlin 1971

Christoph Grundmann, Carnegie-Mellon University, Pittsburgh, PA., USA
Paolo Grünanger, Università di Pavia, Pavia, Italy

ISBN 0-387-05226-7 Springer-Verlag New York · Heidelberg · Berlin
ISBN 3-540-05226-7 Springer-Verlag Berlin · Heidelberg · New York

Preface

From their discovery in the last decade of the nineteenth century on, the nitrile oxides comprised a rather small group of little-known and little-investigated unstable organic derivatives of hydroxylamine. Around 1950, the astonishingly wide variety of their reactions started to become gradually unveiled. This reactivity has made them increasingly important as valuable tools for many syntheses. Simultaneously, the newly discovered transformations raised interesting theoretical questions and their pursuit has generally enriched our understanding of reaction mechanisms of organic chemistry.

At the same time, when nitrile oxides finally had won on their own a position of justified interest in the eyes of the organic chemist, came the discovery that fulminic acid, was in fact the parent member of this series. Thus, suddenly the nitrile oxides had acquired the most illustrious ancestry. Fulminic acid was discovered in 1800, at the dawn of modern chemistry, and since then the riddle of its true structure as well as the proteus-like variability of the molecule in its multitude of reactions has at one time or another attracted some of the most brilliant minds of organic chemistry, among them *Berthollet, Gay-Lussac, Liebig, Wöhler, Berzelius, Gerhardt, Kékulé, Griess, Armstrong, Holleman, Scholl, Nef, H. Wieland.*

The present authors readily confess that they too have succumbed to the fascination obviously emanating from the longest-known member as well as from the younger offspring of the nitrile oxide family. To their surprise, no review of this field has been attempted since Wieland's monograph of 1909, which covered mainly fulminic acid. One of the authors has recently published an article on the preparative aspects of nitrile oxides and a short review of their chemistry. The unexpectedly wide interest that these publications have found among chemists and members of related sciences has encouraged authors and publisher alike to present this monograph as the first comprehensive survey of this area.

The authors did not strive to achieve handbook-like coverage and neutrality. The reader will find the authors' own experience in this field reflected in the elimination of reports of only historical interest or of an erroneous nature, in sometimes critical interpretation of existing data, and in the occasional addition of unpublished facts from their laboratories.

A great help and stimulation for the undersigned has been advice, discussion, and contributions of unpublished data by a number of distinguished colleagues, among them *J. Armand, W. Beck, P. Beltrame, A. Dondoni, F. Eloy, R. Huisgen, A. Quilico, A. Ricca, P. v. R. Schleyer* and *G. Speroni* to whom the authors extend their sincere gratitude.

Christoph Grundmann

Pittsburgh and Pavia, January 1971   *Paolo Grünanger*

Contents

I. Introduction

A. Definition, Nomenclature

Definition. Nitrile oxides, RCNO, are organic compounds which contain the monovalent functional group —CNO bound directly to a carbon atom of the organic moiety of the molecule. Only the first member of this series, fulminic acid or formonitrile oxide, H—CNO is an exception; mainly for this reason its chemistry differs in some respects from that of its homologs.

According to the above definition, the consideration of compounds which may contain a covalent bond between the —CNO group and an element other than carbon, e.g.,

$$(C_6H_5)_3—Si—CNO \quad \text{or} \quad C_6H_5—Hg—CNO,[1]$$

is not included. As with hydrocyanic acid too, the metal salts and the numerous transition metal complex salts of fulminic acid are generally considered part of inorganic chemistry. They will receive only limited coverage, the more so since it is generally established that the inorganic chemistry of fulminic acid—quite in contrast to its organic chemistry— parallels closely that of hydrocyanic acid.[2, 3] Nitrile oxides as transient intermediates have been postulated frequently in reaction mechanisms, especially in reactions leading to 1,2-oxazole (isoxazole) and 1,2,5-oxa-diazole (furazan) derivatives. Such reactions will be discussed in this book only in cases where unambiguous evidence exists for the occurrence of nitrile oxides; for a complete coverage reference is made to recent monographs.[4-7]

1　*Beck, W.:* Angew. Chem. **75**, 872 (1963).

2　*Beck, W.:* Z. Naturforsch. **17**b, 130 (1962).

3　*Beck, W., Lux, F.:* Chem. Ber. **95**, 1683 (1962).

4　*Quilico, A.:* The chemistry of heterocyclic compounds (ed. A.Weissberger), vol. XVII, chap.: Isoxazoles and related compounds, p. 1—176. New York: Interscience 1962.

5　*Behr, L. C.:* The chemistry of heterocyclic compounds (ed. A. Weissberger), vol. XVII, chap.: Furazans, p. 283—320. New York: Interscience 1962.

6　*Barnes, R. A.:* Heterocyclic compounds (ed. R.C. Elderfield), vol. 5, chap.: Isoza-zoles, p. 452—482. New York: J. Wiley & Sons 1957.

7　*Boyer, J.H.:* Heterocyclic compounds (ed. R.C. Elderfield), vol. 7, chap.: Oxadia-zoles, p. 462—508. New York: J. Wiley & Sons 1961.

Nomenclature. The nomenclature of nitrile oxides closely follows that of nitriles, whenever applicable, e. g. when a certain component can be rationally named to include somewhere in the name the term "nitrile", this will be followed by "-oxide". There are a number of occasions, however, where this rule will lead to ambiguity, for instance, when the molecule contains more than one nitrogen function capable of forming an N-oxide. In such, and many other cases, a more succinct and less clumsy name is often arrived at by using the term *fulmido* for the —CNO group (derived from fulminic acid, analogous to *cyano* derived from hydrocyanic acid).

A comprehensive survey of the chemistry of nitrile oxides has never been made. Since 1965, a few shorter review articles have appeared, mostly within the frame of handbooks, stressing either the preparative aspects or the general chemistry of nitrile oxides.[8-12] None of them covers the numerous reactions of nitrile oxides as exhaustively as this book.

8 *Grundmann, C.:* Methoden der organischen Chemie, 4th ed. (ed. E. Müller), vol. 10/3, chap.: Methoden zur Herstellung und Umwandlung von Nitriloxiden, p. 837—870. Stuttgart: G. Thieme 1965.

9 *Grundmann, C.:* Chemistry of nitrile oxides. Fortschr. Chem. Forsch. **7**, 62—127 (1966).

10 *Grundmann, C.:* The chemistry of the cyano group (ed. Z. Rappoport), chap.: Nitrile oxides, p. 791—851. New York: Interscience 1970.

11 *Grundmann, C.:* Syntheses of heterocyclic compounds with the aid of nitrile oxides. Synthesis, 344—359 (1970).

12 *Quilico, A.:* Advances in nitrile oxides chemistry. Experientia **26**, 1169—1183 (1970).

B. History

Some alchemists and pharmacists during the 17th century, among them *Cornelius Drebbel* (1572–1634) and *Johann Kunckel von Löwenstern* (1630–1703) have probably known the mercury and silver salts of fulminic acid (formonitrile oxide),[1a] but from a scientific point of view the history of nitrile oxides begins really with the description of mercury fulminate in 1800 by *Howard*,[1] as the product of the reaction between ethyl alcohol, nitric acid and mercury. It is still interesting—and somewhat amusing too—to note that the British chemist was led to the above combination not by shear Edisonian curiosity, but by serious theoretical considerations.

To translate his reasoning into modern chemical language, what he wanted to do was to synthesize hydrochloric acid. We must remember that chlorine, although prepared by *Scheele* in 1774, was not recognized as an element until 1810, and its relation to hydrochloric acid was equally unclear. From the experience with the better known acids, it was believed that all acids had to contain oxygen (the German name "Sauerstoff" for oxygen, derived from "sauer"=acidic, still bears witness of this old belief), and that hydrochloric acid was therefore a combination of hydrogen, oxygen and the hypothetical element "murium". Chlorine was consequently considered an oxide of murium. *Howard* chooses nitric acid as oxygen donor and alcohol as hydrogen source and continues:

"With this in view, I mixed many substances with alcohol and nitric acid, as I thought might (by predisposing affinity) favor, as well as attract, an acid combination, of the hydrogen of the one, and the oxygen of the other."

He obtained, however, not mercuric chloride, but the similar looking mercuric fulminate; when he tried to liberate hydrogen chloride from it by reaction with concentrated sulfuric acid, he found himself faced to his uttermost surprise by a violent detonation. His further investigation led him to believe that mercuric fulminate was a compound composed of "nitrous etherized gas" (ethyl nitrite), oxalic acid and mercuric oxide.

Not before 1824 did *Gay Lussac* and *Liebig*[2] provide the first correct analysis of silver fulminate. At the same time, *Liebig* recognized that potassium hydroxide removed only one half of the metal giving the complex salt $K[Ag(CNO)_2]$. The same kind of salts were obtained when silver fulminate was reacted with alkali halogenides. This observations together with the fact that fulminic acid is formed from ethyl alcohol, whose nature as a two-carbon compound was early recognized, explain

---

1a *Feldhaus:* Nitrocellulose **4**, 83 (1933). — *Kirk-Othmer:* Encyclopedia of chemical technology, vol. 6, p. 10. New York: Interscience Encyclopedia 1951.

1 *Howard, E.:* Phil. Trans. Roy. Soc. London 204 (1800).

2 *Liebig, J., Gay-Lussac, J. L.:* Ann. Chim. Phys. [2]**25**, 285 (1824).

the otherwise surprising but stubborn attempts, in spite of all other evidence, to ascribe a C_2-formula to fulminic acid which lasted well into the twentieth century.[3,4] The genius of *Liebig*, however, intuitively found the truth when he wrote in the terms of his times:

"... les divers fulminates renferment un principe commun de fulmination, qui est indépendant des bases et qui ne peut étre qu'un composé d'oxigène et de cyanogène ..."[5]

It may be mentioned here only in passing that *Liebig's* analysis of silver fulminate coincided almost with *F. Wöhlers* report on silver cyanate having an identical composition.[6] Out of the ensuing controversy, in which each side first accused the other of incorrect analytical work, developed the concept of isomerism.

With the advent of structural formulas, the somewhat vague ideas of *Gerhardt* were more precisely expressed by *Kékulé* who proposed in 1857 the first structural formula in modern terms, nitro-acetonitrile (*1*) for fulminic acid.[7] The main experimental support for *1* came from the exhaustive chlorination of mercuric fulminate which yielded cyanogen chloride, Cl—CN, and trichloro-nitro-methane (chloropikrin), Cl_3C- $-NO_2$, thus providing obvious evidence for the presence of both a cyano and a nitrogroup in the starting material.[8] Furthermore, *1* seemed to explain well the alleged evolution of HCN from fulminates on acidification (erroneous, as found much later; hydrocyanic and fulminic acid have a very similar odor) and the explosive nature, associated also with many just recently discovered nitrocompounds. For these reasons, *Kékulé's* formula remained unchallenged for more than a quarter of a century, although *P. Griess*, under the impression of the explosivity of the diazonium salts discovered by him, suggested the structure of diazo-acetic acid (*2*) for fulminic acid.[9]

Further progress came when in 1883 simultaneously *Carstanjen* and *Ehrenberg*[10] and *Steiner*[11] found that fulminates are hydrolyzed by concentrated hydrochloric acid with the quantitative formation of

---

3 *Hodgkinson, W.R.:* J. Soc. Chem. Ind. **37**, 190T (1918); — Chem. Abstr. **12**, 2687 (1918).

4 *Bergfeld, L.:* Z. ges. Schiess- u. Sprengstoffw. **37**, 84 (1942); — Chem. Abstr. **37**, 5592^5 (1943).

5 *Liebig, J., Gay-Lussac, J.L.:* Ann. Chim. Phys. [2]**25**, 303 (1824).

6 *Liebig, J.:* Ann. Chim. Phys. [2]**24**, 294 (1823). — *Wöhler, F.:* Ann. Chim. Phys. [2]**27**, 196 (1824).

7 *Kékulé, A.:* Ann. Chem. **101**, 200 (1857).

8 *Kékulé, A.:* Ann. Chem. **105**, 281 (1858).

9 *Griess, P.:* Ann. Chem., Suppl. **1**, 104 (1861).

10 *Carstanjen, E., Ehrenberg, A.:* J. Prakt. Chem. [2]**25**, 232 (1882). — *Ehrenberg, A.* J. Prakt. Chem. [2]**30**, 41 (1884).

11 *Steiner, A.:* Ber. **16**, 1484; 2419 (1883).

hydroxylamine hydrochloride and formic acid.[12] It seems indeed strange that this simple and structurally most revealing transformation of fulminic acid had been so completely overlooked by the elder chemists, and thus delayed the discovery of hydroxylamine (*Lossen*, 1865) by about 50 years. In spite of the simplicity with which the above transformation is expressed by a mono-carbonic formula (Eq. (1)), the subsequent structural proposals

$$HCNO \quad + \quad HCl \quad + \quad 2\,H_2O \quad = \quad H{-}COOH \quad + \quad H_2NOH \cdot HCl \qquad (1)$$

(*3*), (*4*), (*5*) and (*6*) still clung tenaciously to the postulate of fulminic acid being an ethane derivative.

Table I. Some proposed structures for fulminic acid

$O_2N{-}CH_2{-}CN$	$N_2CH{-}COOH$	$HON{=}C{=}C{=}NOH$
1	*2*	*3*
Nitro-acetonitrile	Diazo-acetic acid	Dicarbonyl-dioxime
(*Kékulé*, 1857[7])	(*Griess*, 1861[9])	(*Steiner*, 1883[11])

$\begin{matrix} HC{-}{-}CH \\ \parallel \quad\ \parallel \\ N \quad\ N \\ \mid \quad\ \ \mid \\ O{-}{-}O \end{matrix}$	$\begin{matrix} HC{-}{-}C{-}OH \\ \parallel \quad\ \parallel \\ N \quad\ N \\ \diagdown O \diagup \end{matrix}$	$\begin{matrix} H_2C{-}NO \\ \mid \\ C{\equiv}N{-}O \end{matrix}$
4	*5*	*6*
Glyoxime peroxide	Hydroxy-furazan	Nitroso-acetonitrile oxide
(*Scholl, Holleman*, 1890[13])	(*Divers*, 1884[12,14])	(*Bergfeld*, 1942[4])

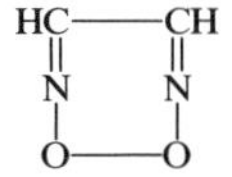

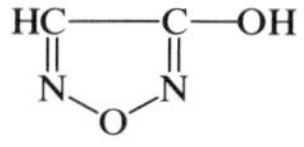

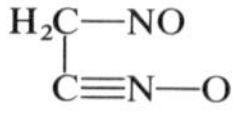

$H{-}O{-}N{=}C\vert \quad \longleftrightarrow \quad HO{-}\overset{\oplus}{N}{\equiv}\overset{\ominus}{C}$	$HC{\equiv}\overset{\oplus}{N}{-}\overset{\ominus}{O} \quad \longleftrightarrow \quad H{-}\overset{\oplus}{C}{=}N{-}\overset{\ominus}{O}$
7 *7a*	*8* *8a*
Carbonyloxime	Formonitrile oxide
(*Nef*, 1893[15])	(*Ley*, 1899;[21] *Pauling*, 1926;[24] *Quilico*, 1946; *Beck*, 1965)

Within the framework of his extended studies on divalent carbon compounds, *Nef*[15] conceived the monomeric structure 7, carbonyl oxime, for fulminic acid which has been the accepted expression until this decade. He felt that the extreme unstability of the free acid and the explosive nature even of its alkali salts (by constrast, the alkali azides are thermally

---

12 *Divers, E., Kawakita, M.*: J. Chem. Soc. **45**, 13 (1884).
13 *Scholl, R.*: Ber. **23**, 3505 (1890). — *Holleman, A. F.*; Ber. **23**, 3742 (1890).
14 *Divers, E.*: J. Chem. Soc. **45**, 19 (1884).
15 *Nef, J. U.*: Ann. Chem. **280**, 291 (1894).

quite stable!) justified a unique and bold departure from the conventional quadrivalent structures of most other carbon compounds. The extreme ease with which fulminic acid adds hydrogen halides (Eq. (2)) seemed only understandable to him as an 1-1-addition to the divalent carbon atom.

$$HON{=}C \quad + \quad H \cdot Hal \quad = \quad HON{=}C\begin{smallmatrix} \diagup Hal \\ \diagdown H \end{smallmatrix} \tag{2}$$

He provided furthermore the first synthesis of fulminic acid from a methane derivative, nitromethane.[16]

Finally, *Wieland* elucidated completely the complex sequences involved in the transformation of ethyl alcohol into fulminic acid,[17] and found additional synthetic routes to fulminic acid starting from methane derivatives.[18] The monomeric formula was further confirmed by *L. Wöhler's* determination of the molecular weight of anhydrous sodium fulminate.[19] A small refinement added much later with the growing understanding of the nature of chemical bonding was the dipolar structure *7a*.[20]

The only other formula, consistent with all chemical data is *8, formonitrile oxide*, which was first suggested by *Ley*[21] and accepted by *Palazzo* who considered fulminic acid a tautomer of structures *7* and *8*.[22]

The developing knowledge of the chemistry of aliphatic and aromatic nitrile oxides provided first by *Wieland*, and since 1920 notably by the Italian school, especially by *Ponzio* and *Quilico* and their students, revealed more and more the complete analogy of many of their reactions with that of fulminic acid. Reading the publications of *Wieland* on fulminic acid,[23a] between 1910 and 1930, the author's awareness of this similarity becomes almost painfully clear, as he tries less and less convincingly to defend *Nef's* structure proposal. The main difficulty for *Wieland*—and many others—to accept *8* were apparently the many reactions of fulminic acid, like Eq. (2), which seemed at that time only explainable as 1,1-additions to a divalent carbon. The Italian school before 1946 did not dismiss *7* outright, but seemed to favour *Palazzo's* suggestion of a tautomeric equilibrium between *7* and *8*. In that year, *Quilico* became convinced by experimental evidence that formula *8* was correct.[23b]

----

16 *Nef, J. U.*: Ann. Chem. **280**, 263 (1894).
17 *Wieland, H.*: Ber. **40**, 418 (1907); **43**, 3362 (1910).
18 *Wieland, H.*: Ber. **42**, 803, 820 (1909).
19 *Wöhler, L.*: Ber. **38**, 1356 (1905).
20 *Lindemann, H., Wiegrebe, L.*: Ber. **63**, 1650 (1930).
21 *Ley, H., Kissel, M.*: Ber. **32**, 1357 (1899).
22 *Palazzo, F. C.*: Atti Accad. Lincei [5] **21** V, 713 (1912).
23a For a summary, cf., *Klages, F.*: Naturwissenschaften **30**, 357 (1942).
23b *Quilico, A., Speroni, G.*: Gazz. Chim. Ital. **76**, 148 (1946).

The formonitrile oxide structure *8* received for the first time theoretical support by *Pauling*[24] who calculated potential energies for all possible structures C—H—N—O, which showed clearly that *8* was favored over *7*. *Pauling* pointed out already that the addition reactions of fulminic acid could be understood as 1,3-additions to the charged structure *8*. This structural problem is naturally nonexistent for the ionized salts of fulminic acid where the structural differences (tautomerism) between *7* and *8* are reduced to mesomeric forms of a hybrid anion.

The last important obstacle for the acceptance of *8*, the above described addition reactions, was finally cleared away by *Huisgen's* concept of the *1,3-dipolar reactivity of nitrile oxides*.[25] This leads to the resonance hybrid *8*↔*8a* for fulminic acid. The 1,3-dipolar structure *8a* explains perfectly the addition reactions such as that of Eq. (2)

$$\overset{\oplus}{HC}=N-\overset{\ominus}{O} \quad + \quad H^{\oplus}Hal^{\ominus} \quad \longrightarrow \quad \overset{H}{\underset{Hal}{>}}C=N-OH \tag{3}$$

which for such a long time stood in the way of formula *8* (Eq. (3)). The final remaining step was an accurate proof for the structure of the free acid. This was done by *Beck* in 1965 by measuring the infrared spectrum of the gaseous fulminic acid,[26] which is unequivocally consistent only with *8*, formonitrile oxide.

The history of the higher homologs of fulminic acid, the nitrile oxides proper, is much shorter. Before the first member of this series was ever prepared, the thermal decomposition of diphenyl-1,2,5-oxadiazole-2-oxide (diphenyl-furoxan, *9*) into two molecules of phenylisocyanate had been postulated to occur via a hypothetic nitrile oxide intermediate[27] (Eq. (4)).

$$\underset{9}{H_5C_6-C\underset{N\diagdown O\diagup N}{\overset{||\quad\quad||}{\quad}}C-C_6H_5} \quad \xrightarrow{\triangledown} \quad [2\,H_5C_6-CNO] \quad \xrightarrow{O} \quad 2\,H_5C_6-N=C=O \tag{4}$$

In the same paper, the reaction of the disodium salt of "nitrobenzalphthalide" with iodine is described, leading to a brown, pungent smelling oil which promptly changed to *9*. It was recently demonstrated that this oil is mainly benzonitrile oxide.[27a]

Benzonitrile oxide (*12*), however, was first identified by *A. Werner* in 1894[28] before he changed his interests to become the father of modern

24 *Pauling, L., Hendricks, S.B.:* J. Am. Chem. Soc. **48**, 641 (1926).
25 *Huisgen, R.:* Proc. Chem. Soc. 357 (1961).
26 *Beck, W., Schuierer, E., Feldl, K.:* Angew. Chem. **77**, 722 (1965).
27 *Gabriel, S., Koppe, M.:* Ber. **19**, 1145 (1886).
27a *Caramella, P., Grünanger, P.:* Unpublished.
28 *Werner, A., Buss, H.:* Ber. **27**, 2193 (1894).

Table II. A. Werner's discovery of benzonitrile oxide

$$H_5C_6-CH=NOH \xrightarrow{Cl_2} H_5C_6-\underset{\underset{\textbf{\textit{10}}}{Cl}}{\overset{\|}{C}}=NOH \xrightarrow{AgNO_2} \nrightarrow H_5C_6-\underset{\underset{\textbf{\textit{11}}}{NO_2}}{\overset{\|}{C}}=NOH$$

From 10 with Na_2CO_3:

$$H_5C_6-C\overset{O}{\underset{}{\diagup\!\!\!\diagdown}}N \quad \text{or} \quad O=N\equiv C-C_6H_5 \qquad\qquad 9+N_2O_3$$
$$\textbf{\textit{12a}} \qquad\qquad\qquad \textbf{\textit{12b}}$$

From 10 with $AgNO_2$:

$$2\,H_5C_6-C\equiv N=O \longrightarrow H_5C_6-\overset{\overset{O-O}{|\quad\ |}}{\underset{}{C}}\underset{\underset{\textbf{\textit{13}}}{}}{\!\!-\!\!}\;C-C_6H_5$$

inorganic chemistry. In an attempt to prepare benzonitrolic acid (*11*) from benzaldoxime, benzhydroximic acid chloride (*10*) was obtained by chlorination of benzaldoxime, but its reaction with silver nitrite did not give *11*, but diphenylfuroxan (*9*), together with nitrous oxides. In following up this unusual reaction, *Werner* demonstrated that *10* is easily transformed by mild bases, for instance sodium carbonate, into an unstable oil with an aggressive odor which soon solidifies to the crystalline furoxan *9*. He postulated this intermediate to be benzonitrile oxide for which he preferred among the two discussed structures *12a* and *12b* the latter one, because it seemed to explain better the spontaneous dimerization to *9* (for which at his time the six-membered formula *13* was generally accepted).

Werner neither purified nor analysed his crude benzonitrile oxide, but supported its assumed structure by hydrolyzing it with acid to hydroxylamine (and benzoic acid) and by converting it with alkali (partially) into benzohydroxamic acid.

In 1907, *Wieland* obtained first benzonitrile oxide in pure form by *Werner's* procedure,[29] determined its molecular weight and discovered some of its addition reactions. Contrary to *Werner*, *Wieland* preferred the cyclic structure *12a*, which, however, was ruled out in 1928 by the comparison of the exaltation of the molecular refraction of benzonitrile oxide with analogous compounds.[30] Since that time, nitrile oxide have always been considered as N-oxides of the corresponding nitriles.

---

29 *Wieland, H.*: Ber. **40**, 1667 (1907).
30 *Auwers, K. v.*: Ber. **61**, 1041 (1928).

The *capacity of nitrile oxides to undergo ring closure reactions* by 1–3 addition over an unsaturated bond was possibly first observed by *Weygand*,[31] but not understood mechanistically. Since 1946, *Quilico* and his students have systematically studied the chemistry both of fulminic acid and its higher homologs in this respect. Finally, since 1961 *Huisgen*[25] has demonstrated that almost all reactions of nitrile oxides can be understood within the general concept of 1,3-dipolar cycloadditions. Recently, the discovery that nitrile oxides can be stabilized by controlled steric hindrance[32] has been of considerable value for the study of the many unexpected and varying reactions of this class of compounds which is still going on at a lively pace while these lines are written.

31 *Weygand, C., Bauer, E.:* Ann. Chem. **459**, 123 (1927).
32 *Grundmann, C., Dean, J.M.:* Angew. Chem. **76**, 682 (1964).

C. General Properties

Structure. Nitrile oxides are isomeric with cyanates (*1*) and isocyanates (*2*). While in the latter, the organic group is connected to the oxygen or the nitrogen atom, respectively, in the nitrile oxides it is the carbon atom which provides the link to the organic group R— of the molecule:

$$R\text{—}O\text{—}C\equiv N \qquad R\text{—}N=C=O \qquad R\text{—}C\underset{O}{\overset{}{\diagdown\!\diagup}}N \qquad R\text{—}C\equiv N\rightarrow O$$

$$(1) \qquad\qquad\quad (2) \qquad\qquad\quad (3) \qquad\qquad (4)$$

The older literature preferred the cyclic structure *3*,[1,2] but optical data and energy level calculations ruled out *3* at a rather early stage in preference of *4*.[3,4] This formula is the most generally accepted simplified expression of the structure of the nitrile oxides which will be used throughout this book. Nevertheless, the structure formulation of a nitrile oxide as a resonance hybrid between the structures *5 a–e* is a more accurate description:[5,6]

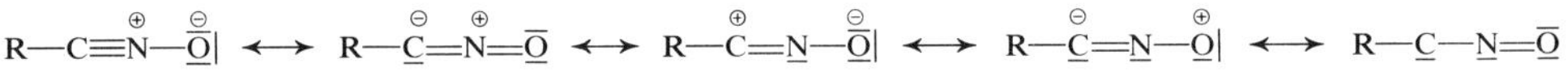

$$R\text{—}C\equiv\overset{\oplus}{N}\text{—}\overset{\ominus}{\underline{\overline{O}}}| \;\longleftrightarrow\; R\text{—}\overset{\ominus}{\underline{C}}=\overset{\oplus}{N}=\overline{O} \;\longleftrightarrow\; R\text{—}\overset{\oplus}{C}=N\text{—}\overset{\ominus}{\underline{\overline{O}}}| \;\longleftrightarrow\; R\text{—}\overset{\ominus}{\underline{C}}=N\text{—}\underline{O}| \;\longleftrightarrow\; R\text{—}\underline{C}\text{—}\underline{N}=\overline{O}$$

$$(5\,a) \qquad\qquad (5\,b) \qquad\qquad (5\,c) \qquad\qquad (5\,d) \qquad\qquad (5\,e)$$

Among the above mesomeric structures, the all-octet-formula *5 a* and *5 b* presumably represent the preferred electron distribution in the ground state, while the sextet-formula *5 c* and *5 d* express best most of the reactions of the nitrile oxides, especially the 1,3-dipolar additions. The resonance form *5 e* with carbene character, however, is probably responsible for the characteristic dimerization reaction of nitrile oxides leading to 1,2,5-oxadiazole-2-oxides (furoxans, cf., Chapter IV, A-3).

The nitrile oxide group is very similar in its electronic structure to the aliphatic diazo- and to the azido group.[5–7] Consequently, there are striking analogies in the chemical behavior of these three classes of compounds. On the other hand, there is a certain, although more superficial, resemblance between the cumulated system of the nitrile oxides and those of the ketenes and isocyanates. But in general, nitrile oxides are far more prone to autocondensation than the latter compounds.

------

1 *Wieland, H.:* Ber. **40**, 1667 (1907).

2 *Wieland, H.:* Die Knallsäure. Sammlung chemischer und chemisch-technischer Vorträge (ed. F. B. Ahrens), vol. 14, p. 385—461. Stuttgart: F. Enke 1909.

3 *Auwers, K. v.:* Ber. **61**, 1041 (1928).

4 *Pauling, L., Hendricks, S. B.:* J. Am. Chem. Soc. **48**, 641 (1926).

5 *Huisgen, R.:* Proc. Chem. Soc. 357 (1961).

6 *Huisgen, R.:* Angew. Chem. **75**, 604 (1963).

7 *Quilico, A., Speroni, G.:* Gazz. Chim. Ital. **76**, 148 (1946). — *Quilico, A., Stagno d'Alcontres, G., Grünanger, P.:* Gazz. Chim. Ital. **80**, 479 (1950); Nature **166**, 226 (1950).

Relation to Nitriles. As nitrogen-containing organic compounds, nitrile oxides are derivatives of hydroxylamine, as nitriles are derivatives of ammonia. The complete *hydrolysis of a nitrile oxide* leads to hydroxylamine and the corresponding carboxylic acid (Eq. (1)) in complete analogy to the corresponding hydrolysis of nitriles to ammonia and carboxylic acids. In both cases, corresponding intermediates occur in the hydrolysis, viz. the hydroxamic acids and the amides.

$$R{-}C{\equiv}N{\longrightarrow}O \xrightarrow{+H_2O} R{-}\underset{\overset{\|}{O}}{C}{-}NHOH \xrightarrow{+H_2O} R{-}COOH \ + \ NH_2OH \qquad (1)$$

As in the case of the nitriles, the hydrolysis can be brought forward as well by mineral acids as by bases.[8,9] While the acid hydrolysis can be controlled to yield either the hydroxamic acid, or hydroxylamine and the carboxylic acid, the alkaline hydrolysis seems not to proceed that smoothly; generally, only moderate yields of the hydroxamic acid have been obtained together with unidentified, often resinous by-products.[10] Nitrile oxides, like amine oxides, are *mildly oxidizing agents.* They liberate iodine from an acidified solution of KI. A number of reducing agents convert them readily into the corresponding nitriles (see Chapter IV, B). Otherwise, however, there are few links with the chemistry of nitriles. The direct oxidation of nitriles to nitrile oxides has not yet been achieved. The fulminate radical ·CNO has been observed spectroscopically in mixtures of cyanogen and ozone subjected to flash photolysis,[11] but it is not likely that such methods will become of practical value.

Earlier attempts in this direction by treatment of benzonitrile with hydrogen peroxide in presence of alkali have led to the formation of small amounts of *benzhydroxamic acid,* identified only by the positive reaction with ferric chloride.[12] It is not likely that the hydroxamic acid was formed via benzonitrile oxide, but by oxidation of benzamide, since hydrogen peroxide catalyzes the alkaline hydrolysis of nitriles to amides.[13]

While many procedures exist by which an amide can be dehydrated to the nitrile, the attempt of an analogous transformation of hydroxamic acids leads always to an α-elimination of water and the subsequent *Lossen*

8 *Steiner, A.:* Ber. **16**, 1484 (1883). *Carstanjen, E., Ehrenberg, A.:* J. Prakt. Chem. [2] **25**, 232 (1882).

9 *Werner, A., Buss, H.:* Ber. **27**, 2193 (1894).

10 *Grundmann, C., Frommeld, H.D.:* J. Org. Chem. **31**, 157 (1966).

11 *McGrath, W.D., Morrow, T.:* Nature **203**, 619 (1964); **204**, 988 (1964).

12 *Olivieri-Mandalá, E.:* Gazz. Chim. Ital. **52** I, 107 (1922).

13 *Radziszewski, B.:* Ber. **18**, 355 (1885). — *McMaster, L., Langreck, F.B.:* J. Am. Chem. Soc. **39**, 103 (1917).

rearrangement of the nitrene (Eq. (2)):

$$R-\underset{\underset{O}{\|}}{C}-NHOH \;\rightleftharpoons\; R-\underset{\underset{OH}{|}}{C}=NOH$$

$$R-N=C=O \leftarrow [R-\underset{\underset{O}{\|}}{C}-N] \;\xleftarrow{\quad -H_2O \quad}\!\!\!\!/\;\longrightarrow\; R-C\equiv N\rightarrow O \qquad (2)$$

$$H-\underset{\underset{O}{\|}}{C}-NHOH \;\xrightarrow{\;Ac_2O\;}\; H-\underset{\underset{O}{\|}}{C}-NH-OAc \;\xrightarrow{\;PCl_5\;}\; \underset{Cl}{\overset{H}{>}}C=N-OAc \qquad (3)$$

$$\xrightarrow{\;AgNO_3\;}\; AgCNO$$

The only exception is the reported conversion of formhydroxamic acid
into silver fulminate by the sequence above (Eq. (3)).[14]

Stability. Nitrile oxides are energy-rich compounds; as derivatives of
fulminic acid all low molecular nitrile oxides and, generally, polynitrile
oxides should be considered as potentially *explosive*. The parent member
of this series, fulminic acid, $H-C\equiv N \rightarrow O$ is, naturally, most dangerous
in this respect. The free undiluted acid polymerizes exothermically below
$-20°$ C;[15] the explosivity of its heavy metal salts is well known, but even
the *alkali salts are highly explosive*.[16-18] At least one of its polymers
Wieland's "trifulmin", is as dangerously explosive as silver fulminate.[19]
Oxalo-bis-nitrile oxide (Cyanogen-bis-N-oxide, $ON\equiv C-C\equiv NO$) is
stable at $-78°$ C, begins visibly to decompose around $-45°$ C and
detonates a few minutes later at that temperature.[20] Dilute ($<5\%$)
*ethereal solutions can spontaneously ignite upon filtration by partial
evaporation of the solvent*, leaving traces of the undiluted compound at the
edges of the paper. The oligomer of cyanogen-bis-N-oxide is not shock-
sensitive, but will detonate over an open flame. When, however, the
tenaciously held solvent is removed at 10^{-2} mm, spontaneous explosion
occurs at $70-80°$ C. Terephthalonitrile oxide (6, R=H) explodes at
$160°$ C,[21] and an explosive decomposition of tetramethyl-terephthalo-
nitrile oxide (6, $R=CH_3$) was observed upon reaction with *p*-diethynyl-
benzene without a diluent.[22]

---

14   *Biddle, H. C.:* Ann. Chem. **310**, 19 (1900); Ber. **38**, 3858 (1905).
15   *Beck, W.:* Private communication.
16   *Ehrenberg, A.:* J. prakt. Chem. [2]**32**, 230 (1885).
17   *Wöhler, L.:* Ber. **38**, 1351 (1905).
18   *Wöhler, L., Weber, A.:* Ber. **62**, 2742 (1929).
19   *Wieland, H.:* Ber. **42**, 803 (1909).
20   *Grundmann, C., Mini, V., Dean, J. M., Frommeld, H.-D.:* Ann. Chem. **687**, 191 (1965).
21   *Eloy, F.:* Private communication.
22   *Grundmann, C.:* Unpublished.

In general, however, the main difficulties in working with nitrile oxides are not caused by their explosivity, but by their rapid *spontaneous polymerization*. Most nitrile oxides dimerize only to the usually stable 1,2,5-oxadiazole-2-oxides (furoxans) (Eq. (4)),

$$2\ R-C\equiv N\longrightarrow O \longrightarrow \qquad (4)$$

7

6

although differently structured oligomers may be obtained under special conditions (see Chapter IV, C-2). Fulminic acid makes an exception, in so far as no stable dimer is formed; furthermore the stable oligomers (trimeric and tetrameric) have a structure unparalleled by its higher homologs (see Chapter IV, A-3a).

All simple aliphatic and most aromatic nitrile oxides are permanently stable as monomers only at temperatures far below 0° C (for specific data, see Table III, pp. 16–20) and are preferably stored at $\sim -70°$ C.

An electronic *influence of substituents on the stability* of aromatic nitrile oxides undoubtedly exists, but it is not very pronounced and is apparently superimposed by other effects.[23, 24] For instance, both electron donor (methyl, methoxy) and acceptor (halogen, nitro-) substituents in the para position seem to stabilize, whereas the nitro group in ortho position makes the nitrile oxide particularly unstable. Among the monochlorobenzonitrile oxides, however, the order of stability is para > ortho > meta.

The relative stability of oxalo-bis-nitrile oxide in dilute solution may be attributed to the effect of the conjugation of the CNO-groups, resulting in the possibility of a higher number of mesomeric structures, while the oxalo-bis-nitrile-mono-N-oxide, $N\equiv C-C\equiv N\rightarrow O$, is as unstable as any of the other lower aliphatic nitrile oxides.[20, 25] The same effect might account for the unusual stability of terephthalo-nitrile-di-N-oxide[26]

23 Beltrame, P., Comotti, A., Veglio, C.: Chem. Commun. 996 (1967).
24 Dondoni, A., Mangini, A., Ghersetti, S.: Tetrahedron Letters 4789 (1966).
25 Grundmann, C., Frommeld, H.-D.: J. Org. Chem. **31**, 4235 (1966).
26 Eloy, F.: Bull. Soc. Chim. Belges **73**, 639 (1964). — Iwakura, Y., Akiyama, M., Shiraishi, S.: Bull. Chem. Soc. Japan **37**, 767 (1964). — Overberger, C.G., Fujimoto, S.: J. Polymer Sci. C4161 (1968).

(6, R=H) while the apparent stability of phenyl-oximino-acetonitrile oxide may be caused by the neighboring oximinogroup, either by steric hindrance or by hydrogen bonding.[27]

A very pronounced effect on *stabilization* can be achieved by substituting aromatic or heterocyclic nitrile oxides in o,o'-positions with substituents of critical spatial requirements which will block sterically the dimerization to furoxans without impairing the ability to react with other unhindered systems. Studies of Stuart-Briegleb models indicate that suitable groups are CH_3, C_2H_5, CH_3—O or CH_3—S, whereas Br, I, NO_2 or SO_2—R are probably too large and F, Cl or OH too small for this purpose.[28,29] In the aliphatic series, a higher degree of steric hindrance is obviously needed; tert-butyl fulmide, (7, $R,R_1 = CH_3$) still dimerizes to the furoxan, although relatively slow, while bis-tert-butyl-acetonitrile oxide (7, $R = (H_3C)_3C$, $R^1 = H$) is permanently stable at room temperature,[30] as well as triphenyl-acetonitrile oxide (7, $R,R^1 = C_6H_5$).[31]

This approach has led to the preparation of a number of nitrile oxides which are stable permanently as monomers and which have been found very useful in the study of reactions of nitrile oxides, especially of those which occur at a slower rate than the dimerization (see Table III, pp. 16–20).

27   *Ponzio, G.:* Gazz. Chim. Ital. **66**, 119, 123 (1936).
28   *Grundmann, C., Dean, J. M.:* Angew. Chem. **76**, 682 (1964); J. Org. Chem. **30**, 2809 (1965).
29   *Grundmann, C., Richter, R.:* J. Org. Chem. **32**, 2308 (1967); **33**, 476 (1968).
30   *Grundmann, C., Datta, S. K.:* J. Org. Chem. **34**, 2016 (1969).
31   *Wieland, H., Rosenfeld, B.:* Ann. Chem. **484**, 236 (1930).

II. Physical Properties

A. General

Physical State. Most nitrile oxides, isolated as individuals, have been obtained as crystalline solids, although the lower homologs of the aliphatic series and benzonitrile oxide melt below room temperature (25°). Only those few known representatives of this group which are prevented from rapid dimerization to furoxans (see Chapter IV, C-2) by steric hindrance are stable in the liquid state and may be distilled under reduced pressure. As compared with the cyano-group, the higher polarity of the fulmido group in the sense of the resonance structures
1a–1c

$$\overset{\oplus}{C}=N-\overset{\ominus}{O} \quad \longleftrightarrow \quad \overset{\oplus}{C}\equiv N-\overset{\ominus}{O} \quad \longleftrightarrow \quad \overset{\ominus}{C}=\overset{\oplus}{N}=O \quad \longleftrightarrow \quad \overset{\ominus}{C}=N-\overset{\oplus}{O}$$

1a   *1b*   *1c*   *1d*

expresses itself in generally higher melting and boiling points (where observable) as those of the corresponding nitriles. Even sterically hindered aromatic nitrile oxides which are the most thermally stable species cannot be investigated conveniently in the gaseous state, since at temperatures where they develop appreciable vapor pressure, rapid isomerization to the isocyanates takes place (see Chapter IV, A). Only the parent member of the series, formonitrile oxide (fulminic acid) has been studied in the gaseous state in high vacuum (see paragraph B).

The following Table III lists melting points and stability of all nitrile oxides which have been prepared by the methods discussed in Chapter III and have been isolated as individuals in pure or approximately pure form.

Table III. Isolated nitrile oxides

Method of Preparation

A. Dehydrogenation of the aldoxime with hypobromite.
B. Dehydrogenation of the aldoxime with N-bromosuccinimide.
C. Dehydrohalogenation of the hydroximic acid chloride.
D. Thermal decomposition of the nitrolic acid.
E. Dehydration of the primary nitroparaffin with phenylisocyanate.
F. Reaction of the organic halide with silver fulminate.

Stability. "Unlimited" means that the physical constants of the compound were unaltered after storage for 30 days at 25°. Any specific time given means the time (at 18°) after which the compound was found to be completely dimerized to the furoxan.[12] More accurate data are known only for 2,6-dichloro-benzonitrile oxide (see Chapter IV, C-2).

No.	Compound	Method of preparation	M.P. [°C]	Stability	References
1	Formonitrile oxide (fulminic acid)	C, D[a]	polymerizes above −15	40 min[b]	[1-6]
2	Acetonitrile oxide	C, D	−5	<1 min	[7-10]
3	2,2-Dimethyl-propionitrile oxide	C	18[c]	2—3 days	[10-12]
4	2-Ethyl-butyronitrile oxide	C	−33	<1 min	[10]
5	Di-tert-butyl-acetonitrile oxide	B	24—25[d]	unlimited	[13]

1 *Beck, W., Feldl, K.:* Angew. Chem. **78**, 746 (1966).

2 *Beck, W.:* Private communication (1966). − *Winnewisser, M., Bodenseh, H. K.:* Z. Naturforsch. **22**a, 1724 (1967).

3 *Birckenbach, L., Sennewald, K.:* Ann. Chem. **512**, 45 (1934).

4 *Huisgen, R., Christl, M.:* Angew. Chem. **79**, 471 (1967).

5 *Kurtz, P.:* Methoden zur Herstellung und Umwandlung von Knallsäure. Methoden der organischen Chemie (ed. *E. Müller*), 4th ed., vol. 8, p. 355—358. Stuttgart: G. Thieme 1952.

6 *Wieland, H., Baumann, A., Reisenegger, C., Scherer, W., Thiele, J., Will, J., Haussmann, H., Frank, W.:* Ann. Chem. **444**, 7 (1925).

7 *Hoshino, T., Mukaiyama, M.:* Japan. Pat. 9855 (1959); — Chem. Abstr. **54**, 7738h (1960).

8 *Mukaiyama, T., Hoshino, T.:* J. Am. Chem. Soc. **82**, 5339 (1960).

9 *Vita Finzi, P., Grünanger, P.:* Chim. Ind. (Milan) **47**, 516 (1965).

10 *Zinner, G., Günther, H.:* Angew. Chem. **76**, 440 (1964).

11 *Grundmann, C.:* Herstellung und Umwandlung von Nitriloxiden. Methoden der organischen Chemie (ed. *E. Müller*) 4th ed. vol. 10/3, p. 837—870. Stuttgart: G. Thieme 1965.

12 *Speroni, G.:* Unpublished, cit. in *A. Quilico:* Isoxazoles and related compounds. The chemistry of heterocyclic compounds (ed. *A. Weissberger*), vol. 17, p. 21, Table II. New York: Interscience 1962.

12a *Dondoni, A., Taddei, F.:* Boll. Sci. Fac. Chim. Ind. Bologna **25**, 155 (1967).

13 *Grundmann, C., Datta, S. K.:* J. Org. Chem. **34**, 2016 (1969).

Table III (continued)

No.	Compound	Method of preparation	M.P. [°C]	Stability	References
6	Oxalo-bis-nitrile oxide	C	explodes at −45	5—6 hours[g]	11, 14, 15
7	2,6,6-Trimethyl-cyclohexene-1-yl-1-nitrile oxide	B	48—49[e]	unlimited	13
8	2,2,6-Trimethyl-cyclohexane-1-nitrile oxide	B	31[f]	unlimited	13
9	Benzonitrile oxide[l]	C, D, E	14—15	30—60 min	8, 11, 16–19
10	2-Chloro-benzonitrile oxide	C	27—28	3—6 days	12
11	3-Chloro-benzonitrile oxide	C	43—44	50—60 min	12, 12a
12	4-Chloro-benzonitrile oxide	C	82—83	10 days	12, 12a, 20
13	2,6-Dichloro-benzonitrile oxide	A, C	86—87	30—35 days	12, 21, 21a
14	2,3,6-Trichloro-benzonitrile oxide	A	93—94	not determined	21a
15	4-Bromo-benzonitrile oxide	C	83—84	not determined	12, 22
16	2-Chloro-6-bromo-benzonitrile oxide	A	102	not determined	21a
17	2-Nitro-benzonitrile oxide	C	76—77 (dec.)	1—2 days	12, 23
18	3-Nitro-benzonitrile oxide	C	82—83	20—25 days	12, 17, 20, 23, 23a
19	4-Nitro-benzonitrile oxide	C	95	>30 days	12, 23, 24
20	4-Methyl-benzonitrile oxide	C	57—58	5—7 days	12, 12a
21	2,6-Dimethyl-benzonitrile oxide	A	79.5—81	unlimited	25

14 *Grundmann, C.:* Angew. Chem. **75**, 450 (1963).

15 *Grundmann, C., Mini, V., Dean, J. M., Frommeld, H.-D.:* Ann. Chem. **687**, 191 (1965).

16 *Wieland, H.:* Ber. **40**, 1667 (1907).

17 *Huisgen, R., Mack, W., Anneser, E.:* Angew. Chem. **73**, 656 (1961).

18 *Quilico, A., Speroni, G.:* Gazz. Chim. Ital. **76**, 148 (1946).

19 *Wieland, H., Semper, L.:* Ber. **39**, 2522 (1906).

20 *Huisgen, R., Mack, W.:* Tetrahedron Letters 583 (1961).

21 *Grundmann, C., Dean, J.M.:* J. Org. Chem. **30**, 2809 (1965); Angew. Chem. **76**, 682 (1964).

21a *Hackmann, J. T., Harthoorn, P. A., Kidd, J.* (to Shell Res. Ltd.): Brit. Pat. 949, 372 (1964); — Chem. Abstr. **60**, 13199a (1964).

22 *Grünanger, P.:* Atti Accad. Naz. Lincei, Rend., Classe Sci. Fis., Mat. Nat. [8]**24**, 163 (1958).

23 *Chang, M.S., Lowe, J.U.:* J. Org. Chem. **32**, 1577 (1967).

23a *Bianchi, G., Frati, E.:* Gazz. Chim. Ital. **96**, 562 (1966).

24 *Eloy, F., Lenaers, R.:* Bull. Soc. Chim. Belges **74**, 129 (1965).

25 *Yamakawa, M., Kubota, T., Akazawa, H.:* Bull. Chem. Soc. Japan **40**, 1600 (1967). — *Yamakawa, M., Kubota, T., Akazawa, H., Tanaka, L.:* Bull. Chem. Soc. Japan **41**, 1046 (1968).

Table III (continued)

No.	Compound	Method of prep-aration	M.P. [°C]	Stability	References
22	2,6-Dimethyl-4-bromo-benzonitrile oxide	A	100—101.5	unlimited	[25]
23	2,4,6-Trimethyl-benzonitrile oxide	A, B, C[i]	114	unlimited	[11,21,25–28]
24	2,4,6-Trimethyl-3,5-dichloro-benzonitrile oxide	C	138	unlimited	[29]
25	2,3,5,6-Tetramethyl-benzo-nitrile oxide	A, C	120	unlimited	[31,26]
26	4-Methoxy-benzonitrile oxide	C	69—70	7—10 days	[30,31,12a]
27	4-Methoxy-2,6-dimethyl-benzo-nitrile oxide	A	69—71	unlimited	[25,32]
28	2,4,6-Trimethoxy-benzonitrile oxide	A	160—170 (dec.)	unlimited	[21]
29	4-Dimethylamino-2,6-dimethyl-benzonitrile oxide	A, B	134—136 (dec.)	unlimited	[25,27,33,34]
30	4-Fulmido-3,5-dimethyl-N-trimethylanilinium iodide	—[k]	188 (dec.)	unlimited	[33]
31	Triphenyl-acetonitrile oxide	F	153—154	unlimited	[35]
32	Oximino-phenyl-acetonitrile oxide	D	112—113	not determined	[36,37]
33	Benzoyl-oximino-phenyl-acetonitrile oxide	C	109—110	not determined	[38]
34	Oximino-p-tolyl-acetonitrile oxide	D	112 (dec.)	not determined	[39]
35	Benzoyl-oximino-p-tolyl-acetonitrile oxide	C	140	not determined	[40]

26 *Califano, S., Moccia, R., Scarpati, R., Speroni, G.:* J. Chem. Phys. **26**, 1777 (1957).
27 *Grundmann, C., Richter, R.:* J. Org. Chem. **33**, 476 (1968).
28 *Just, G., Dahl, K.:* Tetrahedron **24**, 5251 (1968).
29 *Beltrame, P., Veglio, C., Simonetta, M.:* J. Chem. Soc. B 867 (1967).
30 *Dondoni, A., Mangini, A., Ghersetti, S.:* Tetrahedron Letters 4789 (1966).
31 *Rheinboldt, H., Dewald, M., Jansen, F., Schmitz-Dumont, O.:* Ann. Chem. **451**, 161 (1927).
32 *Grundmann, C., Flory, K.:* Unpublished.
33 *Grundmann, C., Richter, R.:* J. Org. Chem. **32**, 2308 (1967).
34 *Grundmann, C., Dean, J.M.:* Angew. Chem. **77**, 966 (1965).
35 *Wieland, H., Rosenfeld, B.:* Ann. Chem. **484**, 236 (1930).
36 *Ponzio, G.:* Gazz. Chim. Ital. **66**, 119 (1936).
37 *Ponzio, G.:* Gazz. Chim. Ital. **66**, 123 (1936).
38 *Ponzio, G.:* Gazz. Chim. Ital. **61**, 570 (1931).
39 *Ponzio, G.:* Gazz. Chim. Ital. **71**, 693 (1941).
40 *Ponzio, G.:* Gazz. Chim. Ital. **61**, 572 (1931).

Table III (continued)

No.	Compound	Method of preparation	M.P. [°C]	Stability	References
36	2-Methoxy-l-naphthonitrile oxide	B	101—103	unlimited	[27]
37	2,6-Dimethoxy-l-naphthonitrile oxide	B	120—122	unlimited	[27]
38	2,6-Dimethoxy-5-bromo-l-naphthonitrile oxide	B	192—194 (dec.)	unlimited	[27]
39	2,7-Dimethoxy-l-naphthonitrile oxide	B	123—124	unlimited	[27]
40	2,7-Dimethoxy-8-bromo-l-naphthonitrile oxide	B	122—124	unlimited	[27]
41	Anthracene-9-nitrile oxide	A	127—128	unlimited[h]	[21,25]
42	10-Methylanthracene-9-nitrile oxide	A, B	165	unlimited[h]	[27]
43	O-Methylpodocarponitrile oxide	—[i]	132	unlimited	[28,41,42]
44	Isophthalo-bis-nitrile oxide	C	92—94 (dec.)	not determined	[43]
45	Terephthalo-bis-nitrile oxide	C	241—242 (dec.) 124—157 (dec.) ~160 (dec.)	not determined not determined not determined	[44] [45,46] [47]
46	2,4,6-Trimethyl-isophthalo-bis-nitrile oxide	A, B	138—139 (dec.)	unlimited	[27,33]
47	4-Dimethylamino-2,6-dimethyl-isophthalo-bis-nitrile oxide	A, B	123—125 (dec.)	unlimited	[27,33]
48	2,3,5,6-Tetramethyl-terephthalo-bis-nitrile oxide	A	169—170 (dec.)	unlimited	[21]
49	5-Methyl-3-phenyl-1,2-oxazole-4-nitrile- oxide	C	83	>30 days	[12,18]
50	2,4,6-Trimethoxy-pyrimidine-5-nitrile oxide	A, B	134—136 (dec.)	unlimited	[27,33]
51	2-Dimethylamino-4,6-dichloro-pyrimidine-5-nitrile oxide	B	159—162 (dec.)	unlimited	[27]

41 *Just, G., Dahl, K.:* Tetrahedron Letters 2441 (1966).
42 *Just, G., Zehetner, W.:* Tetrahedron Letters 3389 (1967).
43 *Iwakura, Y., Akiyama, M., Shiraishi, S.:* Bull Chem. Soc. Japan **38**, 335 (1965).
44 *Iwakura, Y., Akiyama, M., Nagabuko, K.:* Bull. Chem. Soc. Japan **37**, 767 (1964).
45 *Overberger, C.G., Fujimoto, S.:* J. Polymer Sci. B, Polymer Letters **3**, 735 (1965); — J. Polymer Sci. C, 4161 (1968).
46 *Overberger, C.:* Private communication.
47 *Eloy, F.:* Bull. Soc. Chim. Belges **73**, 639 (1964).

Table III (continued)

No.	Compound	Method of preparation	M.P. [°C]	Stability	References
52	2-Dimethylamino-4-chloro-6-methoxy-pyrimidine-5-nitrile oxide	A	154—155	unlimited	[33]
53	2-Dimethylamino-4,6-dimethyl-pyrimidine-5-nitrile oxide	A	178—180 (dec.)	unlimited	[33]
54	1-Fulmido-2-bromo-7,7-dimethylnorbornane	C	123—125	unlimited	[47a]
55	Pentachloro-benzonitrile oxide	C	225—229 (dec.)	unlimited	[47b]

[a] And special methods discussed in Chapter III, B.
[b] Half life in 0.4 N ethereal solution at 0°.
[c] B.p. 61° (15 mm).
[d] B.p. 55—56° (0.2 mm).
[e] B.p. (0.03 mm).
[f] B.p. (0.001 mm).
[g] In approximately 5% solution in methylene chloride at 0°.
[h] The compound is slowly decomposed by light and air.
[i] Dehydrogenation with $Pb(Ac)_4$.
[k] Quaternization of compound No. 29 with CH_3I.
[l] Preparation of $C_6H_5-C^{14}\equiv N \rightarrow O$, see Ref.[55].

Dipole Moment. Quantitative measurements of the dipole moments of several aromatic nitrile oxides have not shown the expected increase over those of the corresponding nitriles. This has been attributed to a redistributions of π electrons from N toward C, resulting in an inhibition of the expected increase in polarity for the nitrile oxide. The reported data are summarized in Table IV.[48, 49, 2, 25]

Solutions. As may be expected from the above-discussed data, the fulmido group imparts upon the molecule about the same solubility characteristics as the cyano group. In alcoholic solvents, however, hydrogen bonding has been demonstrated by I.R. spectroscopy. The frequency shift Δv of the OH absorption of the solvent upon hydrogen bonding to various

47a *Ranganathan, S., Singh, B. B., Panda, C. S.:* Tetrahedron Letters 1225 (1970).
47b *Wakefield, B. J., Wright, D. J.:* J. Chem. Soc. (C) 1165 (1970).
48 *Speroni, G.:* Ric. Sci. **27**, 1199 (1957).
49 *Del Re, G.:* Atti Accad. Naz. Lincei, Rend., Classe Fis., Mat. Nat. [8] **22**, 491 (1957).

Table IV. Dielectric moments of nitrile oxides and nitriles

Compound (in benzene)	μ		μ of the corresponding nitrile (D.)
	found (D.)	calcd. (D.)	
Formonitrile oxide	3.06	—	3.00
Acetonitrile oxide	4.50	—	3.94
Benzonitrile oxide (15°)	4.00	—	3.37
4-Methyl-benzonitrile oxide (25°)	4.58	4.5	4.40
2-Chloro-benzonitrile oxide (25°)	4.78	4.7	4.73
4-Chloro-benzonitrile oxide	2.62	2.6	2.50
2,4,6-Trimethyl-benzonitrile oxide (25°)	4.38	—	4.13
2,4,6-Trimethyl-benzonitrile oxide (25°)	4.39 (in dioxane)	—	—
2,4,6-Trimethyl-benzonitrile oxide (37°)	4.41	—	—
2,3,5,6-Tetramethyl-benzonitrile oxide (25°)	4.33	—	—
2,3,5,6-Tetramethyl-benzonitrile oxide (25°)	4.34 (in dioxane)	—	—

types of N-oxides as proton acceptors is a measure of the ground state basicity of this compound. The available data are summarized in Table V.[50–52]

Table V. Frequency shift of the OH-stretching band in solutions of nitrile and amine oxides

Compound	Δv (in cm^{-1})	
	methanol	phenol
Di-tert-butyl-acetonitrile oxide	120	221
2,4,6-Trimethyl-benzonitrile oxide	78	177, 183 360 (in β-naphthol)
Pyridine-N-oxide	256	—
Trimethylamine-N-oxide	486	648

These results indicate the decreasing *s*-character in nitrogen from aliphatic amine oxides to aromatic nitrile oxides, together with a decreasing electronegativity of the nitrogen and a relatively low basicity of the oxygen. Since the electron density of oxygen is diminished by delocalization, structure *1 c* might represent a preferred resonance form of the ground state.

Essentially the same conclusions have been reached by Japanese workers studying the molecular complexes of 2,4,6-trimethyl-benzonitrile

50 *Schleyer, P. von R., Joris, L.:* Private communication (1967).
51 *Joris, L., Schleyer, P. von R.:* Tetrahedron **24**, 5991 (1968).
52 *Kubota, T., Yamakawa, M., Takasuka, M., Iwatani, K., Akazawa, H., Tanaka, L.:* J. Phys. Chem. **71**, 3597 (1967).

oxide (*2*) or 9-anthronitrile oxide (*3*) with phenol and β-naphthol and the charge-transfer complex of *2* with iodine.[52] Their thermodynamic data, derived from UV and IR spectroscopical investigation of the systems are summarized in Table VI.

Table VI. Molecular complexes of aromatic nitrile oxides

System	K (l/mole)	ΔH (Kcal/mole)	ΔS (eu)
2,4,6-Trimethyl-benzonitrile oxide-Iodine (1:1) (in CCl_4)	2.25 (14.6°)	—3.87	—11.84
2,4,6-Trimethyl-benzonitrile oxide-β-Naphthol (in CCl_4)	15.4 (26.3°)	—4.90	—10.95
2,4,6-Trimethyl-benzonitrile oxide-Phenol (in CCl_4)	3.8 (22.2°)	—	—
9-Anthronitrile oxide-Phenol (in n-Heptane)	3.5 (21.2°) (median of values between ~ 2 and 5)	—	—

Molecular Refraction. The necessary data (refractive index and density) are available only for a few nitrile oxides which are stable enough in the

Table VII. Molecular refraction of nitrile oxides

	Benzonitrile oxide	2,6,6-Trimethyl-*trans*-fulmido-cyclohexane (*4*)	2,6,6-Trimethyl-fulmido-cyclohexene-1 (*5*)
d	$1.2190 \left(\frac{12.2}{2}\right)^{\circ}$ $1.2070 \left(\frac{12.3}{2}\right)^{\circ}$	$0.9739 \left(\frac{26}{4}\right)^{\circ}$	$0.9962 \left(\frac{31}{4}\right)^{\circ}$
n_D	1.5987 1.5986	1.4690	1.5103
MR, found	33.36	48.11	49.74[b]
MR, calcd[a]	33.17	48.21	47.74

[a] From the table of atomic refractions and increments.[53]

[b] Compound *5* shows the exaltation of MR expected for a conjugated aliphatic system.

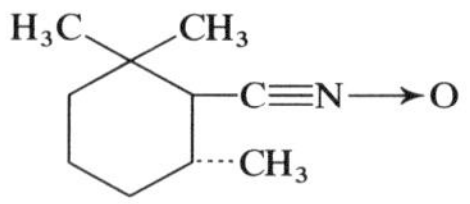

H₃C   CH₃
—C≡N⟶O
····CH₃
4

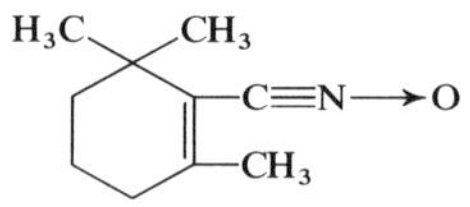

H₃C   CH₃
—C≡N⟶O
—CH₃
5

53 *Asmus, E.:* Refraktometrie. Methoden der organischen Chemie (ed. *E. Müller*), 4th ed., vol. 3/2, p. 407—424. Stuttgart: G. Thieme 1955.

liquid state. The molecular refraction obtained from these data with the Lorenz formula agrees well with the values calculated from the atomic refractions and increments, using for nitrogen the value for nitriles and for oxygen that of carbonyls. This agreement was historically one of the first experimental foundations for the open-chain nitrile oxide formula, $-C \equiv N \rightarrow O$, versus the then preferred ring structure,

The available data are reported in Table VII.[13, 54, 55]

54   *Auwers, K. v.:* Ber. **61**, 1041 (1928).
55   *Alexandrou, N. E.:* J. Org. Chem. **30**, 1335 (1965).

B. Electronic Spectra

Ultraviolet Spectra. With one exception (oxalo-bis-nitrile oxide) quantitative data are only reported for a number of relatively stable aromatic nitrile oxides.[1-3,15] So far as general conclusions are possible, it seems that the CNO group has a similar influence as the CN group in conjugated and aromatic systems. The following conclusions have been drawn from the data of the Table VIII a and VIII b:

1. The $\pi - \pi^*$ band occurred at wavelengths much longer than those of the ^1L_a band of the corresponding nitriles.

2. The $\pi - \pi^*$ band showed a blue shift with an increase in the polarity of the solvents. Quantitative analysis of the solvent effect led to the result of a decreasing dipole moment at the $\pi - \pi^*$ state.

3. The hydrogen bonding ability of the nitrile oxides was found to be weak from analysis of the above solvent effect (see also Chapter III A). Further conclusion pertaining to the electronic structure of nitrile oxides are discussed in the following section (II, C).

Infrared Spectra. Almost all of the isolated nitrile oxides, recorded in Table III, have been characterized by their I.R. spectrum. Beyond that, for many of the less stable nitrile oxides, prepared only in situ, I.R. spectral data, besides of the products of further reactions, have been the sole means of identification of the nitrile oxide intermediate.[4] For some of the more stable species detailed investigations are reported.[1-3,5-7]

Table VIII a. Ultraviolet spectra of nitrile oxides in some solvents

Compound		Solvent							
		n-heptane		CCl_4		CH_3CN		CH_3OH	
		λ_{max} (mμ)	ε	λ_{max} (mμ)	ε	λ_{max} (mμ)	ε	λ_{max} (mμ)	ε
Oxalo-bis-nitrile	I	312	9000	—	—	—	—	—	—
oxide[b,1]		288	10000	—	—	—	—	—	—
	II	256	8000	—	—	—	—	—	—

1 *Grundmann, C., Mini, V., Dean, J.M., Frommeld, H.-D.:* Ann. Chem. **687**, 191 (1965).

2 *Yamakawa, M., Kubota, T., Akazawa, H.:* Bull. Chem. Soc. Japan **40**, 1600 (1967).

3 *Yamakawa, M., Kubota, T., Akazawa, H., Tanaka, L.:* Bull. Chem. Soc. Japan **41**, 1046 (1968).

4 *Wiley, R.H., Wakefield, B.J.:* J. Org. Chem. **25**, 546 (1960).

5 *Califano, S., Moccia, R., Scarpati, R., Speroni, G.:* J. Chem. Phys. **26**, 1777 (1957).

6 *Borello, E., Colombo, M.:* Ann. Chim. (Rome) **46**, 1158 (1956).

7 *Califano, S., Scarpati, R., Speroni, G.:* Atti Accad. Naz. Lincei, Rend., Classe Sci. Fis., Mat. Nat. [8] **23**, 263 (1957).

Table VIII a (continued)

Compound		n-heptane λ_{max} (mμ)	ε	CCl_4 λ_{max} (mμ)	ε	CH_3CN λ_{max} (mμ)	ε	CH_3OH λ_{max} (mμ)	ε
2,6-Dimethyl-benzonitrile oxide[2]	I	293.5	968	294.5	1070	293.5	1230	293.8	1260
		228[a]	970	290[a]	1100	288.5[a]	1200	289[a]	1100
						282.3	1520	282.5	1520
	II	267.4[a]	11100	268.2[a]	12200	263.6[a]	11200	263[a]	11000
		259.7	12700	260.6	13700	255.4	13000	255.3	12900
	III	216.2	27900	—	—	208.8	30100	210.2	28500
		211.6	26800	—	—	—	—	—	—
	IV	195.2	36100	—	—	195.0	34500	194.6	35500
2,4,6-Trimethyl-benzonitrile oxide[2]	I	294.2[a]	667	295.0[a]	803	293.7	863	294.2	970
		—	—	—	—	288.9[a]	970	289.4[a]	1040
	II	269.8[a]	12800	271.1[a]	14200	266.6[a]	12800	267.4[a]	13100
		262.7	15000	263.9	15900	258.7	15000	258.7	15600
	III	219.9	26900	—	—	217.3	24300	217.1	26300
		214.7	26100	—	—	213.0	21300	213.5	26600
	IV	198.4	37300	—	—	197.4	33300	197.9	34400
2,6-Dimethyl-4-bromo-benzonitrile oxide[2]	I	293.2[a]	1350	295[a]	1300	295.6	1030	295.7	1030
	II	274.6[a]	16400	277.4[a]	17900	271.4[a]	17100	271.8[a]	16800
		268.0	19000	268.7	20600	264.0	19400	264.4	19200
	III	216.3	25700	—	—	214.3	26100	214.4	26700
	IV	204.3	31000	—	—	203.0	31600	203.5	31400
2,6-Dimethyl-4-methoxy-benzonitrile oxide[2]	II	267.9	17900	270.2	17900	266.4	18800	266.9	19300
	III	213.7[a]	24700	—	—	215.7	23900	216.0	24200
	IV	202.5	34400	—	—	201.9	31500	201.7	31500
Terephthalo-bis-nitrile oxide[c, 2]	I	300.7	—	301.4	—	290.5	29200	292.1	30800
		298.2	—	—	—	—	—	—	—
		285.8	—	286.6	—	279.4	30000	280.1	30100
	II	222.2	—	—	—	218.1	15500	219.3	15400
		216.8	—	—	—	213.9	15800	214.7	15400
	III	191.5	—	—	—	189.0	41100	189.4	40600
Isophthalo-bis-nitrile oxide[d, 2]	I	306.2	—	—	—	300.7[a]	—	302.2[a]	—
		296.9	—	—	—	291.3[a]	—	292.6[a]	—
	II	268.0[a]	—	—	—	261[a]	—	262[a]	—
		255.8	—	—	—	247.4	—	249.0	—
	III	231.8	—	—	—	217.2 (211.5)	—	212.2[a]	—

Table VIIIa (continued)

Compound		Solvent							
		n-heptane		CCl$_4$		CH$_3$CN		CH$_3$OH	
		λ_{max} (mμ)	ε	λ_{max} (mμ)	ε	λ_{max} (mμ)	ε	λ_{max} (mμ)	ε
9-Anthro-nitrile oxide[2]	I	408.3	12100	411.9	10600	408.1	9230	407.8	9470
		386.5	12900	389.8	11700	386.9	10900	386.7	11300
		367.3	8510	369.5	7840	367.1	7870	366.9	8330
		350.0	4080	352.4[a]	3760	350.0[a]	3940	349.9	4160
	II	259.7	134000	260.4	91300	258.3	133000	257.8	132000
	III	221.0	10500	—	—	220.8	11600	220.7	12100
		217.4	11400	—	—	217.6	12100	217.3	12600
	IV	201.3	16600	—	—	198[a]	15300	199[a]	15000
		198.6	16700	—	—	—	—	—	—
	V	182.9	34400	—	—	—	—	—	—
4-Dimethyl-amino-2,6-dimethylbenzonitrile oxide[3]	I	292.7	22900	—	—	—	—	—	—
	II	229.0	11500	—	—	—	—	—	—
	III	206.8	23000	—	—	—	—	—	—

[a] Shoulder-type band.

[b] In n-hexane.

[c] ε values not given for n-heptane and CCl$_4$ because of poor solubility.

[d] ε values not given, because of questionable purity of sample (may have contained polymeric furoxans) and poor solubility.

Aliphatic and aromatic nitrile oxides are characterized by *two strong absorption bands at around* 2330 cm^{-1} (C≡N stretching) and at around 1370 cm^{-1} (—N=O stretching). Particularly the former band is well suited for the identification of monomeric nitrile oxides.[8] The corresponding nitriles absorb strongly in the same region, but generally ~ 70 cm^{-1} lower. The nitrile oxide band is also usually stronger and broader as the nitrile band which is often weak, but very narrow. The isomeric isocyanates absorb often at close range, but lack the band around 1370 cm^{-1}. Only the two simplest nitrile oxides, fulminic acid and cyanogen-bis-N-oxide, absorb at somewhat different wave numbers,[1, 9] their spectra are given in Table IX together with that of dinitrogen oxide which is isoelectronic with the former and shows a striking similarity with this

8 *Grundmann, C.:* Herstellung und Umwandlung von Nitriloxiden. Methoden der organischen Chemie, 4th ed. (ed. *E. Müller*), vol. 10/3, p. 837—870. Stuttgart: G. Thieme 1965.

9 *Beck, W., Feldl, K.:* Angew. Chem. **78**, 746 (1966); *Winnewisser, B. P., Winnewisser, M.:* J. Mol. Spectry. **29**, 505 (1969).

Table VIIIb. Ultraviolet spectra of Nitrile oxides in some solvents[a]

Compound		ethanol-methanol 4:1		isopentane-methyl-cyclohexane 3:1	
		v, cm^{-1}	ε	v, cm^{-1}	ε
4-Methyl-benzonitrile oxide[15]	I	—	—	35200	1090
		—	—	37730	12130
	II	39680	15140	38600	13680
		—	—	39050	13450
4-Chlorobenzonitrile oxide[15]	I	—	—	34600	830
		—	—	35700	3090
		—	—	36350	8820
	II	38900	17420	37110	10580
		—	—	37880	11730
		—	—	38550	11200
3-Chlorobenzonitrile oxide[15]	I	34130	485	34080	1260
		35200	670	35150	1700
	II	39380	6040	37600	11520
		—	—	38750	11620
4-Methoxy-benzonitrile oxide[15]	I	—	—	33780	1660
	II	38300	21500	38180	18000

[a] At 25°, for spectra at —196° see Ref.[15].

The spectrum of 2,4,6-trimethyl-3,5-dichlorobenzonitrile oxide (solvent not stated) has been reported[16] (λ_{max} in mμ, ε): 306, 1250; 295, 1410; 269, 17000; 2235, ca. 45000.

Table IX. Infrared spectra of fulminic acid, cyanogen-bis-N-oxide and dinitrogen oxide

Compound	C—H— stretching	C≡N— stretching	—N=O stretching	CN—O deformation
H—C≡NO (gaseous) 7	3335	2190	1251	538
ON≡C—C≡NO (in CCl$_4$) 8	—	2190	1235	—[a]
N≡NO (gaseous)	—	2224[b]	1285	589[c]

[a] Could not be observed, because of adsorption of solvent.
[b] N≡N stretching.
[c] N≡NO deformation.

compound.[10,11] The absence of bands around $3600\ \text{cm}^{-1}$ (OH—stretching) and $1400\text{–}1000\ \text{cm}^{-1}$ (OH— deformation) excludes definitely—at least for the gaseous free acid—the old formula, $C\!\stackrel{\leftharpoondown}{=}\!N\!\!-\!\!OH$ for fulminic acid.

A detailed analysis of the I.R. spectrum of acetonitrile oxide, together with the calculation of symmetry force constants, agrees with the above discussed generalizations.[17]

Microwave Spectra. The microwave spectra of fulminic acid and of five of its isotopically substituted species have been studied in the gaseous state in the frequency range from 10 to 46 GHz. The spectrum of the molecules in the ground vibrational state established the linearity of the chain HCNO. The following rotational constants B_0 for the ground state were obtained (Table X).[12] More recently, also the microwave spectra of acetonitrile oxide have been studied, leading to analogous conclusions.[18]

Table X. Microwave ground state vibrations B_0 for various isotopically labelled fulminic acids

Compound	B_0 (in MHz)
$HC^{12}\!\!-\!\!N^{14}\!\!-\!\!O^{16}$	11 469.04
$DC^{12}\!\!-\!\!N^{14}\!\!-\!\!O^{16}$	10 292.51
$HC^{13}\!\!-\!\!N^{14}\!\!-\!\!O^{16}$	11 091.57
$HC^{12}\!\!-\!\!N^{14}\!\!-\!\!O^{17}$	11 151.69
$HC^{12}\!\!-\!\!N^{14}\!\!-\!\!O^{18}$	10 865.34
$DC^{13}\!\!-\!\!N^{14}\!\!-\!\!O^{16}$	10 011.18

X-Ray-Analysis. Crystalline fulminic acid has been investigated and the results have confirmed the conclusion drawn from IR and microwave spectra.[13] (See following paragraph.) An x-ray analysis of the 4-methoxy-2,6-dimethyl-benzonitrile oxide has yielded the following data: (Cu K α radiation)[14]: Monoclinic, $a = 8.36$ Å, $b = 12.76$ Å, $c = 9.01$ Å, $\beta = 110°\ 38'$; $U = 899$ Å^3, $Dm = 1{,}260\ \text{g} \cdot \text{cm}^{-3}$ (at 25°) $Z = 4$, $D_c = 1{,}259\ \text{g} \cdot \text{cm}^{-3}$ (at 25°); space group $P2_1/c$. Details of the molecular structure are summarized in Table XI.

10 *Grosso, R. P., McCubbin, T. K.:* J. Mol. Spectry. **13**, 240 (1964).

11 *Pliva, J.:* J. Mol. Spectry. **12**, 360 (1964).

12 *Winnewisser, M., Bodenseh, H. K.:* Z. Naturforsch. **22**a, 1724 (1967); *Bodenseh, H. K., Winnewisser, M.:* Z. Naturforsch. **24**a, 1966, 1973 (1969).

13 *Beck, W.:* Private communication (1968) (publication in press).

14 *Shiro, M., Yamakawa, M., Kubota, T., Koyama, H.:* Chem. Commun. 1409 (1968).

15 *Di Lonardo, G., Dondoni, A., Mangini, A.:* Boll. Sci. Fac. Chim. Ind. Bologna **26**, 203 (1968).

16 *Beltrame, P., Veglio, C., Simonetta, M.:* J. Chem. Soc. B 867 (1967).

17 *Isner, W. G.: Humphrey, G. L.:* J. Am. Chem. Soc. **89**, 6442 (1967).

18 *Bodenseh, H. K., Morgenstern, K.:* Z. Naturforsch. **25**a, 150 (1970).

Table XI. Molecular structure of 4-methoxy-2,6-dimethylbenzonitrile oxide

Bond	Bond length (in Å)	Bond	Bond length (in Å)
C(1)—C(2)	1.390	C(2)—C(3)	1.401
C(3)—C(4)	1.400	C(4)—C(5)	1.394
C(5)—C(6)	1.385	C(6)—C(1)	1.410
C(1)—C(7)	1.435	C(7)—N(8)	1.147
N(8)—O(9)	1.249	C(2)—C(10)	1.518
C(6)—C(11)	1.517	C(4)—O(12)	1.363
O(12)—C(13)	1.436		

Bond Angles	Degrees	Bond Angles	Degrees
C(1)—C(2)—C(3)	118.9	C(2)—C(3)—C(4)	119.0
C(3)—C(4)—C(5)	121.6	C(4)—C(5)—C(6)	119.9
C(5)—C(6)—C(1)	118.3	C(6)—C(1)—C(2)	122.3
C(2)—C(1)—C(7)	120.2	C(6)—C(1)—C(7)	117.5
C(1)—C(7)—N(8)	173.8	C(7)—N(8)—O(9)	178.3
C(1)—C(2)—C(10)	123.0	C(3)—C(2)—C(10)	118.2
C(1)—C(6)—C(11)	121.4	C(5)—C(6)—C(11)	120.4
C(3)—C(4)—O(12)	122.9	C(5)—C(4)—O(12)	115.5
C(4)—O(12)—C(13)	118.3		

Molecular Geometry of Nitrile Oxides. The conclusions drawn from the measurements of various compounds discussed in the preceding paragraphs are consistent with each other and with theoretical predictions by molecular orbital calculations. The CNO group is almost linear. The atomic distances in formonitrile oxide (fulminic acid) in relation to the coordinate system, and those of 4-methoxy-2,6-dimethylbenzonitrile oxide are depicted in formulas *1* and *2*. Remarkable is the quite short

1

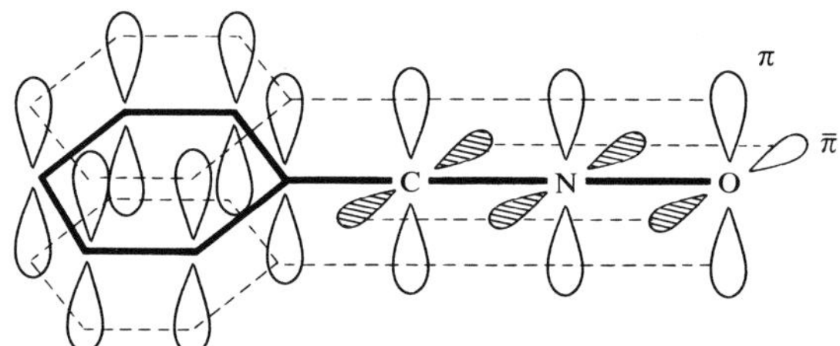

2

(distances in Å)

distance for the N—O— bond in both compounds. As a result from the various cited studies and theoretical calculations the resonance interaction of the two kinds of π systems (π and $\bar{\pi}$) has been represented by Fig. 1.

Fig. 1. Scheme of π-resonance interaction in benzonitrile oxide[2]

III. Preparation

A. General

All known methods for the synthesis of nitrile oxides start with organic systems already containing the C—N—O sequence of the nitrile oxide structure. There is no indication that any possible organic compound containing a C—N single, double or triple bond has ever been converted into a nitrile oxide. Likewise, no example is known where an —N—O— compound has been introduced into an organic residue to generate directly a nitrile oxide.

The most important methods for the preparation of nitrile oxides start with aldoximes, from which by various techniques two hydrogen atoms are abstracted to form the nitrile oxide:

$$R-CH=NOH \xrightarrow{-2H} R-C\equiv N \rightarrow O \tag{1}$$

Less widely applied methods consist in the dehydration of primary nitroparaffins by various procedures:

$$R-CH_2-NO_2 \xrightarrow{-H_2O} R-C\equiv N \rightarrow O \tag{2}$$

Contrary to the chemistry of nitriles, where introduction of a cyano group by reaction of a metal cyanide with an organic halogen compound is a very important synthetic route, the analogous reaction, i.e., the reaction of a metal fulminate with an alkyl or acyl halide, has been realized only in one special case. Most attempts in this direction have usually resulted in a spontaneous rearrangement leading to the isomeric isocyanate.[1–6]

The routes leading to the parent member of the series, formonitrile oxide (fulminic acid), differ in many cases considerably from the ones

1 *Scholl, R.:* Ber. **23**, 3505 (1890).
2 *Holleman, A.F.:* Ber. **23**, 2998 (1890); Rec. Trav. chim. **10**, 70 (1891).
3 *Nef, J.U.:* Ann. Chem. **280**, 339 (1894).
4 *Calmels, G.:* Compt. Rend. **99**, 794 (1884).
5 *Palazzo, F.C., Tamburello, A.:* Gazz. Chim. Ital. **37** I, 1 (1907).
6 *Wieland, H., Höchtlen, A.:* Ann. Chem. **505**, 237 (1933).

suitable for the preparation of the higher homologs. Consequently, they are treated in a special paragraph together with such derivatives of fulminic acid (e.g., metal salts, amides, O-esters) for which no analogs exist among the other nitrile oxides.

A table of all individual nitrile oxides prepared by the methods discussed in the following sections of this chapter is provided in Chapter II, pp. 16–20.

B. Formonitrile Oxide (Fulminic Acid) and Derivatives

Because of its instability, fulminic acid (*6*) has always been prepared first in form of its mercuric or silver salts. Although highly explosive, with the proper precautions both salts can be stored unchanged at room temperature for an extended period of time.

The historically *first discovered preparation of mercuric fulminate*, the reaction of ethanol with concentrated nitric acid in presence of metallic mercury or mercuric nitrate, is still the most convenient route to fulminic acid. From the mercuric salt, all other important derivatives are accessible. In order to start, the reaction needs the presence of catalytic amounts of nitrous acid resp. nitrous oxides. If metallic mercury is used, its dissolution will generate the necessary catalyst.

Wieland uncovered the mechanism of this reaction (Eq. (3)).[1-3]

$$C_2H_5OH \xrightarrow{\quad HNO_3 \quad} \underset{1}{CH_3CHO} \xrightarrow{\quad HNO_2 \quad} \underset{2}{HON{=}CHCHO} \xrightarrow{\quad HNO_3 \quad}$$

$$\underset{3}{HON{=}CHCOOH} \xrightarrow{\quad HNO_3 \quad} \underset{4}{HON{=}C(NO_2)COOH} \xrightarrow{\quad -CO_2 \quad} \qquad (3)$$

$$\underset{5}{HON{=}CHNO_2} \xrightarrow{\quad -HNO_2 \quad} \underset{6}{HC{\equiv}N \rightarrow O} \xrightarrow{\quad Hg^{++} \quad} Hg(CNO)_2$$

The sequence of reactions starts with the oxidation of ethanol to acetaldehyde (*1*) which is then nitrosated to isonitrosoacetaldehyde (*2*). (*2*) is oxidized to isonitroso-acetic acid (glyoxylic acid oxime, *3*) which undergoes further nitration to oxalo-mono-nitrolic acid (*4*). (*4*) decarboxylates to formonitrolic acid (*5*) which converts to formonitrile oxide (*6*) with the loss of nitrous acid. *6* which is not stable under the reaction conditions is immediately trapped by mercuric ion as the stable very little soluble mercuric fulminate which precipitates in the course of the reaction.

This reaction mechanism is essentially supported by the fact that ethanol has been successfully replaced by all the postulated intermediates of Eq. (3), so far as they are known. Accordingly, instead of ethanol acetaldehyde or substances which will generate *1* under the reaction conditions, e.g., paraldehyde or dimethyl-acetal can be used.[4] Further-

1 *Wieland, H.:* Die Knallsäure. Sammlung chemischer und chemisch-technischer Vorträge (ed. F. B. Ahrens), vol. 14, p. 385–461. Stuttgart: F. Enke 1909.

2 *Wieland, H.:* Ber. **40**, 418 (1907).

3 *Wieland, H.:* Ber. **43**, 3362 (1910).

4 *Wöhler, L., Theodorovits, K.:* Ber. **38**, 1345 (1905).

more, either glyoxylic acid oxime (*3*)[3] or formonitrolic acid (*5*) can be substituted for ethanol.

If mercury is replaced by metallic silver or silver nitrate, silver fulminate is procedured accordingly.[5] If properly carried out, the reaction according to Eq. (3) is almost quantitative as far as the conversion of the metal is concerned, but calculated on ethanol the yield is only approximately 6%.[4]

Mercuric fulminate is also obtained from malonic acid and a solution of mercury in an excess of diluted nitric acid,[6] whereby presumably first mesoxalic acid oxime is formed which then decarboxylates to glyoxylic acid oxime (*3*) (Eq. (4)), from which the reaction proceeds according to Eq. (3).

$$HOOCCH_2COOH \xrightarrow{HNO_2} HOOC(C{=}NOH)COOH \xrightarrow{-CO_2} 3 \qquad (4)$$

The mercuric salt of nitromethane decomposes directly into water and mercuric fulminate (Eq. (5)).[7]

$$2\,CH_3NO_2 \xrightarrow{NaOH} 2\,CH_2{=}NO_2Na \xrightarrow{Hg^{++}} (CH_2NO_2)_2Hg \longrightarrow$$

$$\xrightarrow{H_2O} Hg(CNO)_2 \quad + \quad (HO{-}Hg{-}O{-}CH{=}N{-}O)_2Hg \qquad (5)$$
$$7$$

Todays commercial availability of nitromethane, makes this route more interesting than at the time of its discovery, however, the yield is only about 5%, while the majority of the material is converted into a basic (also explosive!) mercury salt of formhydroxamic acid (*7*) which cannot be converted into mercuric fulminate.[8, 9]

Nitromethane can also be converted into fulminic acid by reaction with nitrous acid to form first formonitrolic acid (*5*). Heated with water or nitric acid *5* converts to *6* (as in Eq. (3)) which is trapped with silver nitrate as silver fulminate.[2, 10]

The silver salt of formonitrosolic acid converts into silver fulminate by treatment with nitric acid with the elimination of hyponitrous acid

5   *Nef, J. U.:* Ann. Chem. **280**, 308 (1894).
6   *Angelico, F.:* Atti Accad. Lincei, Rend., Classe Fis., Mat. Nat. **10**, 476 (1901).
7   *Nef. J. U.:* Ann. Chem. **280**, 263 (1894).
8   *Jones, L. W.:* Am. Chem. J. **20**, 1 (1898).
9   *Wöhler, L.:* Ber. **38**, 1351 (1905).
10 *Wieland, H.:* Ber. **42**, 803 (1909).

(Eq. (6)).[11]

$$HC{\overset{NO}{\underset{NOAg}{\big<}}} \quad \xrightarrow{V, HNO_3} \quad AgCNO \quad + \quad (HON)_2 \qquad (6)$$

Under the same conditions, the potassium salt of amino-formonitrosolic acid is likewise converted into silver fulminate[12] (Eq. (7)):

$$H_2NC(NO){=}NOK \quad \xrightarrow{Ag^+, HNO_3} \quad N_2 \quad + \quad H_2O \quad + \quad K^+ \quad + \quad AgCNO \qquad (7)$$

Silver fulminate can also be obtained from formamide oxime by heating with silver nitrate in nitric acid solution [12] (Eq (8)):

$$HON{=}CHNH_2 \quad \xrightarrow{Ag^+, HNO_3} \quad AgCNO \quad + \quad NH_4^+ \qquad (8)$$

The conversion of formhydroxamic acid into silver fulminate has already been discussed in Chapter I, C.

Mercuric Fulminate. *The crude salt*, obtained from the reaction of ethanol and nitric acid is usually of gray to yellowish color caused mainly by mechanical inclusion of finely divided mercury. For its purification, dissolution in aqueous ammonia and reprecipitation with diluted acetic acid[13] or dissolution in sodium thiosulfate solution and reprecipitation with potassium thiocyanate[14] are recommended. An initially *colorless salt*, however, is obtained by addition of 0.01% of Cl^- and Cu^{++} ion to the solution of mercury in nitric acid, used for its preparation.[4] Mercuric fulminate can also be recrystallized from nitric acid or ethanol or pyridine; for its preparation on a laboratory scale several tested and reliable procedures are recorded.[15-17] Mercuric fulminate crystallizes in colorless, heavy (D 4.42[18]) needles, sparingly soluble in water (0.71 g at 12°, 1.74 g at 49° in 1000 ml).[19] Although

11 *Wieland, H., Hess, H.:* Ber. **42**, 4181 (1909).

12 *Wieland, H.:* Ber. **42**, 821 (1909).

13 *Philip, R.:* Z. Ges. Schieß-Sprengstoffw. 7, 109 (1912); — Chem. Abstr. **6**, 3520 (1912).

14 *Wolf, P.:* Z. Ges. Schieß-Sprengstoffw. 7, 272 (1912); — Chem. Abstr. 7, 2684 (1913).

15 *Beckmann, E.:* Ber. **19**, 993 (1886).

16 *Lobry de Bruyn, C.A.:* Ber. **19**, 1370 (1886).

17 *Kurtz, P.:* Methoden der organischen Chemie, 4th ed. (ed. *E. Müller*), vol. 8, chap.: Methoden zur Herstellung und Umwandlung von Knallsäure, p. 355–358. Stuttgart: G. Thieme 1952.

18 *Berthelot, Vieille:* Ann. Chim. [5], **21**, 569 (1880).

19 *Holleman, A.F.:* Rec. Trav. Chim. **15**, 159 (1896).

molecular weight determinations have not been made for lack of a suitable solvent, ample chemical evidence, analogous to that mentioned below for silver fulminate, indicates that mercuric fulminate is indeed a complex salt, $hg^+[hg(CNO)_2]^-$ ($hg = \frac{1}{2}Hg$).

Mercuric fulminate explodes violently upon slight to moderate mechanical impact, on heating over an open flame, and on contact with concentrated sulfuric acid. Because of its explosive properties, it was introduced in 1864 by *A. Nobel* as the first *initial detonating agent for nitroglycerine*, resp. dynamite, and thus helped to shape the revolutionary development of industrial explosives and military weaponry which took place during the second half of the 19th century. Today, other detonators have largely replaced mercuric fulminate to a position of secondary importance, primarily because of its poor storability in warm climates and of the limited availability of mercury, but it is undoubtedly the first, if not so far the only nitrile oxide ever produced commercially on a large scale. There exists a tremendous amount of technical data relating to its use as an explosive for which reference is made to the literature.[20, 21]

Silver Fulminate. Silver fulminate is obtained pure by any of the above mentioned reactions; a reliable laboratory procedure exists,[5, 17] but its very high sensitivity to explosion, even by slight mechanical impacts, requires serious warning. It crystallizes in colorless needlets (D 4.09)[22] frequently of spearhead shape, poorly soluble in water, but considerably more than the silver halogenides, (0.075 g at 13°, 0.18 g at 30° in 1000 ml H_2O).[19] Therefore, silver fulminate can be extracted from a mixture with these by boiling water.[28] Molecular weight determinations of silver fulminate in aniline solution at concentrations below 0.5% give values indicating the monomeric formula, AgCNO, while at a concentration of 2.5% and above the double molecular weight is found. This confirms massive chemical evidence pointing to the complex salt structure, $Ag^+[Ag(CNO)_2]^-$, which was first indicated by Liebig's discovery, that potassium chloride replaces only one half of the total silver content of silver fulminate, with the formation of potassium silver fulminate $K^+[Ag(CNO)_2]^-$.[23]

For numerous other complex salts of fulminic acid, reference is made to the literature.[5, 22, 24]

---

20 *Kirk-Othmer:* Encyclopedia of chemical technology, 2nd ed., vol. 8, p. 581–658. New York: Interscience 1965.

21 Ullmanns Encyklopädie der technischen Chemie (ed. *W. Foerst*), 3rd ed., vol. 16, p. 56–109, München: Urban & Schwarzenberg 1965.

22 *Wöhler, L., Weber, A.:* Ber. **62**, 2742 (1929).

23 *Liebig, J.:* Ann. Chim. Phys. [2] **24**, 294 (1823).

24 *Wöhler, L.:* Ber. **50**, 586 (1917). — *Wöhler, L., Berthmann, H.:* Ber. **62**, 2748 (1929).

Sodium Fulminate. Because of their complex nature, the silver or the mercury salts of fulminic acid are not suited for many applications in organic chemistry, including the generation of the free acid. Therefore, they are often converted into the sodium fulminate by reaction with sodium amalgam. If this reaction is carried out in aqueous suspension, an impure and very explosive sodium fulminate is obtained.[7, 25, 26] A pure and more stable salt results from the use of ethanol or preferably methanol instead of water.[9, 22] The salt can be precipitated by ether from the alcoholic solution and crystallizes in colorless needles. They still explode on heating over an open flame or on contact with concentrated sulfuric acid, but are much less sensitive to mechanical impact than the mercury or silver salts. Sodium fulminate decomposes, either in solution or in solid form rather quickly at room temperature and should be used immediately after preparation. Aqueous solutions are highly dissociated and react strongly alkaline. Molecular weight determination in aqueous solution proved the monomeric formula for the salt.[9] Sodium fulminate is about as toxic as sodium cyanide.[27]

By analogous procedures, the other alkali salts and the earth alkali salts of fulminic acid have been obtained too.[22]

Free Fulminic Acid. An aqueous solution of free fulminic acid (together with hydrobromic acid and probably in equilibrium with formhydroxamyl bromide (Eq. (10)) is obtained by dissolving bromo-hydroximinoacetic acid in water (Eq. (9)).[28]

$$HON{=}\overset{|}{\underset{Br}{C}}{-}COOH \quad \xrightarrow{\ H_2O,\ 0^0\ } \quad H{-}C{\equiv}N{\longrightarrow}O \ + \ CO_2 \ + \ HBr \qquad (9)$$

$$HC{\equiv}N{\longrightarrow}O \ + \ HBr \ \rightleftharpoons \ \overset{H}{\underset{Br}{}}{>}C{=}NOH \qquad (10)$$

An aqueous solution of free fulminic acid can also be obtained by acidification of a solution of sodium fulminate with dilute sulfuric acid preferably at temperatures between $0°$ and $+5°$. To prevent rapid polymerization, the solution of the sodium fulminate should be added to the sulfuric acid, so that there is always an excess of mineral acid present.

At $0°$, the half life time of an aqueous 0.4 N solution, containing at least 0.2 N sulfuric acid is about 90 min.[29]

From such aqueous solutions fulminic acid can be extracted with ether. The half life in 0.4 N ethereal solution at $0°$ is about 40 min.[29] More

25 *Ehrenberg, A.:* J. Prakt. Chem. [2] **32**, 230 (1885).
26 *Scholl, R.:* Ber. **27**, 2818 (1894).
27 *Schischkoff, L.:* Ann. Chem., Suppl. **1**, 109 (1861).
28 *Houben, J., Kauffmann, H.:* Ber. **46**, 4001, 4009 (1913).
29 *Birckenbach, L., Sennewald, K.:* Ann. Chem. **512**, 45 (1934).

convenient, probably, is the conversion of mercuric fulminate into form-hydroxamyl iodide (8) by means of hydroiodic acid and potassium iodide. (Eq. (11)):[30,31]

$$\text{Hg(CNO)}_2 \xrightarrow{\text{HI, KI}} 2\ \underset{\text{I}}{\overset{\text{H}}{>}}\!\!C{=}\text{NOH} + \text{K}_2\text{HgI}_4 \qquad (11)$$
8

Compound 8 is storable below $-20°$, preferably in ethereal solution. Addition of triethylamine to an ethereal, ice cold solution of 8 liberates fulminic acid.

Recently, the free acid has been obtained in a crystalline state by driving out the vapor of 6 with a rapid stream of nitrogen from an aqueous solution, followed by high-vacuum fractionation at $-78°$. Complete removal of the accompagnying HCN is difficult. On warming to about $-15°$ the crystals polymerize rapidly, but do not explode.[32] At 0,5 mm pressure the vapor is stable enough to determine the infrared spectrum. Fulminic acid is volatile with ether vapors and has an odor similar to hydrocyanic acid, but much more aggressive.[33,34]

The triple point of fulminic acid was found to be approximately at $-23.6°$ and 2.6 mm.[35]

All reactions of free fulminic acid have been carried out with dilute aqueous or ethereal solutions.

Analytical Determination. Simple fulminates and soluble complex fulminates, including mercuric fulminate are decomposed quantitatively by thiosulfate according to Eq. (12):

$$2\,\text{MeCNO} + 2\,\text{Na}_2\text{S}_2\text{O}_3 + 2\,\text{H}_2\text{O} = \text{Na}_2\text{S}_4\text{O}_6 + 2\,\text{NaOH}$$
$$+ 2\,\text{MeOH} + \text{C}_2\text{N}_2 \qquad \text{Me} = \text{metal} \qquad (12)$$

The formation of OH^- is only approximately quantitative, but the consumption of $\text{S}_2\text{O}_3^{--}$ corresponds exactly to Eq. (12). Therefore, the fulminate (approx. 0.3 g) is dissolved in 50 ml 0.1 N sodium thiosulfate solution, the solution is then weakly (methyl orange) acidified and titrated back with 0.1 N iodine and starch indicator.

$$1\ \text{ml}\ 0.1\ \text{N}\ \text{Na}_2\text{S}_2\text{O}_3 = 0.004201\ \text{g}\ \text{CNO}^{\ominus}$$

Other complex insoluble fulminates (e.g., silver fulminate) must be converted first into zinc fulminate, by shaking in methanolic suspension with zinc amalgam.[22,36]

30 *Huisgen, R., Christl, M.:* Angew. Chem. **79**, 471 (1967).

31 *Christl, M.:* Ph. D. Thesis, Universität München, 1969.

32 *Beck, W.:* Private communication (1966). — *Winnewisser, M., Bodenseh, H.K.:* Z. Naturforsch. **22**a, 1724 (1967).

33 *Wieland, H., Hess, H.:* Ber. **42**, 1347 (1909).

34 *Wieland, H., Hess, H.:* Ber. **42**, 1353 (1909).

35 *Winnewisser, M.:* Private communication.

36 *Philip, R.:* Z. Ges. Schieß-Sprengstoffw. **7**, 180 (1912); — Chem. Abstr. **6**, 3520 (1912).

Another method utilizes the reaction of fulminate ion with silver nitrate and is especially recommended for the determination of the free fulminic acid in solution. The reaction proceeds according to Eq. (13):

$$
\begin{array}{rclcl}
AgNO_3 + MeCNO &=& MeNO_3 + AgCNO \\
AgCNO + MeCNO &=& Me[Ag(CNO)_2] \\
Me[Ag(CNO)_2] + AgNO_3 &=& MeNO_3 + Ag[Ag(CNO)_2] \\
Me &=& Na \ or \ H
\end{array}
\qquad (13)
$$

6 drops of 0.1 N KI solution are added as indicator and the titration is carried out with 0.1 N AgNO$_3$ solution to the point of beginning turbidity (AgI), signalizing the completion of the reactions of Eq. (13) and an excess of Ag$^+$.[37]

$$1 \ ml \ 0.1 \ N \ AgNO_3 = 0.008403 \ g \ CNO^-$$

The suitability of this method for complex fulminates has not been tested.

Derivatives of Fulminic Acid. In this subparagraph a small number of compounds derived from fulminic acid will be discussed for which no analogy exists within the higher homologs.

Halogen Derivatives. The reaction of fulminic acid or fulminates with fluorine has not yet been reported. Chlorine, bromine, iodine, against which the nitrile oxides proper are remarkably resistant, attack easily mercury or silver fulminate. Within the historical review (Chapter I, B) it has been recorded that the first investigation of this reaction, using a large excess of chlorine, converted mercuric fulminate into cyanuric chloride and trichloro-nitromethane. Later investigations have shown that the above reaction produces primarily a mixture of dichloroformoxime (*9*) and dichlorofuroxan (*10*).[38, 39] The first step in this reaction is obviously the formation of the transient chloro-formonitrile oxide (*11*), which will either stabilize itself by dimerization to *10*, or by addition of hydrogen chloride to *9*. (Eq. (14))

$$Hg(CNO)_2 + 2\,Cl_2 \longrightarrow HgCl_2 + 2\,Cl\!-\!C\!\equiv\!N\rightarrow O$$

11

$$
\begin{array}{ccc}
Cl\!-\!\underset{\underset{N}{\|}}{C}\!-\!\underset{\underset{N}{\|}}{C}\!-\!Cl & \qquad & Cl_2C\!=\!NOH \\
\end{array}
\qquad (14)
$$

10   *9*

---

37 *Birckenbach, L., Sennewald, K.:* Ann. Chem. **512**, 38 (1934).
38 *Birckenbach, L., Sennewald, K.:* Ann. Chem. **489**, 7 (1931); — Ber. **65**, 546 (1932).
39 *Endres, G.:* Ber. **65**, 65 (1932).

In an aqueous medium, especially in presence of hydrochloric acid the formation of *9* is favored (40%), while in a nonpolar solvent, e.g., ethyl-chloride 25% of *10* are obtained. In aqueous systems and even in the anhydrous state *9* attains an equilibrium with *11* and hydrogen chloride.[40] The bromination reaction proceeds entirely analogously, in aqueous systems about equal quantities (25-30%) of dibromo-furoxan (*10*, Br for Cl) and dibromoformoxime (*9*, Br for Cl) are produced.[38-45] Under any conditions, during the chlorination or the bromination at least 45-55% of the applied metal fulminate are destroyed by oxidation. The reaction of mercury or silver fulminate with iodine, however, is much smoother and gives over 90% of only the diiodo-furoxan (*10*, I for Cl).[38, 42] Free fulminic acid forms predominantly the dichloro-, resp. dibromo-formoximes.[38, 46] Dichloro-furoxan is an oil, b.p. 63–64/12 mm, while dibromo- (m.p. 50°) and diiodo-furoxan (m.p. 89°, dec.) are solids. All three compounds are volatile with steam and possess an extremely irritating lachrymatory odor, their halogen seems to have a considerable positive character and is not prone to nucleophilic substitution, although only the dibromo-furoxan has been investigated in some detail.[45]

Carbonyloxime-Derivatives. Attempts to alkylate fulminic acid were made be a number of chemists around the turn of the century after *Nef's* formula, |C=NOH, had been accepted in the hope of stabilizing the elusive acid in form of its O-esters, |C=N—O—R (*12*). All these efforts have failed so far in producing *12*, but have resulted in the formation of the corresponding isocyanates, if any identifiable products were isolated (see Chapter IV, A-2). Treatment of sodium fulminate with dimethyl sulfate gave a very small yield of a crystalline, trimeric, explosive methyl derivative, $(CH_3)_3(CNO)_3$, m.p. 149°, of unknown structure, which might be derived from *12*.[47] In the reaction of silver fulminate with triphenylmethyl chloride, C-alkylation occurred (see Chapter III, C-3). The reaction of free fulminic acid with diazomethane has apparently not been studied.

In an attempt to prepare the alkyl-carbonyloximes by a different route, methyl formhydroxamate was obtained from O-methylhydroxylamine and formic acid. Treatment of this amide with phosphorus pentachloride yielded methoximino-formyl chloride (*13*, R = CH₃), dehydrohalogenation with solid potassium hydroxide in ether produced an intense,

40   *Seher, A.:* Ber. **83**, 400 (1950).
41   *Kékulé, A.:* Ann. Chem. **105**, 281 (1858).
42   *Sell, E., Biedermann, R.:* Ber. **5**, 89 (1872).
43   *Holleman, A.F.:* Ber. **26**, 1403 (1893).
44   *Scholl, R., Brenneisen, M.:* Ber. **31**, 642 (1898).
45   *Wieland, H.:* Ber. **42**, 4192 (1909).
46   *De Paoloni, I.:* Gazz. Chim. Ital. **60**, 700 (1930).
47   *Palazzo, F.C., Tamburello, A.:* Gazz. Chim. Ital. **37** I, 1 (1907).

evil odor, resembling that of isonitriles, indicating that initially the O-ether might have been formed as a transient intermediate, but all attempts to isolate it failed[48] (Eq. (15)):

$$CH_3-O-NH_2 \xrightarrow{\text{HCOOH}} CH_3-O-NH-CHO \xrightarrow{\text{PCl}_5}$$

$$R-O-N=CHCl \xrightarrow{\text{KOH}} \not\longrightarrow CH_3-O-N=C|$$

13

$$\Big\downarrow \text{MeX}$$

(15)

$$n\text{-}C_4H_9-O-N=C\diagup^{Me}_{\diagdown Cl} \longrightarrow n\text{-}C_4H_9OMe \quad +$$

14

$$N\equiv C-Cl \xrightarrow{2\,C_4H_9OH} HN=C\diagup^{OC_4H_9}_{\diagdown OC_4H_9}$$

15

16

A reinvestigation of this reaction sequence, using O-n-butylhydroxyl-amine proceeded well to the substituted formylchloride (13, R=n—C$_4$H$_9$), especially when PCl$_5$ was replaced by phosgene in the last step. Dehydrochlorination with the strong bases potassium tert-butoxide or lithium-n-butyl, proceeded apparently first to the metalated intermediate 14. This, however, did not cleave into metal halide and 12, but under attack on oxygen to metal butylate and cyanogen chloride (15), which could be identified by IR spectroscopy, but reacted soon further in a known manner[49] with the alkoxide to form di-n-butyl-iminocarbonate (16), the isolable end product.[50]

The o-phenanthroline iron complex of fulminic acid is alkylated with triethyl-oxonium fluoroborate to a diethyl derivative, in which spectro-scopical evidence indicates that the ethyl groups are bound to oxygen. In this complexed form, true alkyl derivatives of carbonyloxime are possibly obtained[32, 51] (Eq. (16)).

$$(Phen)_2Fe(CNO)_2 \xrightarrow{(C_2H_5)_3O^{\oplus}\cdot BF_4^{\ominus}} (Phen)_2Fe(C=N-O-C_2H_5)_2$$

(16)

$$Phen \quad = \quad$$

---

48 *Biddle, H. C.:* Am. Chem. J. **33**, 60 (1905); **35**, 349 (1906).

49 *Nef. J. U.:* Ann. Chem. **287**, 265 (1895).

50 *Grundmann, C., Richter, R.:* Unpublished; see also *Schulze, K. W.:* Diss. Technische Hochschule, Stuttgart, 1967.

51 *Beck, W., Schuirer, E.:* Chem. Ber. **95**, 3048 (1962).

In 1934, *E. Müller* obtained from the reaction of diazomethane with sodium or triphenylmethyl sodium a very explosive sodium compound CHN_2Na, which on hydrolysis yielded partly diazomethane back (~ 30–40%) besides of another very unstable compound CH_2N_2 which was tentatively characterized by its reactions and termed "isodiazomethane".[52] The preparation of isodiazomethane was improved by using lithium methyl and by decomposing the lithium salt with weakly acidic buffers, whereby little diazomethane is formed.[53, 54] A number of possible structures were discussed for "isodiazomethane", but not until 1968 spectroscopical evidence (NMR and IR) proved that isodiazomethane was N-isocyanamine (*17*), or the amide of the carbonyl oxime form of fulminic acid, fulmamide[55, 56] (Eq. (17)).

$$H_2CN_2 \xrightarrow{\text{LiCH}_3} CHLiN_2 \xrightarrow{\text{pH4}-5} H_2N{-}N{=}C| \longleftrightarrow H_2N{-}\overset{\oplus}{N}{\equiv}\overset{\ominus}{C}$$

$$\textit{17}$$

$$\text{R}{-}\text{I, Ba(OH)}_2, \text{DMF} \downarrow$$

$$\underset{R}{\overset{R}{>}}N{-}NH{-}CHO \xrightarrow[\text{Et}_3N]{\text{COCl}_2 \text{ or POCl}_3} \underset{R \quad \textit{18}}{\overset{R}{>}}N{-}N{=}C| \qquad (17)$$

Fulmamide is a colorless oil with strong isonitrile-like odor, which explodes at 15° and even in ethereal, diluted solutions is stable for only a limited time. Alkali rearranges quickly to diazomethane. In its reactions fulmamide shows no similarity at all with fulminic acid, but behaves as a typical isonitrile (isocyanamine). This lack of analogy between *17* and fulminic acid is possibly the strongest purely chemical evidence against *Nef's* formula.

Fulmamide is dialkylated by alkyl iodide ($R = CH_3$ or C_2H_5) in dimethylformamide in presence of barium hydroxide to the N-dialkyl-fulmamides (*18*),[56] a group of compounds which had been obtained earlier by a different route,[57] i.e., the reaction of N,N-dialkyl-N'-formyl hydrazides with phosgene or phosphorus oxychloride in presence of trialkylamines (Eq. (17)).

With the exception of the lowest homolog (*18*, $R = CH_3$) the dialkyl isocyanamines (N-dialkyl-fulmamides) are stable liquids which can be

52 *Müller, E., Disselhoff, H.:* Ann. Chem. **512**, 250 (1934). — *Müller, E., Kreutzmann, W.:* Ann. Chem. **512**, 264 (1934).

53 *Müller, E., Ludsteck, D.:* Chem. Ber. **87**, 1887 (1954).

54 *Müller, E., Kästner, P., Rundel, W.:* Chem. Ber. **98**, 711 (1965).

55 *Müller, E., et al.:* Ann. Chem. **713**, 87 (1968).

56 *Müller, E., Beutler, R., Zeeh, B.:* Ann. Chem. **719**, 72 (1968).

57 *Bredereck, H., Föhlisch, B., Walz, K.:* Ann. Chem. **686**, 92 (1965); **688**, 93 (1965); — Angew. Chem. **74**, 388 (1962). — *Föhlisch, B., Bredereck, H., Walz, K.:* Angew. Chem. **76**, 580 (1964).

distilled in vacuum. They possess the typical disagreeable odor of isocyanides, their IR spectrum shows the $N\equiv C-$ band at ~ 2090 cm^{-1}. At
room temperature, the dialkyl fulmamides decompose spontaneously
within several days. In the case of the N,N-diethyl compound, the reaction
products were found to be glyoxylic acid nitrile-N,N-diethylhydrazone
and acetaldehyde ethyl imine. A synchronous mechanism is suggested
(Eq. (18)).

$$\text{(18)}$$

Like the parent compound, *18* react as isonitriles and not as derivatives
of fulminic acid.

C. Other Nitrile Oxides

1. From Aldoximes

Dehydrogenation of Aldoximes. Aliphatic, aromatic and heterocyclic aldoximes have been dehydrogenated to nitrile oxides in alkaline solution (in the form of their oximate anions) by potassium ferric cyanide or, preferably, hypohalites[1-3]:

$$R\text{---}CH=\!NO^{\ominus} \quad + \quad OHal^{\ominus} \quad \longrightarrow \quad R\text{---}C\equiv N\longrightarrow O \quad + \quad OH^{\ominus} \quad + \quad Hal^{\ominus} \quad (1)$$

The most preferable reagent is an alkaline solution of sodium hypobromite. The hypobromite dehydrogenation is a very fast reaction occurring at temperatures around $0°\,C$, with good to excellent yields. Alkali hypoiodite offers no preparative advantage since the reaction is slower and the yields are much inferior.

The steric configuration of the oxime (*syn* or *anti*) was found without influence in such cases were alkali provides a facile rearrangement to the *syn* form, e.g., with mesitaldoxime. 4-Dimethylamino-2,6-dimethylbenzaldehyde forms rather stable *syn-* and *anti*-oximes and in this case only the *syn*-oxime yielded the nitrile oxide *1* whereas the *anti*-oxime was brominated to *2*. By analogy, it is inferred that the reaction of equation 1, occurs only with the *syn*-oximes, and that the reaction begins with an attack of the oxidant on the negatively charged oxygen atom of the oximate ion which is, as models easily demonstrate, less hindered than the *anti*-configuration. These conclusions have been corroborated by the recent studies of the lead tetraacetate oxidation of aldoximes (see below).

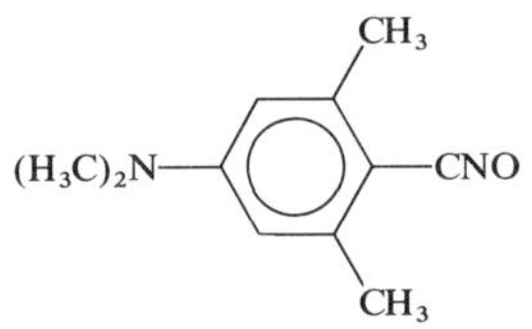

(structure *1*: 4-dimethylamino-2,6-dimethylbenzonitrile oxide, $(H_3C)_2N$—ring with CH_3, CH_3 and —CNO)   (structure *2*: $(H_3C)_2N$—ring with CH_3, CH_3, Br and $CH=NOH$ group, C—H, N—OH)

1 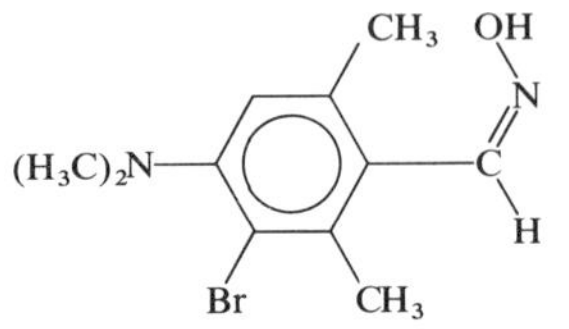   *2*

The reaction of sodium hypochlorite with aromatic aldoximes, which has been studied much earlier,[4] has sometimes yielded a mixture of a minor amount of the desired nitrile oxide together with a dimeric compound resulting from the abstraction of only one hydrogen atom from each molecule of aldoxime. The different route taken by the oxidation in the case of hypochlorite has been attributed to the higher

---

1 *Grundmann, C., Dean, J. M.:* Angew. Chem. **76**, 682 (1964); J. Org. Chem. **30**, 2809 (1965).

2 *Grundmann, C., Richter, R.:* J. Org. Chem. **32**, 2308 (1967).

3 *Grundmann, C., Datta, S. K.:* J. Org. Chem. **34**, 2016 (1969).

4 *Ponzio, G., Busti, G.:* Gazz. Chim. Ital. **36**II, 338 (1906).

oxidation potential of this reagent. In the earlier literature, these compounds were formulated as aldoxime peroxides, but more recently the structures of an oxime anhydride N-oxide (3)[5,6] or an aldazine bis-N-oxide (4)[7] have been proposed:

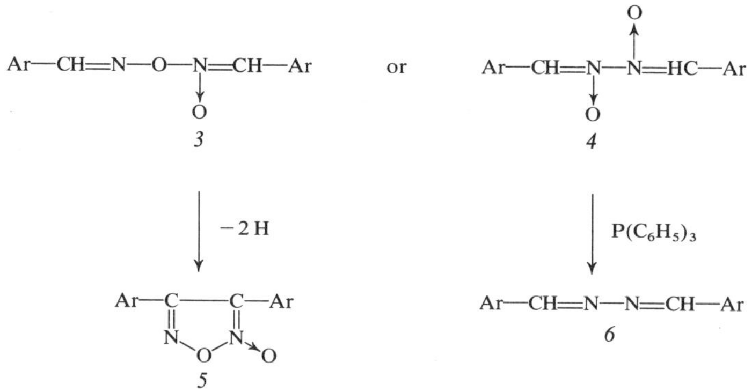

$$Ar—CH{=}N—O—N{=}CH—Ar \qquad or \qquad Ar—CH{=}N—N{=}HC—Ar$$

$$3 \qquad\qquad\qquad 4$$

$$\Big\downarrow {-2\,H} \qquad\qquad\qquad \Big\downarrow {P(C_6H_5)_3}$$

$$Ar—C{\equiv\!\!\!}C—Ar \qquad\qquad Ar—CH{=}N—N{=}CH—Ar$$

$$5 \qquad\qquad\qquad 6$$

It seems difficult to decide between 3 and 4 on the basis of chemical reactions alone. Further oxidation of these compounds leads to the corresponding diaryl-1,2,5-oxadiazole-1-oxides (furoxans, 5) which is easily understood on the basis of 3, but not to reconcile with structure 4 without accepting a unprecedented rearrangement. On the other hand, these compounds are smoothly reduced by triphenyl phosphine to the diaryl aldazines 6, a reaction more compatible with 4, since trialkyl and triaryl phosphines as well as trialkyl-phosphites are known to be specific deoxidants for N-oxides, e.g., azoxy-compounds, nitrile oxides, or furoxans.[8–10] The I.R. spectra of these dehydro-oximes favor structure 3, because of the strong similarity with the spectra of the corresponding furoxans 5.[6]

Compounds 3, resp. 4, decompose under mild conditions, e.g., refluxing in chloroform, into one mole of oxime and one mole of nitrile oxide which can be trapped in situ by a suitable dipolarophile[10a] (see also Section D and Chapter V).

The hypobromite oxidation of aldoximes has been especially useful in cases where the nitrile oxides could not be prepared by the older route via the hydroximic acid chlorides because of side reactions during the

5 *Wieland, H., Semper, L.:* Ber. **39**, 2522 (1906).

6 *Kropf, H., Lambeck, R.:* Ann. Chem. **700**, 18 (1966).

7 *Horner, L., Hockenberger, L., Kirmse, W.:* Chem. Ber. **94**, 290 (1961).

8 *Grünanger, P., Langella, M.:* Atti Accad. Naz. Lincei, Rend., Classe Fis., Mat. Nat. [8] **36**, 387 (1964).

9 *Grundmann, C., Frommeld, H.D.:* J. Org. Chem. **30**, 2077 (1965).

10 *Grundmann, C.:* Chem. Ber. **97**, 575 (1964).

10a *Grundmann, C., Kite, G.F.:* Unpublished results.

chlorination step.[1,11] The synthesis fails, however, in cases where the aldoxime contains other functional groups which are alkali-labile or unstable toward the oxidizing agent. Also polyfunctional nitrile oxides are usually obtained in poor yields.

A milder and more selective dehydrogenation of aldoximes has been achieved with N-bromosuccinimide in presence of alkali alkoxides or tertiary bases.[12] This modification allows the preparation, e. g., of amino-substituted aromatic and heterocyclic nitrile oxides as well as of polyfunctional nitrile oxides in satisfactory to very good yields, and is probably the most generally applicable procedure for the synthesis of nitrile oxides.

Aldoximes can also be dehydrogenated to nitrile oxides be means of lead tetraacetate.[13-15] Only *syn*-oximes lead to nitrile oxides, the oxidation has to be carried out at $-78°$ otherwise different products are obtained, identical with those produced from *anti*-oximes (see Eq. (3)). The reaction is believed to proceed via the lead organic compound 7 which through the conformer 7a, assisted by hydrogen bonding, disintegrates into nitrile oxide, lead acetate and acetic acid. (Eq. (2).) If at this point, the acetic acid is neutralized with pyridine or, preferably, triethylamine, the nitrile oxide can be isolated or identified by typical 1,3-dipolar cycloaddition reactions (see Chapter V). Otherwise, it will react with the acetic acid to the acetyl-hydroximic acid 8 which isomerizes under the reaction conditions to the acetyl hydroxamate 9.

$$\begin{array}{ccccc}
\underset{H}{\overset{R}{\diagdown}}C{=}N{-}OH & \xrightarrow[-78°]{Pb(OAc)_4} & \underset{H}{\overset{R}{\diagdown}}C{=}N{-}O{-}Pb(OAc)_3 & + & AcOH \\
 & & 7 & &
\end{array}$$

$$Pb(OAc)_2 \; + \; HOAc \; + \; R{-}C{\equiv}N{\rightarrow}O \; \xleftarrow{-78°} \; 7a$$

$$[R{-}\underset{OAc}{\overset{|}{C}}{=}NOH] \longrightarrow R{-}\underset{O}{\overset{\|}{C}}{-}NH{-}OAc$$
$$8 \qquad\qquad\qquad 9$$

(2)

11 *Grundmann, C.*: Herstellung und Umwandlung von Nitriloxiden, Methoden der Organischen Chemie, 4th ed. (ed. *E. Müller*), vol. 10/3, p. 841–870. Stuttgart: G. Thieme 1965.

12 *Grundmann, C., Richter, R.*: J. Org. Chem. **33**, 476 (1968).

13 *Just, G., Dahl, K.*: Tetrahedron Letters 2441 (1966).

14 *Just, G., Dahl, K.*: Tetrahedron **24**, 5251 (1968).

15 *Just, G., Zehetner, W.*: Tetrahedron Letters 3389 (1967).

Anti-aldoximes, when treated with lead tetraacetate at $-78°$, probably also form a lead organic compound of type *7*, since no precipitation of lead diacetate can be observed, the compound seems to be stable under these conditions. A conformational equilibrium analogous to $7 \leftrightarrow 7a$ does not provide a low energy path to product formation. Above $-70°$ unassisted homolysis occurs apparently leading to iminoxy radicals *10*, which are quite stable at $-55°$, (green color), but are rapidly consumed at higher temperatures by dimerization to aldazine-bis-N-oxides (*4*) in the case of R = aryl, while in the aliphatic series the end products are the dimeric nitroso-acetates[16] (*11*), probably formed by reaction of *10* with acetoxy radicals, and subsequent rearrangement (Eq. (3)).

$$(3)$$

A satisfactory explanation for the different behavior of aliphatic and aromatic iminoxy radicals has not yet been found.

Dehydrohalogenation of Hydroximic Acid Halides. Although this route does not start directly from the aldoximes, the necessary precursors to the nitrile oxide, the hydroximic acid halides, are most conveniently

---

16 *Kropf, H., Lambeck, R.*: Ann. Chem. **700**, 1 (1966).

prepared from the corresponding aldoximes. Thus, it must be considered also as a synthesis starting with the aldoximes. For all practical purposes, only the hydroximic acid chlorides *13* are used (Eq. (4)):

$$R\text{—}CH\text{=}NOH \quad \xrightarrow{\text{Cl}_2 \text{ or NOCl}} \quad [R\text{—}\underset{\underset{12}{|}}{\overset{}{C}}H\text{—}NO] \quad \longrightarrow \quad R\text{—}\underset{\underset{13}{|}}{\overset{}{C}}\text{=}NOH \quad (4)$$

with Cl substituents on compounds *12* and *13*.

The halogenation of aldoximes,[17, 18] which is generally carried out either in an inert solvent, such as chloroform and carbon tetrachloride, or in acetic acid, water or hydrochloric acid, seems—at least in some cases— to proceed via the intermediate of a geminal chloro-nitroso-compound *12* as indicated by the transient blue-green color of the reaction mixture. Aliphatic hydroximic acid chlorides are preferably prepared in ether at $-60°$ C, via *12* (or its dimer) which rearranges to *13* within one hour.[19]

Another route to hydroximic acid chlorides employs nitrosyl chloride as chlorinating agent[20].

This variation is sometimes advantageous when the use of free chlorine leads to unwanted side reactions, e. g., the direct chlorination of thiophene-2-aldoxime gave 5-chloro-thiophen-2-hydroximyl chloride, while nitrosyl chloride at -15 to $-10°$ yielded 65 % of the unsubstituted hydroximyl chloride.[21]

Finally, hydroximic acid halides have been obtained by reaction of concentrated hydrochloric or hydrobromic acid on certain nitrolic acids,[22–25] e. g., (Eq. (5)):

$$ROOC\text{—}\underset{\underset{NO_2}{|}}{\overset{}{C}}\text{=}NOH \quad + \quad HHal \quad \longrightarrow \quad ROOC\text{—}\underset{\underset{Hal}{|}}{\overset{}{C}}\text{=}NOH \quad (5)$$

$$R = C_2H_5, \text{ H; Hal} = \text{Cl, Br}$$

Nitrolic acids, however, can be converted in most cases directly into nitrile oxides (see Chapter III, C-2), so this method offers little advantage.

There are indications that at least the direct chlorination of aromatic aldoximes provides a mixture of *syn-* and *anti-* isomers of the hydroximic

17 *Henecka, H., Kurtz, P.:* Methoden zur Herstellung und Umwandlung von funktionellen N-Derivaten der Carboxyl-gruppe. Methoden der Organischen Chemie (ed. *E. Müller*), 4th ed., vol. 8, p. 691–692. Stuttgart: G. Thieme 1952.

18 *Metzger, H.:* Herstellung und Umwandlung von Oximen. Methoden der organischen Chemie (ed. *E. Müller*), 4th ed., vol. 10/4, p. 98–101, 206–208. Stuttgart: G. Thieme 1968.

19 *Casnati, G., Ricca, A.:* Tetrahedron Letters 327 (1967).

20 *Rheinboldt, H., Dewald, M., Jansen, F., Schmitz-Dumont, O.:* Ann. Chem. **451**, 161 (1927).

21 *Iwakura, Y., Uno, K., Shiraishi, S., Hongu, T.:* Bull. Chem. Soc. Japan **41**, 2954 (1968).

22 *Houben, J., Kauffmann, H.:* Ber. **46**, 2821 (1913).

23 *Houben, J.:* Ber. **46**, 4001 (1913).

24 *Jovitschitsch, M.:* Ber. **28**, 1213 (1895); **35**, 151 (1902); **39**, 784 (1906).

25 *Semper, L., Lichtenstadt, L.:* Ann. Chem. **400**, 302 (1913).

acid halides *13*. In some cases, e.g., $R = C_6H_5$, or o-ClC_6H_4, using this mixture (which analyses correctly for *13*!) directly in the dehydrochlorination reaction (Eq. (6)) produced nitrile oxides in moderate to good yields (40–60%), while the isolation of a crystalline constant melting *13* gave an essentially quantitative yield of the nitrile oxide. This has been interpreted as an indication that only one stereoisomer is capable of neat nitrile oxide formation. If it is permissible to draw analogies to the dehydrogenation of aldoximes (see preceding paragraph) one would assume that it is the *syn* isomer of *13* which provides the nitrile oxide.[26, 27] This assumption seems to be supported by NMR-studies of the equilibrium between hydroximic acid chlorides and nitrile oxides (+ HCl).[27a] But these so far isolated observations need further study.

Hydroximic acid chlorides are permanently stable at room temperature and are therefore conveniently storable precursors of the generally unstable nitrile oxides, which can be generated from them when needed almost instantaneously by action of base, the reaction being a neat dehydrochlorination (Eq. (6)):[11]

$$ \underset{\underset{Cl}{|}}{R\!-\!C}\!=\!NOH \quad \xrightarrow{\text{Base}} \quad R\!-\!C\!\equiv\!N\!\longrightarrow\!O \quad + \quad \text{Base}\cdot HCl \qquad (6)$$

Any base, inorganic or organic, in either an aqueous or anhydrous medium may be used. Earlier investigators preferred aqueous sodium carbonate and by this method, the first nitrile oxide, benzonitrile oxide, was prepared.[28] More recently, the reaction of the hydroximic acid chloride, dissolved or suspended in an inert organic solvent, e.g., anhydrous ether, with one equivalent of a tertiary organic base, preferably triethylamine, has been recommended.[29] Using this technique at temperatures as low as $-40°$ C, even the most unstable aliphatic nitrile oxides have been successfully prepared.[11, 30] Tertiary amines as dehydrohalogenating agents may, however, not be applicable in all cases, since they seem to form with some hydroximic acid chlorides rather stable addition compounds, which are apparently quaternary amidoximidinium salts[31–33] (see also Chapter VI, G).

26 *Babad, H.:* The Ott Chemical Company, Muskegon, Michigan, private communication.

27 *Grundmann, C., Dean, J.M.:* Unpublished.

27a *Guetté, J.P., Armand, J., Lacombe, L.:* Compt. Rend. **264**, C, 1509 (1967).

28 *Werner, A., Buss, H.:* Ber. **27**, 2193 (1894).

29 *Huisgen, R., Mack, W.:* Tetrahedron Letters 583 (1961).

30 *Zinner, G., Günther, W.:* Angew. Chem. **76**, 440 (1964).

31 *Wieland, H., Höchtlen, A.:* Ann. Chem. **505**, 237 (1933).

32 *Quilico, A., Gaudiano, G., Ricca, A.:* Gazz. Chim. Ital. **87**, 638 (1957).

33 *Grundmann, C., Mini, V., Dean, J.M., Frommeld, H.D.:* Ann. Chem. **687**, 191 (1965).

But the main limitations of this synthesis lie in the first step of the process, i.e., the chlorination of the aldoximes. Unsaturated aldoximes add chlorine to the double bond, e.g., from 1-oximino-2-methyl-butene-2 (*14*) 2,3-dichloro-2-methyl-butane-hydroximic acid chloride (*15*) and from 1-oximino-2,2-dimethyl-pentene-4 (*16*) 4,5-dichloro-2,2-dimethyl-pentanehydroximic acid chloride (*17*) are obtained[34]

$$CH_3 \cdot CH{=}C(CH_3)CH{=}NOH \xrightarrow{Cl_2} CH_3 \cdot CHCl \cdot CCl(CH_3) \cdot CCl{=}NOH$$

$$\textit{14} \qquad\qquad\qquad\qquad\qquad\qquad \textit{15}$$

$$H_2C{=}CH \cdot CH_2 \cdot C(CH_3)_2CH{=}NOH \xrightarrow{Cl_2} ClCH_2 \cdot CHCl \cdot CH_2C(CH_3)_2CCl{=}NOH$$

$$\textit{16} \qquad\qquad\qquad\qquad\qquad\qquad \textit{17}$$

$$H_3C{-}\underset{\underset{O}{\|}}{C}{-}CH{=}NOH \xrightarrow{Cl_2} ClCH_2{-}\underset{\underset{O}{\|}}{C}{-}CH{=}NOH$$

$$\textit{18} \qquad\qquad\qquad\qquad\qquad\qquad \textit{19}$$

Methylglyoxaldoxime (*18*) is not chlorinated to the expected hydroximic chloride, but to the isomeric *19*.[34a] Salicylaldoxime (*20*) and 2-methoxybenzaldoxime (*21*) cannot be transformed into the corresponding hydroximic acid chlorides, but products of additional chlorination in the benzene ring are obtained, i.e., 3,5-dichloro-2-hydroxybenzhydroximic acid chloride (*22*) and 5-chloro-2-methoxy-benzhydroximic acid chloride (*22a*).[34] Likewise, the chlorination of the oximes of mesitylaldehyde (*23*) or 2,3,5,6-tetramethyl-benzaldehyde leads to inseparable mixtures of the desired hydroximic acid chlorides and products of further chlorination of the aromatic ring and of the methyl groups.[1] But by using three moles of chlorine *23* is neatly converted into 2,4,6-trimethyl-3,5-dichlorobenz-hydroximic acid chloride[35] (*24*). For a further example of such side reactions see p. 48.

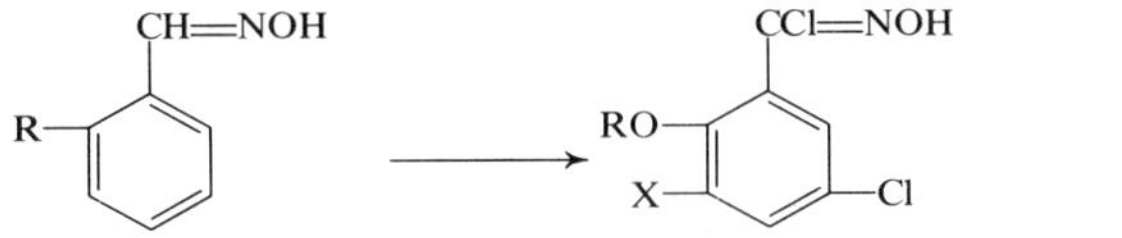

20, R = OH
21, R = OCH₃

22, R = H,   X = Cl
22a, R = CH₃, X = H

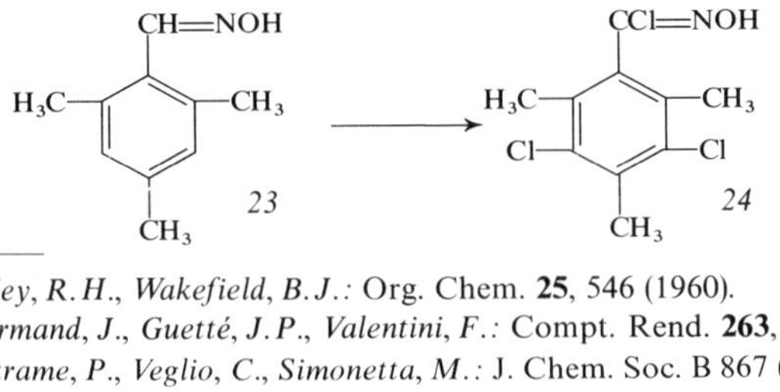

23

24

34   *Wiley, R. H., Wakefield, B. J.:* Org. Chem. **25**, 546 (1960).
34a   *Armand, J., Guetté, J. P., Valentini, F.:* Compt. Rend. **263**, C, 1388 (1966).
35   *Beltrame, P., Veglio, C., Simonetta, M.:* J. Chem. Soc. B 867 (1967).

Nonetheless, the route via the hydroximic acid chlorides is the one most widely applied in the preparation of nitrile oxides and it has been successfully used in the synthesis of aliphatic, aromatic, and heterocyclic nitrile oxides. It has been found especially valuable for the technique of generating nitrile oxides in situ (see Chapter III, D).

2. From Primary Nitroparaffins

Via the Conversion to Hydroximic Acid Chlorides. Primary nitroparaffins, in the form of their aci-salts, react with hydrogen chloride under anhydrous conditions to form hydroximic acid chlorides (Eq. (7)),[36] the conversion of which into nitrile oxides has been discussed in the preceding paragraph:

$$R-CH_2NO_2 \xrightarrow{\ ^{\ominus}OH\ } R-CH=N-O^- \xrightarrow{\ 2\,HCl\ } R-\underset{\underset{Cl}{|}}{C}=NOH \qquad (7)$$

$$+ \quad H_2O \quad + \quad Cl^{\ominus}$$

$$R = CH_3,\ C_6H_5,\ C_6H_5CO,\ C_2H_5OOC$$

Hydroximic acid chlorides are generally more easily accessible by the routes discussed above, and there are now better methods to convert primary nitroparaffins into nitrile oxides. These are discussed below.

Via the Decomposition of Nitrolic Acids. Nitrolic acids (*24a*) are obtained by reaction of primary nitroparaffins with nitrous acid, or, less frequently, by the nitration of aldoximes (Eq. (8)).

$$\begin{array}{c} R-CH_2NO_2 \quad + \quad HNO_2 \\[1.2em] R-CH=NOH \quad + \quad N_2O_4 \end{array} \searrow\!\!\!\!\nearrow\ R-\underset{\underset{NO_2}{|}}{C}=NOH \longrightarrow R-C\equiv N\longrightarrow O \ +\ HNO_2 \qquad (8)$$

$$24a$$

All nitrolic acids easily lose the elements of nitrous acid with the formation of nitrile oxides. Sometimes this decomposition occurs spontaneously at room temperature; generally, it is induced by gentle heating.

The applicability of this reaction has been only slightly investigated, but benzonitrile oxide (R = C_6H_5) and mesitonitrile oxide (R = 2,4,6(CH_3)$_3C_6H_2$) were obtained in this manner.[5, 10a, 11]

---

36 *Steinkopf, W., Jürgens, B.*: J. prakt. Chem. [2] **84**, 686 (1911).

The three known isomers of phenylglyoxime (25; syn, anti, amphi-(antiphenyl)) are oxidized by dinitrogen tetroxide, presumably via the intermediate nitrolic acid, to the same 4-phenyl-furoxan (26; m.p. 111–112°) which is isomerized by mild treatment with alkali to oximino-phenylacetonitrile oxide (27).[44] Most of the earlier work on 26 and 27 is erroneous, at least in interpretation.[37-43] None of the previous investigators started with a pure isomer of 25; additional difficulties arose from the fact that both compounds have almost identical melting points and that 26 gives the same reactions with HCl or aniline as would be expected from a true nitrile oxide (see Chapter VI, F and G). The final structure proof for 26 and 27 rests therefore mainly on spectroscopical data.[44,45]

$$C_6H_5\text{—}C(NOH)\text{—}CH\text{=}NOH \xrightarrow{\ \ N_2O_4\ \ } [C_6H_5\text{—}C(NOH)\text{—}\underset{NO_2}{\overset{|}{C}}\text{=}NOH] \longrightarrow$$

25

$$\xrightarrow{\ -HNO_2\ } C_6H_5\text{—}\underset{\underset{O}{N\diagdown\ \diagup N\to O}}{\overset{|}{C}\text{——}\overset{|}{C}H} \xrightarrow{\ ^-OH\ } C_6H_5\text{—}\underset{NOH}{\overset{\|}{C}}\text{—}C\text{≡}N\to O$$

26   27

Homologs and their corresponding o-benzoyl derivatives have been obtained analogously.[40,43]

Via the Dehydration by Isocyanates. The modes of formation of nitrile oxides from primary nitroparaffins, as discussed above, represent in the final analysis a dehydration of the nitroparaffin. Recently, a method has become available which achieves this reaction in one step under rather mild conditions using phenylisocyanate as the dehydrating agent in the presence of catalytic amounts of triethylamine.[46-49] The original

37 *Scholl, R.:* Ber. **23**, 3505 (1890).

38 *Wieland, H., Semper, L.:* Ann. Chem. **358**, 36 (1908).

39 *Wieland, H.:* Ann. Chem. **424**, 107 (1921).

40 *Ponzio, G.:* Gazz. Chim. Ital. **66**, 119, 127, 134 (1936); **53**, 379 (1923).

41 *Avogadro, L.:* Gazz. Chim. Ital. **53**, 824 (1923).

42 *Ponzio, G., Ruggerio, G.:* Gazz. Chim. Ital. **56**, 733 (1926).

43 *Ponzio, G.:* Gazz. Chim. Ital. **61**, 561, 570, 572 (1931); **71**, 693 (1941).

44 *Burakevich, J.V., Lore, A.M., Volpp, G.P.:* Abstracts of the 5th Middle Atlantic Regional Meeting of the Am. Chem. Soc., April 1970, Newark Del., p. 70, 73.

45 *Borello, E., Colombo, M.:* Ann. Chim. (Rome) **46**, 1158 (1956).

46 *Hoshino, T., Mukaiyama, M.:* Japan Pat. 9855 (1959); — Chem. Abstr. **54**, 7738h (1960).

47 *Mukaiyama, T., Hoshino, T.:* J. Am. Chem. Soc. **82**, 5339 (1960).

48 *Eloy, F.:* Bull. Soc. Chim. Belges **73**, 793 (1964).

49 *Vita-Finzi, P., Grünanger, P.:* Chim. Ind. (Milan) **47**, 516 (1965).

authors have suggested the following mechanism (Eq. (10)):

$$
\begin{aligned}
R{-}CH_2{-}NO_2 &\xrightarrow{(C_2H_5)_3N} R{-}CH{=}NO_2^{\ominus} \longrightarrow \\[4pt]
\xrightarrow{C_6H_5{-}N{=}C{=}O}\ & R{-}CH{=}\underset{\downarrow O}{N}{-}O{-}\underset{\parallel O}{C}{-}\overset{\ominus}{N}{-}C_6H_5 \longrightarrow \\[4pt]
\xrightarrow{H^{\oplus}}\ & R{-}CH{=}\underset{\downarrow O}{N}{-}O{-}\underset{\parallel O}{C}{-}NH{-}C_6H_5 \longrightarrow \\[4pt]
\longrightarrow\ & R{-}C{\equiv}N \rightarrow O \ +\ HOOC{-}NH{-}C_6H_5 \longrightarrow \\[4pt]
\xrightarrow{C_6H_5N{=}C{=}O}\ & C_6H_5{-}NH{-}\underset{\parallel O}{C}{-}NH{-}C_6H_5 \ +\ CO_2
\end{aligned}
\tag{10}
$$

The reaction proceeded well with nitroethane, 1-nitropropane, and phenylnitromethane and has since been applied successfully to a number of other nitroparaffins. With nitromethane, however, the reaction took a slightly different course, since nitromethane reacts with phenylisocyanate[50, 52] (see Chapter V, B, p. 102).

Also phosphorus oxychloride has been used for the dehydration of the nitronic acid anions derived from primary nitroalkanes.[51] In some cases, similar conversions as with phenylisocyanate have been obtained, but, in general, this variation seems to give less satisfactory results. It is, however, claimed that this reaction does not provide an organic by-product (symm. diphenylurea), but gives only inorganic water-soluble by-products (HCl and H_3PO_4) which are more easy to separate from the nitrile oxide. An attempt to use acetic anhydride as dehydrating agent gave very poor results. Diketene has been claimed also as a dehydrating agent.[53]

None of the investigators of the above reactions attempted to isolate the nitrile oxides thus prepared, but proved their existence either by dimerization to 1,2,5-oxadiazol-1-oxides (see Chapter IV, C-2) or by 1,3-dipolar cycloaddition with olefins to isoxazoles (see Chapter V, B). It is likely, however, that this method could also be used to prepare individual nitrile oxides of sufficient stability. Since, especially in the lower aliphatic series, the primary nitrocompounds are often more easily accessible today than the corresponding hydroximic acid chlorides, there is no doubt about the usefulness of this method.[54]

---

50 *Paul, R., Tchelitscheff, S.:* Bull. Soc. Chim. France 140 (1963).

51 *Bachmann, G. B., Strom, L. E.:* J. Org. Chem. **28**, 1150 (1963).

52 *Christl, M., Ph. D. Thesis:* Universität München, 1969.

53 *Tsushima, S., Tsujikawa, T., Aki, O.:* Japan. Pat. 24902 (1968); — Chem. Abstr. **70**, 87344 (1969).

54 *Grundmann, C.:* Fortschr. Chem. Forsch. **7**, 62 (1966).

3. From Fulminates

Early attempts to react metal salts of fulminic acid with alkyl or acyl halides were aimed at preparing derivatives of fulminic acid of structure *28*, a still unknown class of compounds (see Chapter III, B). The only identifiable products obtained were the isomeric isocyanates *30* or products derived from these (Eq. (11)) (see Chapter IV, A).[31, 37, 55]

$$R\text{—Hal} \ + \ MeCNO \ \longrightarrow\!\!\!\!\!/ \ R\text{—O—N}\!\!=\!\!Cl$$
$$28$$

$$[R\text{—C}\!\equiv\!N \longrightarrow O] \longrightarrow R\text{—N}\!\!=\!\!C\!\!=\!\!O$$
$$29 \qquad\qquad 30$$

$$R = \text{Alkyl or Acyl,} \ Me = Ag, Hg, Na$$

(11)

The reaction of triphenylmethylchloride with silver fulminate leading in good yield to triphenylacetonitrile oxide $(29, \ R = C(C_6H_5)_3)$ is the only case in which a nitrile oxide has ever been isolated in this type of reaction.[56] If tritylchloride was replaced by diphenylmethyl bromide, the only product which could be isolated was diphenylmethyl isocyanate[31] 2,4,6-Trinitro-chlorobenzene (picryl chloride)—which has a chlorine atom at least as reactive—failed to react at all with silver fulminate, possibly because of excessive steric hindrance.[57]

Steric hindrance, which is undoubtedly present to a large degree in triphenylacetonitrile oxide stabilizes nitrile oxides toward dimerization, but does not prevent rearrangement to isocyanates (see Chapter IV, A). Thus, there is at present no explanation for the apparent stability of this compound under the reaction conditions, but the fact that it is formed may be considered as an indication that nitrile oxides are also the primary reaction products in those reactions where earlier investigators isolated only isocyanates.

55 *Calmels, G.:* Compt. Rend. **99**, 794 (1884). — *Holleman, A.F.:* Ber. **23**, 2998 (1890); — Rec. Trav. Chim. **10**, 65 (1891). — *Nef. J. U.:* Ann. Chem. **280**, 339 (1894).
56 *Wieland, H., Rosenfeld, B.:* Ann. Chem. **484**, 236 (1930).
57 *Grundmann, C., Frommeld, H.D.:* Unpublished.

D. Preparation in situ

The development of the chemistry of the nitrile oxides has been as much advanced as it has been hindered by their extreme reactivity, especially with themselves by polymerization and rearrangement. Only two recent innovations have dealt successfully with this predicament and are to a considerable degree responsible for the fast development of this area in the last 10 years. One is the employment of sterically hindered nitrile oxides, discussed in Chapter I, C, which, however, imposes automatically certain structural requirements. The other is the in situ preparation of the nitrile oxide, a technique first consciously and successfully applied by *Huisgen*.[1-4]

The nitrile oxide is generated from a stable precursor in the presence of a partner with whom it will react about as fast as with itself. In order to minimize the tendency toward autocondensation—in most cases dimerization to the furoxan (see Chapter IV, C-2)—it is advisable to generate the nitrile oxide very slowly, so that the stationary concentration is rather low; this serves to impede polymerization. At the same time, the reaction partner (dipolarophile) should be present in as high a concentration as possible.

Practically all reactions of free fulminic acid studied before, however, were in fact in situ reactions. In this case, the formonitrile oxide was most conveniently generated by acidification of its sodium salt. An improved recent method uses the formhydroximic iodide[5] (Chapter III, B, Eq. (11)).

For the higher homologs, *hydroximic acid chlorides* are most frequently used as precursors of nitrile oxides to be generated in situ. The acid chloride is dissolved or suspended in ether together with the dipolarophile and one mole of triethylamine, or another suitable tertiary amine, is added gradually at 0 to 20° C. In cases where the dipolarophile itself might react with the base, ethereal solutions of the acid chloride and of triethylamine are added simultaneously in equivalent amounts to the solution of the dipolarophile, or the nitrile oxide is first generated at −20° C and then the reaction partner is added while the reaction mixture warms slowly to room temperature.

The same technique has been applied to in situ reaction of nitrile oxides generated *from primary nitroparaffins* by dehydration with phenyl-

1 *Huisgen, R., Mack, W., Anneser, E.:* Angew. Chem. **73**, 656 (1961).

2 *Huisgen, R., Mack, W.:* Tetrahedron Letters 583 (1961).

3 *Huisgen, R.:* Angew. Chem. **75**, 604 (1963).

4 *Grundmann, C.:* Herstellung und Umwandlung von Nitriloxiden. Methoden der Organischen Chemie, 4th ed. (ed. *E. Müller*), vol. 10/3, p. 841–870. Stuttgart: G. Thieme 1965.

5 *Huisgen, R., Christl, M.:* Angew. Chem. **79**, 471 (1967).

isocyanate in the presence of catalytic amounts of triethylamine.[6-8] Here, a solution of the nitro-compound and triethylamine in a suitable inert solvent, e. g., benzene, is added slowly to the solution of the dipolarophile and two moles of phenylisocyanate.

Hydroximic acid chlorides seem generally to be in equilibrium with the corresponding nitrile oxides (see also Chapter VI, F).[9,10] It is obvious that the equilibrium is far on the left side even in dilute solutions and at elevated temperatures. But when the dissociation is carried out in an inert solvent with no solubility for hydrogen chloride, such as boiling toluene, a stationary very low concentration of nitrile oxide is maintained. Thus, in presence of a suitable scavenger for the nitrile oxide, the thermal dissociation of hydroximic chlorides provides a very elegant way of reacting a nitrile oxide in situ.[11-15]

Thermal *decomposition of nitrolic acids* (see this Chapter, Eq. (8)) has also occasionally been used for the generation of unstable nitrile oxides in situ, for example, acetyl fulmide,

$$CH_3—CO—C\equiv N \longrightarrow O,$$

or carbethoxy-fulmide,[16,17,17a]

$$C_2H_5OOC—C\equiv N \longrightarrow O.$$

But the formation of nitrile oxides in situ by thermal degradation of nitrolic acids is most probably responsible for the production of a variety of heterocyclic compounds, mostly isoxazoles and furoxans, which have been observed as the final products resulting from the action of concentrated nitric acid or higher oxides of nitrogen on unsaturated

6 *Mukaiyama, T., Hoshino, T.:* J. Am. Chem. Soc. **82**, 5339 (1960).

7 *Eloy, F.:* Bull. Soc. Chim. Belges **73**, 793 (1964).

8 *Vita-Finzi, P., Grünanger, P.:* Chim. Ind. (Milan) **47**, 516 (1965).

9 *Souchay, P., Armand, J.:* Compt. Rend. **256**, 4907 (1963). — *Souchay, P., Armand, J., Guetté, J.P., Valentini, F.:* Compt. Rend. **262**, [C] 985 (1966).

10 *Armand, J.:* Bull. Soc. Chim. France, 882 (1966). — *Armand, J., Guetté, J.P., Valentini, F.:* Compt. Rend. **263**, [C] 1388 (1966). — *Armand, J., Souchay, P., Valentini, F.:* Bull. Soc. Chim. France 4585 (1968).

11 *Eloy, F., Lenaers, R.:* Bull. Soc. Chim. Belges **72**, 719 (1963).

12 *Lenaers, R., Eloy, F.:* Helv. Chim. Acta **46**, 1067 (1963).

13 *Arbasino, M., Grünanger, P.:* Ric. Sci. **34**, [II A], 561 (1964).

14 *Vita-Finzi, P., Arbasino, M.:* Ric. Sci. **35**, [II A], 1484 (1965).

15 *Sasaki, T., Yoshioka, T.:* Bull. Chem. Soc. Japan **40**, 2604 (1967); **41**, 2206 (1968); **42**, 258 (1969).

16 *Quilico, A., Simonetta, M.:* Gazz. Chim. Ital. **76**, 200 (1946); **77**, 586 (1947).

17 *Jovitschitsch, M.:* Ber. **28**, 1213 (1895).

17a *Biekert, E., Kössel, A.:* Ann. Chem. **662**, 93 (1963).

compounds (see also Chapter V, A, pp. 86–88). It is likely that nitrolic acids will form under such conditions and that the nitrile oxides resulting from them can either add to the unsaturated compound still present, or dimerize to the furoxans. A remarkable example of how uniformly such reactions may proceed is the formation of 3,3'-diisoxazole (*31 a*), from acetylene and a mixture of nitrogen monoxide and nitrogen dioxide in ethyl acetate at 60° C under 10–15 atm pressure[18] (Eq. (12)).

$$HC{\equiv}CH \ + \ 2\,NO \ + \ 2\,NO_2 \ \longrightarrow \ HON{=}\underset{O_2N}{C}{-}\underset{NO_2}{C}{=}NOH \ \longrightarrow$$

$$\xrightarrow{-2\,HNO_2} \quad O{\leftarrow}N{\equiv}C{-}C{\equiv}N{\rightarrow}O \quad \longrightarrow \tag{12}$$

31

$$\xrightarrow{+2\,C_2H_2} \quad \text{31 a}$$

31 a

Although the proposed reaction scheme involves the highly unstable cyanogen-bis-N-oxide(oxalo-bis-nitrile oxide) (*31*) as an intermediate, yields of 60–70 % of *31 a* can be obtained.

In general, however, reactions of this type are beyond the scope of this book; moreover, they are amply covered in several recent summaries.[19–23]

18 *Cramer, R., McClellan, W.R.:* J. Org. Chem. **26**, 2976 (1961).

19 *Barnes, R.A.:* Isoxazoles. Heterocyclic compounds (ed. *R.C. Elderfield*), vol. 5, p. 452–483. New York: John Wiley & Sons 1957.

20 *Boyer, J.H.:* Oxadiazoles. Heterocyclic compounds (ed. *R.C. Elderfield*), vol. 7, p. 462–540. New York: John Wiley & Sons 1961.

21 *Quilico, A.:* Isoxazoles and related compounds. The chemistry of heterocyclic compounds (ed. *A. Weissberger*), vol. 17, p. 1–176. New York: Interscience Publishers 1962.

22 *Behr, L.C.:* Oxadiazoles and related compounds. The chemistry of heterocyclic compounds (ed. *A. Weissberger*), vol. 17, p. 283–319. New York: Interscience Publishers 1962.

23 *Kochetkov, N.K., Sokolov, S.D.:* Recent developments in isoxazole chemistry. Advances in heterocyclic chemistry (ed. *A.R. Katritzky*), vol. 2, p. 365–422. New York: Academic Press 1963.

E. Functional Groups in Nitrile Oxides

The wide variety of functional groups which react spontaneously with
nitrile oxides (as discussed in Chapters V and VI) restrict markedly the
chances of preparing nitrile oxides with functional groups. Since the
$C\equiv N\rightarrow O$ group will easily react with acids, bases, reducing agents,
and is furthermore prone to hydrolysis as well as polymerization, the
chances are slim to introduce another functional group into the molecule,
once the nitrile oxide group has been formed. Indeed, no such attempt
has been recorded so far in the literature. All successful routes to func-
tionally substituted nitrile oxides have started by introducing the desired
group into a suitable precursor and generating the nitrile oxide in the last
step of the synthesis.

Functional groups which are a priori compatible with a nitrile oxide
are halogen (aromatic as well as aliphatic), the nitro group (only realized
in aromatics so far, in o-position $-NO_2$ seems, however, to have a
marked destabilizing effect on $C\equiv N\rightarrow O$, cf., data of Table III, p.17),
the $-SO_2$ group in sulfones, the $-COOR$ group (but not the free acid,
nor its anion), the $R-O-R$ group (in aliphatic and aromatic ethers)—
to some extend (see below)—the tertiary amino group (tested so far as
$-N=$ in various heterocyclic rings and as $Ar-N(CH_3)_2$), and, finally,
the nitrile oxide group itself.

Since the preparation of the first and simplest difunctional nitrile
oxide, oxalo-bis-nitrile oxide (*31*), and the demonstration of its potential
as a building block for new polymers,[1, 2] the synthesis of other difunctional nitrile oxides has attracted a number of chemists. Aside from *31*,
the terephthalo-bis-nitrile oxide (*32*) and isophthalo-bis-nitrile oxide (*33*)
have been obtained. All three nitrile oxides were synthesized from the
aldoximes via the hydroximic acid chlorides. While no difficulties were
reported in the preparation of *33*,[3] in the synthesis of *32*, the first attempts
lead to polymers only,[4] but later investigations demonstrated that with
proper precautions a product can be isolated that consists largely of the
monomer.[5, 6] The wide differences in melting points reported for *32* (see
Table III, p.19) are attributed partly to the various states of crys-
tallinity of the different preparations and partly to polymerization during
slow heating. Since the molecular weight determinations of the purest
samples are still unsatisfactory, these differences may also be caused by
a varying amount of oligomers already present in freshly prepared *32*.

1 *Grundmann, C.*: Angew. Chem. **75**, 450 (1963).
2 *Grundmann, C., Mini, V., Dean, J. M., Frommeld, H. D.*: Ann. Chem. **687**, 191 (1965).
3 *Iwakura, Y., Akiyama, M., Shiraishi, S.*: Bull. Chem. Soc. Japan **38**, 335 (1965).
4 *Wiley, R. H., Wakefield, B. J.*: J. Org. Chem. **25**, 546 (1960).
5 *Eloy, F.*: Bull. Soc. Chim. Belges **73**, 639 (1964).
6 *Overberger, C. G., Fujimoto, S.*: J. Polymer Sci., B, Polymer Letters **3**, 735 (1965).

Two sterically hindered bifunctional nitrile oxides *34* and *35* which were prepared by the hypobromite oxidation of the corresponding aldoximes, are, however, well characterized compounds, stable indefinitely at room temperature.[7, 8] An attempt to prepare the sterically hindered trifunctional nitrile oxide *36*, proceeded well to the corresponding trisoxime, but its dehydrogenation with either hypobromite or N-bromosuccinimide failed, presumably because not all of the oximino-groups had the required *syn* configuration, resp. would not rearrange to it under the applied conditions.[9]

$$O \leftarrow N \equiv C - C \equiv N \rightarrow O$$

31

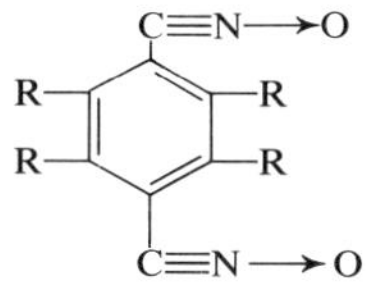

32, R = H
34, R = CH$_3$

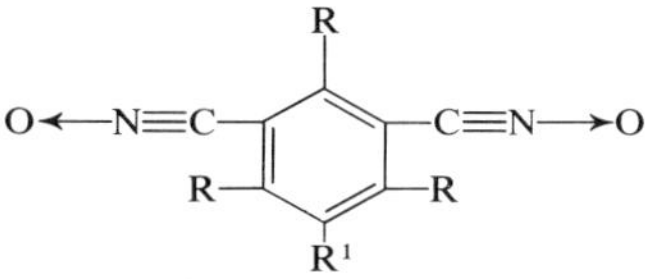

33, R and R^1 = H
35, R = CH$_3$, R^1 = H
36, R = CH$_3$, R^1 = C≡N → O

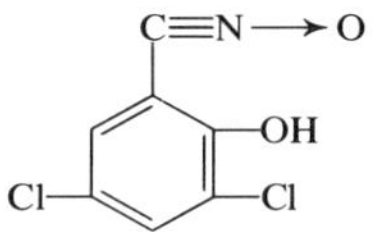

37

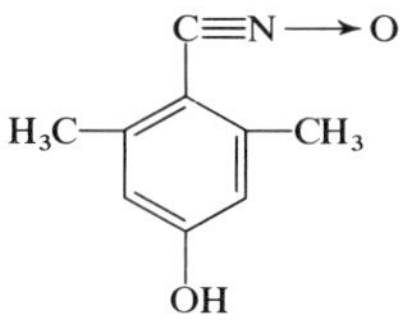

38

Since aromatic nitrile oxides do not react readily with alcoholic or phenolic groups unless catalyzed by strong bases or acids, (see Chapter VI, D), it should be possible to synthesize nitrile oxides containing these functional groups. The nitrile oxide *37*, obtained from the corresponding hydroximic acid chloride, was isolated only as a dilute solution in CCl$_4$ and identified by its infrared spectrum and by subsequent dimerization to the furoxan.[4]

An attempt to generate *38* from the corresponding aldoxime by dehydrogenation with hypobromite failed because of the preferred attack of the oxidant on the ring.[9] Once the phenolic group is alkylated it is entirely compatible with the nitrile oxide function as demonstrated by the preparation of a number of methoxy-substituted nitrile oxides listed in Table III, pp. 18–20. Other methoxylated aromatic nitrile oxides have been characterized only in solution by IR spectroscopy.[4] The compati-

7 *Grundmann, C., Dean, J. M.:* Angew. Chem. **76**, 682 (1964); — J. Org. Chem. **30**, 2809 (1965).

8 *Grundmann, C., Richter, R.:* J. Org. Chem. **33**, 476 (1968).

9 *Grundmann, C., Dean, J. M., Richter, R.:* Unpublished.

bility of the oximino group, $>$C=NOH, with the nitrile oxide function is proven by several examples, discussed already in paragraph C-2 of this chapter.

The stability of nitrile oxides with free carboxylic groups depends largely on the inherent acidity of the carboxyl function, since the attack of the latter on the C$\equiv$N$\rightarrow$O group starts with protonation to the conjugate acid. Thus, sterically hindered aromatic nitrile oxides can be recrystallized unchanged from hot acetic acid,[10] but formic acid causes hydrolysis to the corresponding hydroxamic acid.[11] (See also Chapter VI, C and F.) This explains the failure to convert 4-formoximino-3,5-dimethylphenoxyacetic acid (*39*) into the corresponding nitrile oxide by hypobromite oxidation; only polymeric products were obtained since the phenoxyacetic acids are generally strong organic acids.[9]

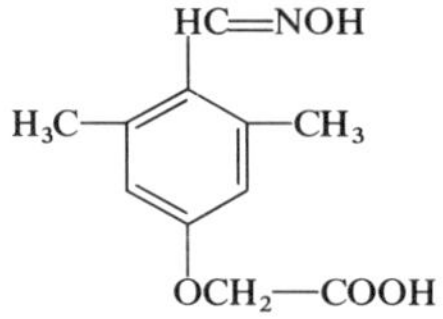

39

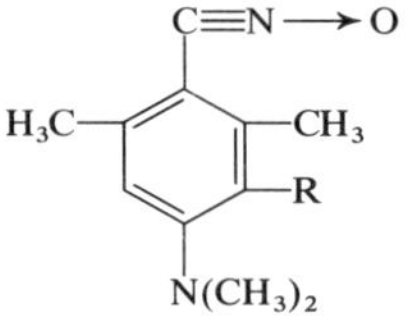

40, R = H
41, R = C$\equiv$N $\rightarrow$ O

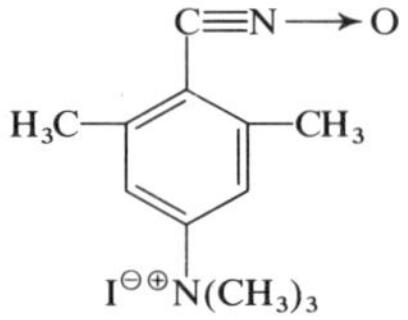

42

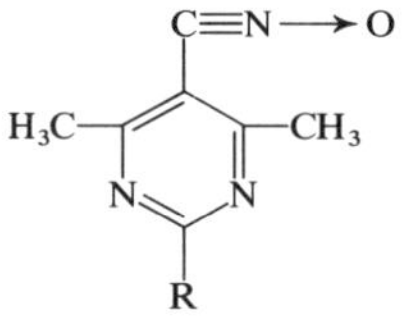

43, R = OCH$_3$
44, R = N(CH$_3$)$_2$

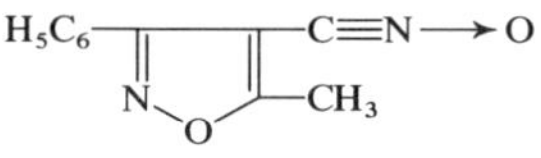

45

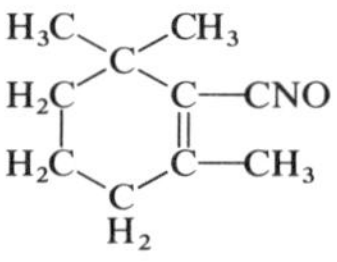

46

The ease with which nitrile oxides react with primary and secondary amines makes it unlikely that these functional groups will ever be found capable of existing together with a nitrile oxide group in a monomeric molecule, but tertiary amines seem to form stable adducts only in certain cases (see also Chapter VI, G). Thus, the 4-dimethylamino-2,6-dimethyl-benzonitrile oxide (*40*) and the 4-dimethylamino-2,6-dimethylisophthalo-bis-nitrile oxide (*41*) could be prepared by oxidation of the corresponding aldoximes with hypobromite or, preferably with N-bromosuccinimide.[8, 12, 13] The nitrile oxide *40* was converted with methyl iodide in

10 *Grundmann, C., Frommeld, H. D.:* J. Org. Chem. **31**, 157 (1966).

11 *Overberger, C. G.:* Private communication.

12 *Grundmann, C., Richter, R.:* J. Org. Chem. **32**, 2308 (1967).

13 *Yamakawa, M., Kubota, T., Akazawa, H., Tanaka, I.:* Bull. Chem. Soc. Japan **41**, 1046 (1968).

benzene into the quaternary ammonium iodide *42*, the first example of a water-soluble nitrile oxide. While *42* could be recrystallized unchanged from 95 % ethanol, its aqueous solutions are only stable at 0° for several days. At 25°, quick decomposition occurs, the nitrile oxide being reduced to the nitrile by the iodide ion.

Attempts to prepare pyrrolo- or pyridino-nitrile oxides have so far failed,[9, 14] but in the pyrimidine series a number of nitrile oxides, stabilized by steric hindrance, e. g., *43* and *44* were obtained by hypobromite or N-bromosuccinimide dehydrogenation of the corresponding aldoximes.[8, 12] The isoxazolo-nitrile oxide *45* was prepared by dehydrohalogenation of the hydroximic acid chloride.[15]

Nitrile oxides react generally with ethylenic double bonds (see Chapter V, B), but the reaction rate decreases sharply with increasing substitution.[16, 17] Thus, it was possible to obtain the sterically hindered, α,β unsaturated nitrile oxide *46* which was permanently stable at room temperature.[18]

The above discussion relates mainly to isolated or isolable nitrile oxides, however, if the nitrile oxide is generated *in situ* (see Section D), it may be reacting faster with an offered partner (e.g., a dipolarophile) then with functional groups present on the nitrile oxide molecule itself. Thus, a few more functional groups may be compatible with the nitrile oxide function. (For examples see the tables of Chapter IX).

14 *Chang, M.S., Matuszko, A.J.:* J. Org. Chem. **28**, 2260 (1963).

15 *Quilico, A., Speroni, G.:* Gazz. Chim. Ital. **76**, 148 (1946).

16 *Grundmann, C.:* Fortschr. Chem. Forsch. **7**, 62 (1966).

17 *Grundmann, C., Frommeld, H.D., Flory, K., Datta, S.K.:* J. Org. Chem. **33**, 1464 (1968).

18 *Grundmann, C., Datta, S.K.:* J. Org. Chem. **34**, 2016 (1969).

IV. Chemical Reactions Involving Only the Nitrile Oxide

A. Isomerization to Isocyanates

Experimental. Heating a nitrile oxide above its limits of thermal stability initiates generally two different, competing reactions: a) The dimerization to 1,2,5-oxadiazoles (furoxans) which are stable compounds with exception of the parent member; therefore, the polymerization of fulminic acid takes a different course (see Sections C-1 and C-2). b) The rearrangement to isocyanates. (Eq. (1))

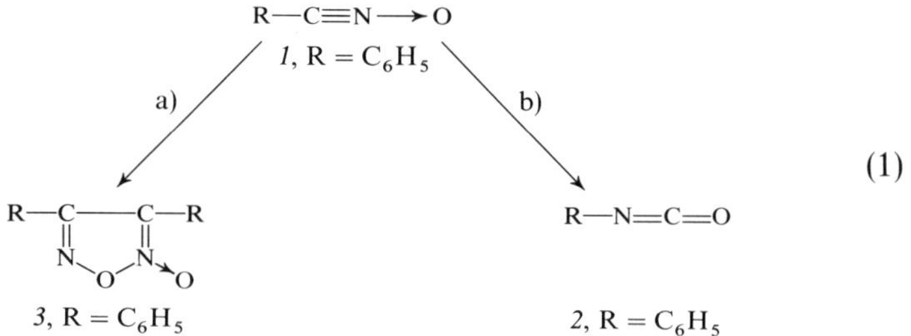

$$\tag{1}$$

From the few qualitative studies available, it seems that reaction pathway a) predominates with increasing thermal instability of the nitrile oxide, thus reaction b) has never been observed with lower aliphatic nitrile oxides. On the other hand, aromatic nitrile oxides of moderate stability will at room temperature follow route a) almost quantitatively, while on rapid heating route b) will be favored to some extent. This behavior strongly suggests quite different activations energies for both reactions. Quantitative studies are sorely needed.

Heating benzonitrile oxide (*1*) in xylene solution quickly to 110° results in ~10% conversion to phenylisocyanate (*2*) while the rest dimerizes to diphenylfuroxan (*3*).[1, 16] When distilled under atmospheric pressure, the latter is also converted into *2*, presumably after initial depolymerization to *1*.[2, 3]

1 *Wieland, H.:* Ber. **42**, 803, 4207 (1909).
2 *Gabriel, S., Koppe, M.:* Ber. **19**, 1145 (1886).
3 *Auwers, K., Meyer, V.:* Ber. **21**, 784, 804 (1888); **22**, 705, 716 (1889).

If the apparently either kinetically or thermodynamically preferred dimerization is made difficult or nearly impossible by proper steric hindrance around the CNO-group, aliphatic as well as aromatic nitrile oxide follow pathway b) to isocyanates with excellent to quantitative yields.[4, 5] At temperatures between 110 and 140°, this reaction is complete in less than one hour. The isocyanates thus formed were mostly identified, following reaction with aniline, as substituted diaryl ureas. The sterically hindered nitrile oxides Nos. 5, 7, 8, 21, 23, 25, 27, 46, 48 of Table III have thus been rearranged.

The polymeric nitrile oxides which are obtained from nitrolic acids by reaction with weak aqueous alkali (see Section C-2) undergo the same thermolysis as the dimers, the furoxans. Presumably without initial depolymerization to the monomeric nitrile oxides, isocyanates are obtained in good yields.[1, 6]

As mentioned briefly in Chapter III, C-3, the reaction of silver and mercury fulminate with alkyl or acyl halides did not produce the expected nitrile oxides—with one exception discussed loc. cit.—but led instead to the corresponding isocyanates (Eq. (2), p. 64).

Silver fulminate and ethyl iodide in boiling ether resulted after some time in destructive explosions; but when the reaction mixture was kept < 10° for three weeks, ethyl isocyanate and its trimer 4 were obtained in fair yield.[7] Acetyl chloride reacted in ligroin or ether with mercury fulminate to acetyl isocyanate in 50–55 % yield; isocyanic acid was obtained as a minor by-product (10 %) and, after reaction with ethanol, was identified as ethyl carbamate (5).[8] This is the only case on record where the expected isomerization of fulminic to isocyanic acid has been observed; under most conditions, the polymerization is apparently much faster. Analogously, when benzoyl chloride was reacted with mercury fulminate at ∼10° for 5–7 days, and the reaction mixture subsequently decomposed with water, symm. dibenzoyl-urea (6) was obtained as the expected hydrolysis product of the initially formed benzoylisocyanate.[9] 6 was also obtained under similar conditions from silver fulminate and benzoyl bromide. Finally, the reaction of silver fulminate with benzhydryl bromide in benzene yielded benzhydryl isocyanate (7).[10] A similar rearrangement is obviously occuring in the long ago reported reaction

4 *Grundmann, C., Dean, J. M.:* Angew. Chem. **76**, 682 (1964); — J. Org. Chem. **30**, 2809 (1965).

5 *Grundmann, C., Datta, S. K.:* J. Org. Chem. **34**, 2016 (1969).

6 *Wieland, H.:* Ber. **42**, 816 (1909).

7 *Nef, J. U.:* Ann. Chem. **280**, 339 (1894).

8 *Scholl, R.:* Ber. **23**, 3505 (1890).

9 *Holleman, A. F.:* Ber. **23**, 2998 (1890); — Rec. Trav. Chim. **10**, 70 (1891).

10 *Wieland, H., Höchtlen, A.:* Ann. Chem. **505**, 237 (1933).

of cupric fulminate with hydrogen sulfide which gave thiocyanic acid.[11] Besides, the older literature cites numerous examples where urea, resp. its derivatives have been observed in the alkaline decomposition of metal fulminates.[12]

a) C_2H_5—I $+$ AgCNO $\longrightarrow$ C_2H_5—N=C=O $+$

$$\text{(cyclic trimer)}\quad \mathbf{4}$$

b) CH_3—COCl $+$ Hg(CNO)$_2$ $\longrightarrow$ CH_3—$\overset{\displaystyle O}{\underset{\displaystyle \parallel}{C}}$—N=C=O $+$

H—N=C=O $\xrightarrow{\;C_2H_5OH\;}$ C_2H_5—O—$\overset{\displaystyle}{\underset{\displaystyle \parallel O}{C}}$—NH$_2$ (2)

$$\mathbf{5}$$

c) C_6H_5—COCl $+$ Hg(CNO)$_2$ $\longrightarrow$ [C_6H_5—C(=O)—N=C=O] $\longrightarrow$

$\xrightarrow{\;H_2O\;}$ C_6H_5—C(=O)—NH—C(=O)—NH—C(=O)—C_6H_5

$$\mathbf{6}$$

d) $(C_6H_5)_2CH$—Br $+$ AgCNO $\longrightarrow$ $(C_6H_5)_2CH$—N=C=O

$$\mathbf{7}$$

There is some evidence for the assumption that the isocyanate isolated in the above-cited reactions might have resulted from the subsequent rearrangement of the initially formed nitrile oxides. There is, however, another possibility. All isocyanates where obtained either from silver or mercuric fulminate. It is well known that in the analogous reaction of metal cyanides with alkyl halides, the "normal" products (i.e., the nitriles) are formed in good yields only from the alkali cyanides, whereas silver cyanide gives in the same reaction, often almost exclusively, the "anormal" products, the isocyanides. It should be, therefore, interesting to study again the reaction of alkyl halides with alkali fulminates, and to trap the transient nitrile oxides either by 1,3-dipolar cycloaddition to a suitable dipolarophile or to identify them as their stable dimers, the furoxans.

11 *Gladstone, J. H.:* Ann. Chem. **66**, 1 (1848).

12 *Wieland, H.:* Die Knallsäure. Sammlung chemischer und chemisch-technischer Vorträge (ed. *F. B. Ahrens*), vol. 14, p. 385–461. Stuttgart: F. Enke 1909.

Mechanism. The mechanism of the *thermal* rearrangement of nitrile oxides to isocyanates is unknown. It is, however, tempting to speculate that it might occur via the sequence of Eq. (3), formation of an oxazirene 8 which opens to the acylnitrene 9, followed by the well known migration of R to the stable isocyanate:

$$R\!-\!C\!\equiv\!N\!\longrightarrow\!O \quad\longrightarrow\quad R\!-\!C\overset{O}{\diagup\!\diagdown}N \quad\longrightarrow\quad R\!-\!\overset{O}{\overset{\|}{C}}\!-\!\overline{N} \quad\longrightarrow\quad R\!-\!N\!=\!C\!=\!O$$

$$8 \qquad\qquad 9$$

$$\Big\downarrow C_6H_{10} \tag{3}$$

$$C\overset{O}{\diagup\!\diagdown}N \qquad\qquad R\!-\!\overset{}{\underset{O}{\overset{\|}{C}}}\!-\!N$$

$$8a \qquad\qquad\qquad 10$$

The INDO calculation of the fulminate-isocyanate rearrangement indicates a cyclic intermediate $(8a)$.[13] The actual formation of (8), however, as a discrete step of the overall reaction would violate the *Woodward-Hoffman* rules.[14]

Attempts to trap 9, by thermal isomerization of 2,4,6-trimethyl benzonitrile oxide (11) in cyclohexene have, however, failed to produce any of the expected acylaziridine 10 or of compounds derived from it. Similar experiments with other nitrile oxide have been likewise unsuccessful. Also, attempts to monitor the thermal isomerization of 11 by means of either ultraviolet or infrared spectroscopy have so far failed to produce evidence for the transient existence of the oxazirene 8.[15]

The possibility that the *thermal* rearrangement occurs via an ionic mechanism (ion pair or separated ions) (Eq. (4)) can probably be

$$R\!-\!C\!\equiv\!\overset{(+)}{N}\!-\!\overline{\underline{O}}|^{(-)} \quad\longrightarrow\quad [R^{(+)}\cdot[CNO]^{(-)}] \quad\longrightarrow\quad R\!-\!N\!=\!C\!=\!O \tag{4}$$

excluded. The thermal rearrangement of several nitrile oxides was investigated in which R represented bicyclic carbon systems, e. g., of the nonbornyl or camphenilyl type, whose propensity as carbonium ions to undergo a Wagner-Meerwein skeletal rearrangement is well known. The obtained isocyanates, respective their derivatives (symmetrical ureas), showed in the mass spectrometer a disintegration pattern compatible only with the unrearranged skeleton R. Furthermore, two

13 *Holsboer, F.J., Beck, W.:* Chem. Commun. 262 (1970). — *Beck, W.:* Chem. Ber. **95**, 341 (1962).

14 *Schmitz, E.:* Private communication.

15 *Grundmann, C., Richter, R.:* Unpublished; *Grünanger, P.:* Unpublished.

isotopically tagged benzonitrile oxides, *11a* and *11b* were mixed and subjected to thermal rearrangement. The formed isocyanate was isolated as the phenyl-o-tolyl-urea and subjected to analysis by mass spectrography. The obtained disintegration pattern shows only the species *12a* and *12b*, formed by intramolecular reaction, and excludes *13a* or *13b*, which would result from cross-over of the tag (intermolecular reaction) with 98% certainty.[16]

$$C_6H_5\text{---}\overset{13}{C}\equiv N\rightarrow O \quad + \quad p\text{---}D\text{---}C_6H_4\text{---}C\equiv N\rightarrow O$$

$$\textit{11a} \qquad\qquad\qquad\qquad \textit{11b}$$

1. V
2. $o\text{-}CH_3\text{---}C_6H_4\text{---}NH_2$

$$C_6H_5\text{---}NH\text{---}\overset{13}{\underset{O}{C}}\text{---}NH\text{---}C_7H_7 \qquad\qquad p\text{---}D\text{---}C_6H_4\text{---}NH\text{---}\underset{O}{C}\text{---}NH\text{---}C_7H_7$$

$$\textit{12a} \qquad\qquad\qquad\qquad\qquad \textit{12b}$$

$$C_6H_5\text{---}NH\text{---}\underset{O}{C}\text{---}NH\text{---}C_7H_7 \qquad\qquad p\text{---}D\text{---}C_6H_4\text{---}NH\text{---}\overset{13}{\underset{O}{C}}\text{---}C_7H_7$$

$$\textit{13a} \qquad\qquad\qquad\qquad\qquad \textit{13b}$$

14 $\xrightarrow{h\nu}$ *15* $\xleftarrow[-N_2]{h\nu}$ *16*

17 $\xrightarrow{h\nu}$ *18*   +   *19*

---

16 *Grundmann, C., Kochs, P.:* Angew. Chem. **82**, 637 (1970).

But there is strong evidence that the *photochemical* rearrangement of nitrile oxides to isocyanates proceeds along the pathway of equation 3.[17] Photolysis of the sterically hindered stable nitrile oxides *17* or o-methyl-podocarponitrile oxide (*14*) with ultraviolet light (peak intensity 2537 Å) in 0.4% solution in pentane, hexane or methanol gave yields of 20–50% of the five-membered lactones *15* and *18* which are obviously formed by insertion of the intermediate nitrene into a sterically favored methyl group. In the photolysis of *17* an equimolecular amount of the mesityl-isocyanate *19* (isolated from irradiation in methanol as the methyl carbamate) was obtained, while in the case of *14*, no other products than *15* were identified. The possibility that *15* is actually formed via the acyl-nitrene was strongly supported by the synthesis of *15* through the photo-lysis of the corresponding acyl azide *16*.[18]

17 *Just, G., Zehetner, W.:* Tetrahedron Letters 3389 (1967).
18 *ApSimon, J. W., Edwards, O. E.:* Can. J. Chem. **40**, 896 (1962).

B. Deoxygenation to Nitriles

As other N-oxides, all nitrile oxides are easily deoxygenated, whereby the corresponding nitrile is formed (Eq. (5)).

$$R—C\equiv N\rightarrow O \quad\longrightarrow\quad R—C\equiv N \tag{5}$$

Earlier investigators used acetic acid and zinc dust or tin and hydrochloric acid.[1, 2] Deoxygenation is also effected by isocyanides which are converted into isocyanates,[3, 4] although the reaction is slow and complications arise with unstable nitrile oxides by the competing dimerization to furoxans (see Section C-2). Dicobalt-octacarbonyl, $Co_2(CO)_8$ reduces mesitonitrile oxide already at room temperature to the corresponding nitrile, the reducing agent is not only CO, but also Co^0.[10]

The most effective method is the reaction of the nitrile oxide with a trivalent phosphorus compound. Trialkyl- and triaryl-phosphines or trialkoxy phosphines (trialkyl phosphites) are equally satisfactory.[5, 6] The reaction is fast and quantitative, often already at $0°$, and also very specific. The reagent will not attack the dimeric nitrile oxide *3* (furoxan) at a comparable rate, although under more stringent conditions the latter are deoxygenated too to the corresponding 1,2,5-oxadiazoles (Eq. (6))[7]

$$R—C{=\!=\!=}C—R \;+\; PX_3 \quad\longrightarrow\quad R—C{=\!=\!=}C—R \;+\; O{=}PX_3 \tag{6}$$

This reaction is therefore especially suited for the analytical determination of nitrile oxides, and has been found useful even with the very sensitive oxalo-bis-nitrile oxide (cyanogen-bis-N-oxide).[7, 8] More recently the method has been used to study the kinetics of the dimerization of nitrile oxides. (See Section C-2)

This synthesis of nitriles has only occasionally been used for preparative purposes,[9] its main value lies in the analysis and structure elucidation of nitrile oxides.

1 *Wieland, H.:* Ber. **40**, 1667 (1907); — *Wieland, H., Rosenfeld, B.:* Ann. Chem. **484**, 236 (1930).

2 *Ponzio, G.:* Gazz. Chim. Ital. **53**, 379 (1923); **61**, 561 (1931); **66**, 127 (1936). — *Avogadro, L.:* Gazz. Chim. Ital. **53**, 824 (1923).

3 *Olofson, R. A., Michelman, J. S.:* J. Am. Chem. Soc. **86**, 1863 (1964).

4 *Vita-Finzi, P., Arbasino, M.:* Tetrahedron Letters 4645 (1965).

5 *Grünanger, P., Langella, M.R.:* Atti Accad. Naz. Lincei, Rend., Classe Fis. Mat. Nat. [8] **36**, 387 (1964).

6 *Grundmann, C., Frommeld, H.-D.:* J. Org. Chem. **30**, 2077 (1965).

7 *Grundmann, C.:* Chem. Ber. **97**, 575 (1964).

8 *Grundmann, C., Mini, V., Dean, J. M., Frommeld, H.-D.:* Ann. Chem. **687**, 191 (1965).

9 *Yamakawa, M., Kubota, T., Akazawa, M.:* Bull. Chem. Soc. Japan **40**, 1600 (1967).

10 *Beck, W., Mielert, A., Schier, E.:* Z. Naturforsch. **24**b, 936 (1969).

C. Polymerization

This section is subdivided into the discussion of polymerization reactions of fulminic acid (formonitrile oxide), the parent member of the series, and of those of the higher homologs. This is convenient, since the most important polymerization reaction, the dimerization, stops with the latter, while the fulminic acid dimer is only of transient existence and leads through its capability for various further reactions to types of compounds not encountered with other nitrile oxides. On the other hand, there are a few polymerization reactions in which fulminic acid behaves quite analogous to the higher homologs. These will consequently be discussed within the second part of this section.

1. Polymerization of Fulminic Acid

No reaction of fulminic acid has been studied over the past 100 years more thoroughly than its polymerization.[1,2] *Scholvien* obtained first true polymers, a trimer and a tetramer, named *metafulminuric acid* and *isocyanilic acid*, by the spontaneous, exothermic polymerization of aqueous or ethereal solutions of fulminic acid.[3] Much earlier *Liebig* had obtained from fulminic acid a compound which was considered to be a trimer, and named *fulminuric acid*.[4] But subsequent investigations have demonstrated, that fulminuric acid is the result of a series of complex transformations of the originally formed polymers and just accidentally has the empirical formula of a trimeric fulminic acid. The earlier investigators were unaware that mechanistic considerations require the initial formation of one or several dimers from which the higher oligomers derive.

The Dimer. In order to explain the formation of isocyanilic acid, *Wieland* possibly was the first one to suggest that the immediate precursor was a dimer, oximino-acetonitrile oxide[5] (*19a*). While working with *Nef's* formula (*20b*) for fulminic acid, he had to postulate a not very plausible rearrangement. With fulminic acid now established as formonitrile oxide (*20a*) this reaction is easily understood as a nucleophilic addition of its dissociated form to the 1,3-dipolar resonance structure. (Eq. (7)). This

1 *Wieland, H.:* Die Knallsäure, Sammlung chemischer und chemisch-technischer Vorträge (ed. *F.B. Ahrens*), vol. 14, p. 385–461. Stuttgart: E. Enke 1909.

2 *Klages, F.:* Naturwissenschaften **30**, 351 (1942).

3 *Scholvien, L.:* J. Prakt. Chem. [2] **32**, 461, 487 (1885).

4 *Liebig, J. v.:* Ann. Chem. **95**, 282 (1855).

5 *Wieland, H., Baumann, A., Reisenegger, C., Scherer, W.; Thiele, J., Will, J., Haussmann, H., Frank, W.:* Ann. Chem. **444**, 7 (1925).

is perfectly analogous to the addition of mineral and organic acids to *1*
(see also Chapter VI, F).

$$H-\overset{\oplus}{C}=N-\overset{\ominus}{O} \quad + \quad O\leftarrow N\equiv \overset{\ominus}{C} \quad + \quad \overset{\oplus}{H} \quad \longrightarrow \quad O\leftarrow N\equiv C-CH=NOH$$

20a $\qquad\qquad\qquad\qquad\qquad\qquad\qquad\qquad\qquad\qquad$ *19a*

$$\xrightarrow{\text{HCl}} \quad HON=CH-CCl=NOH \quad \xleftarrow{\text{HCl}}$$

20 $\qquad\qquad\qquad\qquad\qquad\qquad\qquad\qquad$?

$HON\overset{\leftharpoonup}{\equiv}C \qquad\qquad\qquad\qquad\qquad\qquad\qquad HON=C=C=NOH$

20b $\qquad\qquad\qquad\qquad\qquad\qquad\qquad\qquad\qquad$ *19b*

$$(7)$$

The first experimental evidence for a dimeric intermediate in the
polymerization of fulminic acid came from kinetic and thermodynamic
studies, demonstrating that—at least in aqueous solutions more than
0.2 normal in mineral acid—the reaction was clearly (pseudo-) second
order (K = 0.036 moles/l min at 0°).[6] The dimer was also trapped from
partially polymerized solutions with the aid of hydrogen halides in form
of the halogeno-glyoximes *20*, after the remaining monomer had been
removed with silver nitrate.[7] These results, however, were interpreted as
support for the symmetrical structure *19b*, dicarbonyl-dioxime, for the
dimer. Today, this reaction is better understood via structure *19a* as the
normal 1,3-addition of hydrogen halide to a nitrile oxide.

Convincing proof for the existence of the dimer as the nitrile oxide
19a was furnished later by the reaction of free fulminic acid with acetylene
or mono-substituted acetylenes.[8, 9] In a typical 1,3-dipolar cycloaddition
20a gave the expected isoxazoles *21*. At same time considerable amounts
of the corresponding isoxazole-3-aldoximes *22* were always obtained
which are formed by the analogous reaction of *19a*. Since *21* does not
react with fulminic acid, the dimerization of *20a* to *19a* and the dipolar
cycloaddition must occur at comparable rates (Eq. (8)).

$$(8)$$

---

6 *Birckenbach, L., Sennewald, K.*: Ann. Chem. **512**, 45 (1934).
7 *Sennewald, K., Birckenbach, L.*: Ann. Chem. **520**, 201 (1935).
8 *Quilico, A., Panizzi, L.*: Gazz. Chim. Ital. **72**, 155 (1942).
9 *Quilico, A., Stagno D'Alcontres, G.*: Gazz. Chim. Ital. **79**, 654, 703 (1949).

Table XII. Reactions of meta-fulminuric acid

(Configurational formulas are those of *Wieland et al.* and may be subject to revision.)

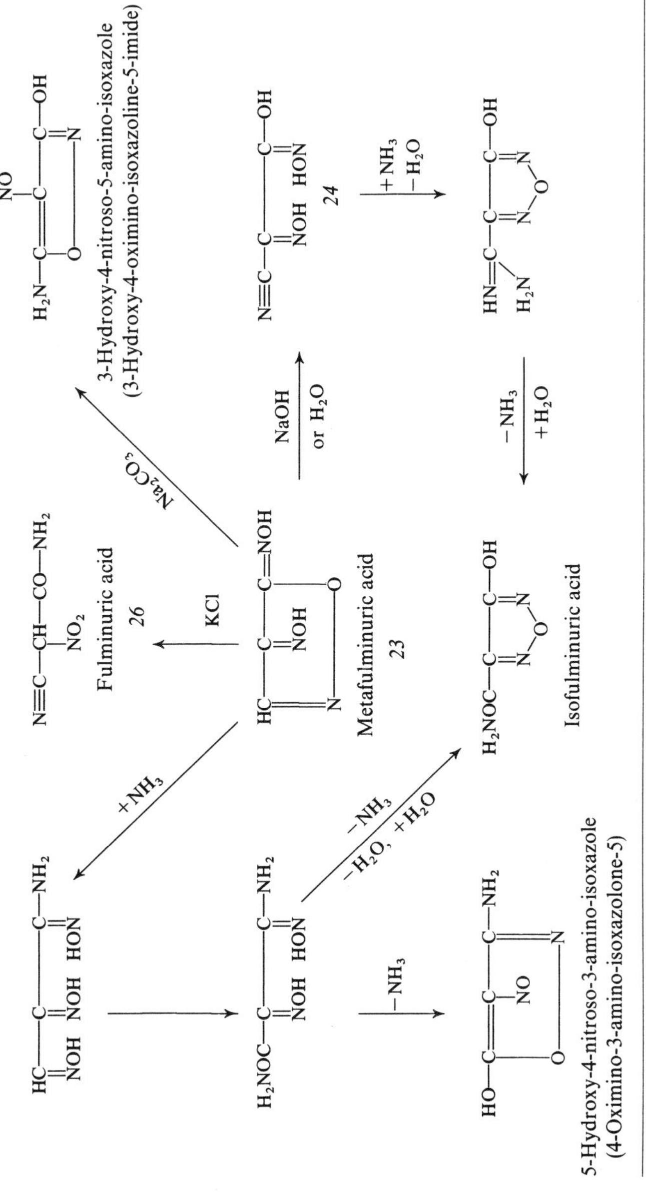

Metafulminuric Acid. Metafulminuric acid is 4,5-dioximino-isoxazoline-2 (*23*). This structure was proven by the alkaline cleavage to oximino-cyano-acet-hydroxamic acid (*24*),[10] a reaction common to all isoxazole derivatives, unsubstituted in position 3[11] (Table XII).

Formula *24* has been corroborated by further degradation and by qualitative reactions. Although there is no reason to challenge the structure *23* for metafulminuric acid, it should be mentioned that neither the acid nor the malonic acid derivative *24* have ever been synthesized. The mechanism of the formation of *23* can be visualized either (a) as a 1,3-dipolar addition of fulminic acid to the dimer *19b*, dicarbonyldioxime, or (b) as a further nucleophilic addition of fulminic acid to the dimer *19a*, oximino-acetonitrile oxide, to an open chain trimer *25*, which then closes the ring to *23* by an 1,3-addition of the oximino group to the nitrile oxide function of *25* (Eq. (9)).

$$
\text{(9)}
$$

Metafulminuric acid, usually under the influence of mild alkaline reagents, undergoes a series of interesting isomerizations and transformations which involve mostly hydrolytic ring opening and reclosure in a different position.[12] These reactions are summarized in Table XII.

Fulminuric Acid. Although fulminuric acid *26* was originally obtained directly from fulminic acid (by boiling mercuric fulminate with potassium chloride),[4] the reactions of Table XII demonstrate that it is a secondary product originating from metafulminuric acid.[12] The structure, nitro-cyanacetamide, was established both by successive degradation and by synthesis long before the controversy over the parent compound could

10 *Nef, J. U.:* Ann. Chem. **280**, 291 (1894). — *Wieland, H., Hess, M.:* Ber. **42**, 1346 (1909). — *Ulpiani, C.:* Gazz. Chim. Ital. **46**I, 1 (1916).

11 *Hill, H. B., Hale, W. J.:* Am. Chem. J. **29**, 253 (1903).

12 *Wieland, H., Baumann, A.:* Ann. Chem. **392**, 196 (1912).

be settled and certainly did not help to solve the latter.[13, 14] The mechanism of its formation may involve first the cleavage of *23* to *24*, which is also effected by boiling water alone.[7] The dioxime *24* then closes the ring to the 1,2,5-oxadiazole (furazan) *27*, a reaction for which many analogies do exist. But as the final step we would have to assume a hydrolytic cleavage of *27* between O and N of the ring in a manner to create the nitronic acid *28* which then tautomerizes to fulminuric acid (Eq. (10)). It should be emphasized, however, that the whole sequence is highly speculative, but earlier attempts in explaining the formation of *26* are completely unacceptable by today's knowledge of the chemistry of nitrile oxides.

$$(10)$$

Isocyanilic Acid. Isocyanilic acid (*29*) is formed only to a very minor extent in the spontaneous polymerization of aqueous solutions of fulminic acid, if no mineral acid is present. In strongly acidic solution (>2N) the yield can be raised to 15–20% of the applied monomer. The rest forms mostly metafulminuric acid, resp. its primary hydrolysis product *24*.[5, 7] The difficult accessibility of *29* prevented extended studies until it was found that *29* can be synthesized conveniently from nitromethane[5, 15] (Eq. (11a)). A further unambiguous synthesis starts from glyoxime (Eq. (11b)). It is interesting to note that both pathways produce first oximino-acetonitrile oxide *19a*, identical with the—then only

13 *Steiner, A.:* Ber. **9**, 779 (1876).
14 *Seidel, P.:* Ber. **25**, 431, 2756 (1892). — *Conrad, M., Schulze, A.:* Ber. **42**, 735 (1909).
15 *Steinkopf, W.:* J. Prakt. Chem. [2] **81**, 193 (1910).

Table XIII. Reactions of isocyanilic acid

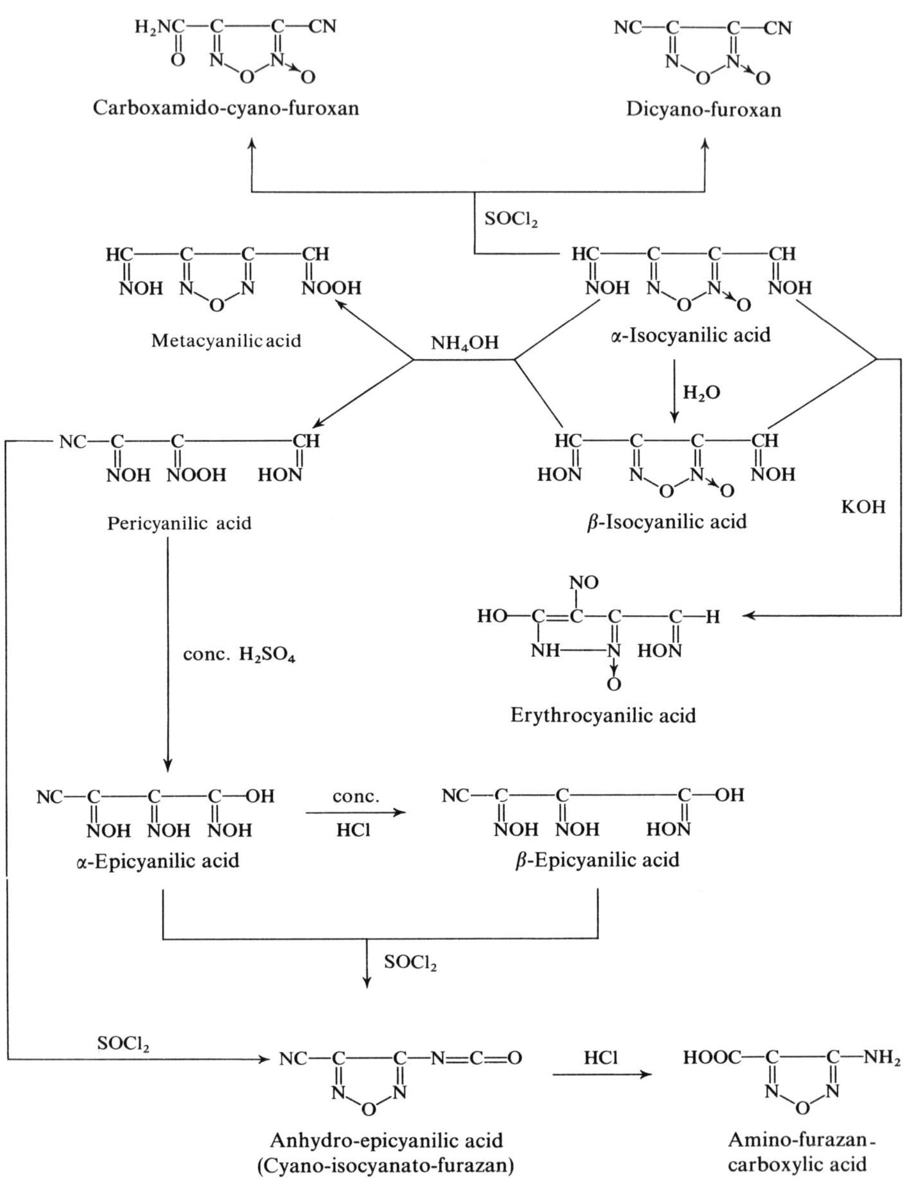

postulated—dimer of fulminic acid which stabilizes itself by the well known furoxan-dimerization reaction of nitrile oxides (see Chapter IV, C-2).

a) $2\,CH_3\!-\!NO_2$ $\xrightarrow[-H_2O]{KOH}$ $HON\!=\!CH\!-\!CH_2\!-\!NO_2 \longrightarrow$
Methazonic acid

$\xrightarrow[-H_2O]{conc.\ H_2SO_4}$ $HON\!=\!CHC\!\equiv\!N\!\rightarrow\!O$ $\xrightarrow{dimerization}$ $HON\!=\!CH\!-\!C\underset{N}{\overset{\parallel}{}}\quad\underset{N}{\overset{\parallel}{}}C\!-\!CH\!=\!NOH$

19a

29 Isocyanilic acid

$$\uparrow\ Na_2CO_3 \mid -HCl$$

$$(11)$$

b) $HON\!=\!CH\!-\!\underset{Cl}{\overset{|}{C}}\!=\!NOH$ $\xleftarrow{Cl_2}$ $HON\!=\!CH\!-\!CH\!=\!NOH$

Isocyanilic acid is a very reactive substance; a number of its transformations have been studied by *Wieland* and his associates.[5, 16-18] The results are summarized in Table XIII. Configurational formulas are those preferred by the original investigators and may be subject to revision. After it was found that isocyanilic acid is transformed by short heating with water into a stereoisomer (*syn-anti* isomerism of the oximinogroups), designated as β-isocyanilic acid, the parent compound was named α-isocyanilic acid.

Crystalline fulminic acid converts to a solid polymer of unknown structure (molecular weight ~ 1500, in $CHCl_3$) on warming slowly above $-15°$.[34]

2. Polymerization of Other Nitrile Oxides

Dimerization to Furoxans (1,2,5-oxadiazole-2-oxides). This is the most frequently observed reaction of nitrile oxides (Eq. (12)). It occurs during the formation of nitrile oxides both in an acidic (thermal decomposition of nitrolic acids) and in an alkaline environment (dehydrohalogenation of hydroximic acid chlorides), and it is the normal reaction of nitrile oxides during storage under neutral conditions at room temperature. The rate of this dimerization is unmeasurably fast (already at $0°\,C$) for the lower aliphatic nitrile oxides, while the half life of most aromatic nitrile oxides at room temperature is of the order of minutes to days. Bulky neighboring substituents seem to enhance the stability, and even the 2,2-dimethyl-propionitrile oxide (tert-butylfulmide) has a half life

16 *Wieland, H., Frank, W., Kitasato, Z.*: Ann. Chem. **475**, 42 (1929).
17 *Wieland, H., Kitasato, Z., Fromm, F.*: Ann. Chem. **475**, 54 (1929).
18 *Wieland, H., Kitasato, Z., Utzino, S.*: Ann. Chem. **478**, 43 (1930).

of at least two orders of magnitude higher than the few other known, not sterically hindered, aliphatic nitrile oxides. For qualitative observations of the rate of this reaction, see Table III, pp. 16–17. Nitrile oxides which contain substituents of sufficient spatial requirements in the vicinity of the CNO group do not form furoxans at a rate comparable with that of the isomerization to the isocyanates (see Section A, and Chapter I, C).

However, heating 2,4,6-trimethyl-benzonitrile oxide in an inert solvent for a day to 65–70°—where the rate of the isomerization reaction is still low—has produced a small yield of the extremely sterically hindered dimesitylfuroxan (*31*, R = 2,4,6-trimethylphenyl). The furoxan structure of the dimer was rigidly proven by degradation and independent synthesis.[19]

An equimolar mixture of p-chloro and p-methoxy-benzonitrile oxide in carbon tetrachloride gives the four possible isomers in nearly equal amounts.[19a]

Two alternative mechanisms have been proposed for the dimerization of *unhindered* nitrile oxides.

Kinetic studies on several aromatic nitrile oxides[20–22] have demonstrated that the reaction takes place with a clean second order kinetics, at least in dilute solutions (0.004–0.03 mol/l). Results are summarized in Table XIV a. The rate is increased by electron-withdrawing substituents and decreased by electron-releasing substituents, a Hammett type relationship holding with $\rho = +0.86$. The last example of Table XIV a does not fit into this relation, because here the rate appears to be dominated by steric rather than electronic influences. The reaction rate is only slightly affected by the polarity or solvating power of the medium, and activation parameters are substantially unmodified in the various solvents, being characterized by a negative value of the activation entropy (ca. -20 e.u.) (Table XIV b).

The overall features of the kinetics strictly parallel those reported for the cycloaddition of nitrile oxides to double or triple bond dipolarophiles, whose mechanism has been outlined as a concerted one-step process (see Chapter V, A). Therefore it would be obvious to postulate the dimerization as a 1,3-dipolar cycloaddition, where a more or less polarized transition state could be involved, because of the non-syn-

19 *Grundmann, C., Frommeld, H.-D., Flory, K., Datta, S. K.:* J. Org. Chem. **33**, 1464 (1968).

19a *Panattoni, C., Clemente, D. A., Bandoli, G., Battaglia, A., Dondoni, A.:* Chem. Commun. 60 (1970).

20 *Dondoni, A., Mangini, A., Ghersetti, S.:* Tetrahedron Letters 4789 (1966).

21 *Reid, Jon B.:* The Ott Chemical Company, Muskegon, Michigan, private communication.

22 *Barbaro, G., Battaglia, A., Dondoni, A.:* J. Chem. Soc. (B) 588 (1970).

Table XIV a. Rate constants for the dimerization of benzonitrile oxides

Compound	Temperature	$K \cdot 10^3$ [c] ($l\,mol^{-1}\,sec^{-1}$)	References
4-Methoxybenzonitrile oxide[a]	$40° \pm 0.1°$	2.37	[20]
4-Methylbenzonitrile oxide[a]	$40° \pm 0.1°$	3.06	[20]
Benzonitrile oxide[a]	$40° \pm 0.1°$	4.15	[22]
4-Chlorobenzonitrile oxide[a]	$40° \pm 0.1°$	6.78	[22]
3-Chlorobenzonitrile oxide[a]	$40° \pm 0.1°$	8.53	[20]
2,6-Dichlorobenzonitrile oxide[b]	$40° \pm 2°$	0.32	[21]
2,6-Dichlorobenzonitrile oxide[b]	$25° \pm 2°$	0.072	[21]
2,6-Dichlorobenzonitrile oxide[b]	$0° \pm 2°$	0.029	[21]

[a] In CCl_4; the reaction was followed by infrared absorption measurements of the $C\equiv N$ stretching band (around $2290\,cm^{-1}$) and by volumetric analysis based on the fast reaction of nitrile oxides with n-butylamine (see Chapter VI, G). — [b] In toluene; the data were obtained by quenching samples into excess trimethylphosphite and analyzing the formed nitrile (see Chapter IV, B) by gas-liquid partition chromatography versus an external nitrile standard. — [c] Standard deviation $\pm 3-4\%$.

Table XIV b. Solvent effect on rate constants and activation parameters for dimerization of p-chlorobenzonitrile oxide [22]

Solvent	$K \cdot 10^3$ ($l\,mole^{-1}\,sec^{-1}$)[a]			E_a[b]	$\log A$[c]	$S^{\neq}$[d]
	$t\,°C = 25$	40	50			
CCl_4	1.81	6.78	16.0	16.7	9.5	-18
$1,2\text{-}Cl_2C_2H_4$	0.231	0.849	2.24	17.1	8.8	-20
$CHCl_3$	0.177	0.747	1.80	18.1	9.5	-17
CH_3OH	0.647	2.51	5.01	17.2	9.4	-18
C_2H_5OH	0.593	2.18	4.85	16.0	8.5	-22
$i\text{-}C_3H_7OH$	0.780	2.72	5.62	15.1	8.0	-24
1,4-Dioxane	0.460	1.68	3.74	16.0	8.4	-22
CH_3CN	0.657	$-$	$-$	$-$	$-$	$-$

[a] Standard deviation $\pm 3\text{-}4\%$.
[b] Accurate to $\pm 0.4\,kcal.\,mole^{-1}$.
[c] Accurate to ± 0.4 unit.
[d] Accurate to ± 1.5 unit.

chronous formation of the σ-bonds (Eq. (12a)).[22] A zwitterionic discrete intermediate (Eq. (12b)) is on the contrary far less probable.

However, as pointed out first by *Huisgen*,[23] this mechanism violates the principle of maximum gain in σ-bond energy, most generally found valid in all the many other types of 1,3-dipolar cycloadditions, although at least one other exception[24] is known. To satisfy this principle, the dimerization should lead to the isomeric 1,2,4-oxadiazole-4-oxides *32*

23 *Huisgen, R.:* Angew. Chem. **75**, 742 (1963).
24 *Kadaba, P. K.:* Tetrahedron **22**, 2453 (1966).

instead of the furoxans *31*. Excluding structural elements common to both isomers, the value $\Delta H°$ for *31* amounts to 136 Kcal $(1\,C—C(83) + 1\,N—O(53))$ whilst $\Delta H°$ for *32* calculates to 158 Kcal $(1\,C—N(73) + 1\,C—O(85))$. Besides, the 1,2,4-oxadiazole-4-oxides are well known compounds[25] which are accessible by various routes, some starting with nitrile oxides (see below), but they have never been observed, even in traces, as products of the spontaneous dimerization of the nitrile oxides.

$$
\begin{array}{c}
\text{(reaction scheme, Eq. (12))}
\end{array}
$$

(12)

It has therefore been suggested that the formation of the furoxan might occur in a two-step reaction via the dimerization of the nitrile oxide to a 1,2-dinitroso-ethylene *33* involving the mesomeric structure *30b* of the nitrile oxide with carbene character (Eq. (12c)). *33* will then immediately stabilize itself by a simple regrouping of electrons as the furoxan *31*. Such a mechanism would naturally not interfere with the rule of maximal σ-bonding which is only applicable to true 1,3-dipolar cycloaddition.[19, 23] This explanation is supported by NMR studies of benzofuroxan *34* which indicate that a mobile equilibrium exists between *34* and a modest amount of 1,2-dinitrosobenzene *35*,[26, 27] although the aromaticity of the

25 *Eloy, F.:* Fortschr. Chem. Forsch. **4**, 807 (1965).
26 *Katritzky, A.R., Øksne, S., Harris, R.K.:* Chem. Ind. (London) 990 (1961).
27 *Diehl, P., Christ, H.A., Mallory, F.B.:* Helv. Chim. Acta **45**, 504 (1962).

benzene ring should play here an important role:

$$
\underset{34}{\text{[structure]}} \;\rightleftharpoons\; \underset{35}{\text{[structure]}}
$$

(13)

$$
\underset{R=C_6H_5,\ C_2H_5}{\text{[structure]} \;\rightleftharpoons\; \text{[structure]} \;\rightleftharpoons\; \text{[structure]}}
$$

Similarly, NMR and kinetic studies of the isomerization of unsymmetrical substituted furoxans have brought evidence for the transient existence of 1,2-dinitroso-ethylene intermediates[28] (Eq. (13)).

Furthermore, for sterical reasons, the formation of dimesityl-furoxan from 2,4,6-trimethylbenzonitrile oxide can be more easily rationalized via Eq. (12c).[19]

On the whole, however, at present some more experimental evidence seems desirable in order to decide definitively between the concerted process and the "carbene" route.

A regeneration of nitrile oxides from 3,4-disubstituted furoxans seems in general only possible at temperatures where the former rearrange immediately to isocyanates. Monosubstituted furoxans, however, seem to depolymerize more easily.[29] An attempt to distill under vacuum 3-carbethoxy-furoxan-4-carboxylic acid 36 resulted—apart from the formation of gaseous decomposition products—in an approximately 50% yield of 3,4-dicarbethoxy-furoxan 37, a reaction which can be understood by assuming that initial decarboxylation to 3-carbethoxy-furoxan is followed by cleavage to fulminic acid (which undergoes further degradation) and carbethoxy-fulmide which in turn dimerizes to 37 (Eq. (14)).[30, 31]

Furoxans attached to five-membered rings are especially strained, e. g. the camphor derivative 37a. When heated above its melting point, it decomposes violently (normally, furoxans are quite stable thermally) with intermediate formation of the bis-nitrile oxide 37b which could be trapped by cycloaddition with phenylacetylene as the isoxazole 37c.[31a]

28 Mallory, F. B., Cammarata, A.: J. Am. Chem. Soc. **88**, 61 (1966).
29 Wieland, H., Semper, L., Gmelin, E.: Ann. Chem. **367**, 52 (1909).
30 Grundmann, C.: Fortschr. Chem. Forsch. **7**, 62 (1966).
31 Bouveault, L., Bongert, A.: Bull. Soc. Chim. France [3] **27**, 1164 (1902).
31a Altaf-ur-Rahman, Boulton, A. J.: Chem. Commun. 73 (1968); — Private communication.

Photolysis of diphenyl-1,2,5-oxadiazole (diphenylfurazan) seems to involve as first step a cleavage into benzonitrile oxide and benzonitrile.[31b]

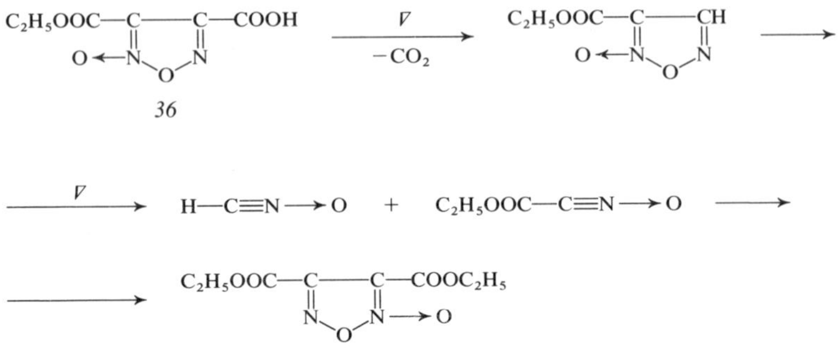

$$C_2H_5OOC-C \cdots C-COOH \quad \xrightarrow[-CO_2]{\nu} \quad C_2H_5OOC-C \cdots CH \longrightarrow$$

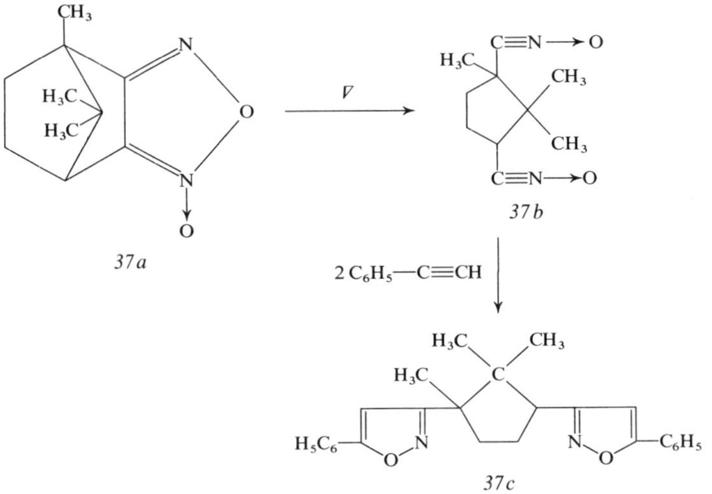

36

$$(14)$$

$$\xrightarrow{\nu} \quad H-C{\equiv}N{\rightarrow}O \quad + \quad C_2H_5OOC-C{\equiv}N{\rightarrow}O \longrightarrow$$

$$\longrightarrow \quad C_2H_5OOC-C \cdots C-COOC_2H_5$$

37

37a $\xrightarrow{\nu}$ *37b*

$$2\,C_6H_5-C{\equiv}CH$$

37c

Aside from their formation from nitrile oxides, furoxans are obtained by a number of other routes, some involving nitrile oxides as hypothetical transient intermediates, for which reference is made to recent surveys of the chemistry of furoxans.[32,33]

Oxalo-bis-nitrile oxide (cyanogen-bis-N-oxide) *38*, polymerizes in dilute solution at $-25°$ to a solid orange to red explosive oligomer which is very probably a poly-furoxan (*39*, $n=5$–8). The end groups

31b *Mukai, T., Oine, T., Matsubara, A.:* Bull. Chem. Soc. Japan **42**, 581 (1969).

32 *Behr, L. C.:* Oxadiazoles and related compounds. The chemistry of heterocyclic compounds (ed. *A. Weissberger*), vol. 17, p. 283–320. New York: Interscience 1962.

33 *Boyer, J. H.:* Oxadiazoles. Heterocyclic compounds (ed. *R. C. Elderfield*), vol. 7, p. 462–507. New York: J. Wiley & Sons 1961.

are formed by addition of water which cannot be completely eliminated from _38_.[35] Analogous polymers, containing alternating benzene and furoxan rings (_40_), are obtained by the spontaneous or thermally induced polymerization of isophthalo- or terephthalo-bis-nitrile oxide.[36–38]

$$O \longleftarrow N{\equiv}C{-}C{\equiv}N \longrightarrow O$$

_38_

_39_

_40_

The degree of polymerization of _40_ is not known, the thermal stability of such polymers is definitely inferior to analogous types containing other kinds of heterocyclic rings.

Dimerization to 1,2,4-Oxadiazole-4-oxides and 1,4,2,5-Dioxadiazines. Benzonitrile oxide which dimerizes in neutral medium exclusively to diphenyl-furoxan (_41_) furnishes two other dimers under the catalytic influence of aprotic or protic acids. As Eq. (15) shows, the presence of boron trifluoride etherate (where the _Lewis_ acid is strongly coordinated) does not affect the course of the fast dimerization to furoxan _41_; in the case of mesitonitrile oxide, however, where the dimerization to furoxan is practically suppressed at room temperature, the _Lewis_ acid slowly coordinates with the nitrile oxide to yield the boron complex of 3,5-dimesityl-1,2,4-oxadiazole-4-oxide (_43_), which can be transformed with $KMnO_4$ in acetone into _45_. Benzonitrile oxide dimerizes to _42_ when treated with gaseous BF_3 in a non-coordinating solvent such as hexane and with a ratio 2:1. Treatment with an excess of boron trifluoride in hexane yields a different dimer, _i.e._ 3,6-diaryl-1,4,2,5-dioxadiazine (_46_) from benzo-, resp. _47_ from mesito-nitrile oxide.[39] The dimer _46_ has

34 _Beck, W.:_ Private communication (1969).

35 _Grundmann, C., Mini, V., Dean, J.M., Frommeld, H.-D.:_ Ann. Chem. **687**, 191 (1965).

36 _Iwakura, Y., Akiyama, M., Shiraishi, S.:_ Bull. Chem. Soc. Japan **38**, 335 (1965).

37 _Overberger, C.G., Fujimoto, S.:_ J. Polymer Sci. B, Polymer Letters **3**, 735 (1965).

38 _Eloy, F.:_ Bull. Soc. Chim. Belges **73**, 639 (1964).

39 _Morrocchi, S., Ricca, A., Selva, A., Zanarotti, A.:_ Chim. Ind. (Milan) **50**, 558 (1968); — Gazz. Chim. Ital. **99**, 165 (1969).

previously been obtained by the action of pyridine on benzonitrile oxide.[40]

$$ \tag{15} $$

$$ \text{42, } \quad Ar = C_6H_5 \quad : \quad 44 $$
$$ \text{43, } \quad Ar = 2,4,6\text{-}(CH_3)_3C_6H_2 \quad : \quad 45 $$

46, Ar = C_6H_5
47, Ar = 2,4,6-$(CH_3)_3C_6H_2$

3,5-Diphenyl-1,2,4-oxadiazole-4-oxide (*44*) is also formed by treatment of benzonitrile oxide in ether with dry hydrogen chloride[40] or by spontaneous decomposition of benzohydroximic acid chloride in a closed vessel.[41] The action of both aprotic and protic acids in small amounts can be rationalized on the base of a common mechanism.[39] (Eq. (16)). When hydrogen chloride is used, the intermediate *48* (X = H, Ar = C_6H_5) is stabilized as a hydrochloride.

$$ Ar-C\equiv N \rightarrow O \; + \; H^{\oplus} \; \rightleftharpoons \; Ar-\overset{\oplus}{C}=N-OH \tag{16} $$

$$ X = BF_3^{\ominus} \text{ or } H \qquad 48 $$

<hr>

40 *Speroni, G., Bartoli, M.:* Sopra gli ossidi di benzonitrile. Nota VIII. Florence: Stabilimento Tipografico Marzocco 1952.

41 *Wieland, H.:* Ber. **40**, 1667 (1907).

Other Polymers. If nitrous acid is abstracted from nitrolic acids by means of mild aqueous alkalies, e. g. dilute sodium carbonate or ammonia, polymeric nitrile oxides are obtained which differ widely in their properties from the types discussed previously.[42] (Eq. (17)). These polymers are, contrary to furoxans and 1,2,4-oxadiazole-4-oxides, energy-rich, very reactive compounds. Similar polymers have been obtained from benzonitrile oxide under the catalytic influence of trimethylamine[40] and by the spontaneous dehydrohalogenation of cyano-formhydroximic chloride with water.[43,44]

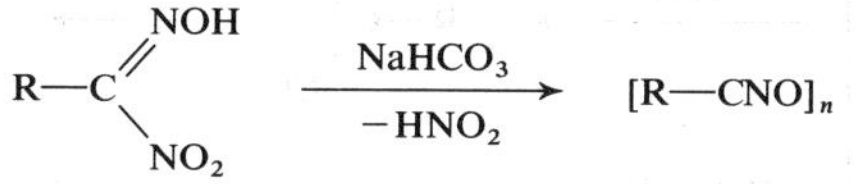

$$R-C\underset{NO_2}{\overset{NOH}{\diagdown}} \quad \xrightarrow[-HNO_2]{NaHCO_3} \quad [R-CNO]_n \qquad (17)$$

It is unlikely that the formation of these polymers occurs via the monomeric nitrile oxides, since at least benzonitrile oxide is sufficiently stable to be detected in substance under the conditions employed. The polymers are obtained as solids, some amorphous, some definitely crystalline and are characterized, contrary to the furoxans, by their thermal instability. The parent member, named by *Wieland* "trifulmin", is dangerously explosive. The high sensitivity and the poor solubility of these products made a determination of their molecular weight impossible.

The originally proposed structure of s-triazine-tris-N-oxides (*49*) had to be abandoned in view of our recent knowledge of the chemistry of s-triazines and the failure to obtain s-triazines from them by chemical reduction. At least in the case of *49* (R = C_6H_5) the applied reducing agent should have led to 2,4,6-triphenyl-s-triazine which is entirely stable under these conditions. But all attempts to reduce these polymers resulted only in the formation of the corresponding monomeric nitriles. With the structure *49* thus having become untenable, a dipolar chain structure (*50*) has been suggested,[40,45] but rigid proof is still missing as well as an idea about the average molecular weight.

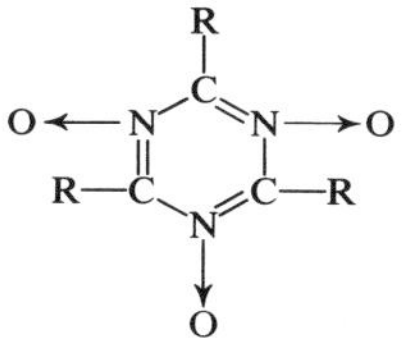

$$\overset{⊕}{C}=N-O-[-C=N-O-]_n-C=N-O^⊖$$

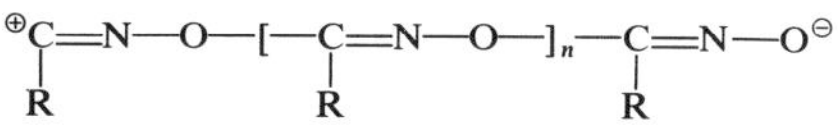

50

R = H, CH_3, C_6H_5, C_2H_5OOC, CN

49

42 *Wieland, H.:* Ber. **42**, 803, 816 (1909).

43 *Houben, J., Kauffmann, H.:* Ber. **46**, 2821 (1913).

44 *Grundmann, C., Frommeld, H.-D.:* J. Org. Chem. **31**, 4235 (1966).

45 *Speroni, G.:* Sitzungsber. d. 14. Internat. Kongr. f. reine u. angewandte Chemie, Zürich 1955, No. 52, p. 30.

These polymers behave in their reactions in many instances like the parent monomeric nitrile oxides. Controlled thermal decomposition by heating in an inert solvent led to the isomeric isocyanates, concentrated hydrochloric acid caused conversion to hydroximic acid chlorides and 1,2,4-oxadiazole-4-oxides, while amines form the corresponding ureas (Eq. (18)).[42]

$$
50 \quad
\begin{array}{l}
\xrightarrow{V} \quad R\!-\!N\!=\!C\!=\!O \\[2ex]
\xrightarrow{HCl} \quad R\!-\!C\!\!\begin{array}{c}\nearrow^{NOH}\\ \searrow_{Cl}\end{array} \longrightarrow 44 \\[2ex]
\xrightarrow{H_2N\!-\!R'} \quad R\!-\!NH\!-\!CO\!-\!NH\!-\!R'
\end{array}
\qquad (18)
$$

None of the above depolymerization reactions seems to involve the monomeric nitrile oxide as an intermediate.[46]

---

46   *Grundmann, C., Kite, G.F.:* Unpublished.

V. Addition Reactions Leading to Cyclic Structures: 1,3-Dipolar Cycloadditions

As can be expected from their dipolar structure (see resonance hybrid *5a—e*, Chapter I, C), nitrile oxides are highly reactive compounds and may lead either to open-chain products through a 1,3-addition reaction with nucleophiles, or to heterocyclic products through a 1,3-dipolar cycloaddition with unsaturated bonds, both electron-rich and electron-poor.

Reagents, containing functional groups apt to react with nitrile oxides under both schemes, usually enter only one of the two. Sometimes, however, 1,3-addition and cycloaddition can be concurrent within the same type of compounds, as for instance with arylacetylenes (see Section C), unsaturated carboxylic acids (see Chapter VI, F) and amines (see Chapter VI, G). In this case, prevailing of either type of addition depends on the nature of the substituents. A kinetic comparison between amines and alkenes or alkynes led to conclude that, at least under the experimental conditions employed, the nucleophilic reactivity of the former type of compound prevails over the dipolarophilic activity of the latter ones.[1]

The two types of addition reactions will be discussed in two separate chapters.

---

1 *Caramella, P., Vita Finzi, P.:* Chim. Ind. (Milan) **48**, 963 (1966).

A. General

Historical. The reaction between nitrile oxides and triple bond derivatives is now one of the principal methods of synthesis of the isoxazole ring.[1,2]

An analogous reaction starting from hydroximic acid chlorides and aryl acetylides was long before known[3] (Eq. (1)). *Quilico*[4] first proved

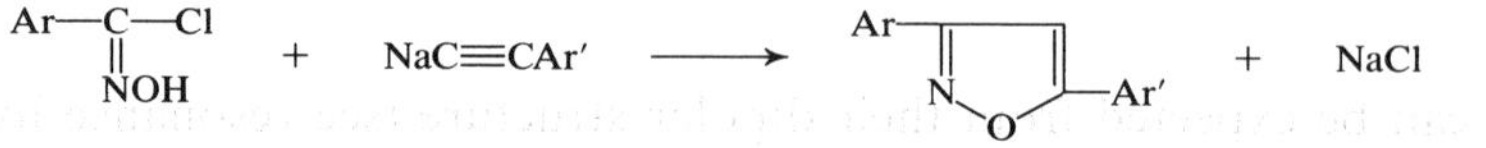

$$
\underset{\text{NOH}}{\text{Ar—C—Cl}} \;+\; \text{NaC} \equiv \text{CAr}' \;\longrightarrow\; \text{(isoxazole)} \;+\; \text{NaCl} \qquad (1)
$$

that the nitrile oxide should be regarded as an intermediate both in the above reaction and in the reactions of hydroximic acid chlorides with the sodium salts of β-dicarbonyl and similar compounds,[5–7] according to the mechanism exemplified in Eq. (2). Furthermore, the same author

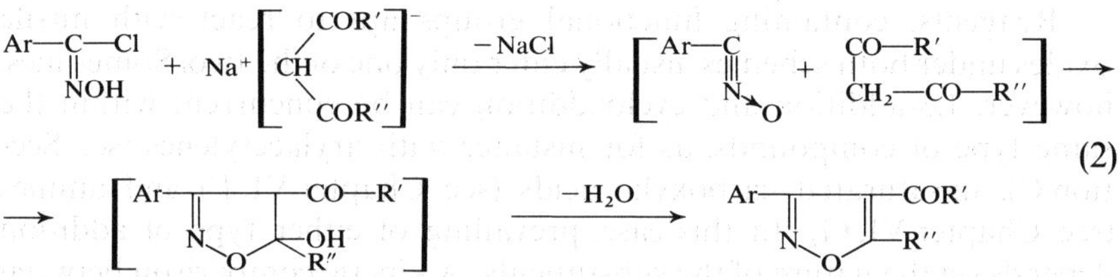

$$
\underset{\text{NOH}}{\text{Ar—C—Cl}} + \text{Na}^{+} \left[\underset{\text{COR}''}{\overset{\text{COR}'}{\text{CH}}} \right]^{-} \xrightarrow{-\text{NaCl}} \left[\underset{N\to O}{\text{Ar—C}} + \underset{\text{CH}_2\text{—CO—R}''}{\text{CO—R}'} \right] \longrightarrow
$$

$$
\longrightarrow \left[\underset{N\ O}{\text{Ar}} \overset{\text{CO—R}'}{\underset{\text{R}''}{\text{OH}}} \right] \xrightarrow{-\text{H}_2\text{O}} \underset{N\ O}{\text{Ar}} \overset{\text{COR}'}{\underset{\text{R}''}{}}
\qquad (2)
$$

discovered the parallelism of behaviour between fulminic acid and nitrile oxides in the reaction with acetylenic derivatives,[8–12] and consequently pointed out[4,13] that fulminic acid should be considered the simplest term of the series, *i.e.* formonitrile oxide. Thus, the formation of several isoxazole and furoxan derivatives, known to arise in the complex reaction between fuming nitric acid and various unsaturated compounds, could be ascribed to the generation of intermediate nitrile oxides and their subsequent cycloaddition reactions.

1 *Quilico, A.:* The chemistry of heterocyclic compounds (ed. *A. Weissberger*), vol. XVII, chap.: Isoxazole and related compounds, p. 1–176. New York: Interscience 1962.

2 *Kochetkov, N. K., Sokolov, S. D.:* Advances in heterocyclic chemistry (ed. *A. R. Katritzky*), vol. II, chap.: Recent developments in isoxazole chemistry, p. 365–422. London: Academic Press 1963.

3 *Weygand, C., Bauer, E.:* Liebigs Ann. Chem. **459**, 123 (1927).

4 *Quilico, A., Speroni, G.:* Gazz. Chim. Ital. **76**, 148 (1946).

5 *Quilico, A., Fusco, R.:* Rend. Ist. Lombardo Sci. Lettere **69**, 439 (1936).

6. *Fusco, R.:* Rend. Ist. Lombardo Sci. Lettere **70**, 225 (1937).

7 *Quilico, A., Fusco, R.:* Gazz. Chim. Ital. **67**, 589 (1937).

8 *Quilico, A., Speroni, G.:* Gazz. Chim. Ital. **69**, 508 (1939).

9 *Quilico, A., Speroni, G.:* Gazz. Chim. Ital. **70**, 779 (1940).

10 *Quilico, A., Panizzi, L.:* Gazz. Chim. Ital. **72**, 155 (1942).

11 *Quilico, A., Stagno d'Alcontres, G.:* Gazz. Chim. Ital. **79**, 654 (1949).

12 *Quilico, A., Stagno d'Alcontres, G.:* Gazz. chim. ital. **79**, 703 (1949).

13 *Quilico, A., Simonetta, M.:* Gazz. Chim. Ital. **76**, 200 (1946).

A few examples should be mentioned here. The presence of isoxazole-3-carboxylic acid (1)[14, 15] in the reaction mixture from acetylene and fuming nitric acid has been explained through the following scheme, (Eq. (3))[13] an extension of *Wieland's* classical sequence[16] for the synthesis of fulminic acid.

$$
\begin{array}{c}
CH \\
\parallel \\
CH
\end{array}
\xrightarrow{HNO_2}
\begin{array}{c}
CH-CHO \\
\parallel \\
NOH
\end{array}
\xrightarrow{Oxid.}
\begin{array}{c}
CH-COOH \\
\parallel \\
NOH
\end{array}
\longrightarrow
$$

$$
\xrightarrow{HNO_3}
\begin{array}{c}
HOOC-C-NO_2 \\
\parallel \\
NOH
\end{array}
\xrightarrow{-HNO_2}
HOOC-C\equiv N\rightarrow O
\longrightarrow
$$

$$
\xrightarrow{+CH\equiv CH}
HOOC\text{-isoxazole } (1)
\qquad (3)
$$

The formation of 3,3'-diisoxazolylketone (3)[17] as major and of furoxan 4[18] as minor product in the reaction mixture from acetylene, acetone and fuming nitric acid can be easily interpreted by assuming nitrile oxide 2 as intermediate (Eq. (4)).[19]

$$
CH_3COCH_3
\xrightarrow[HNO_3]{HNO_2}
\begin{array}{c}
CH_3COCNO_2 \\
\parallel \\
NOH
\end{array}
\xrightarrow{-HNO_2}
CH_3CO-C\equiv N\rightarrow O
\longrightarrow
$$

$$
\xrightarrow{+CH\equiv CH}
CH_3CO\text{-isoxazole}
\xrightarrow[HNO_3]{HNO_2}
\text{isoxazole-}CO-C(NO_2)=NOH
\longrightarrow
$$

$$
\xrightarrow{-HNO_2}
\text{isoxazole-}CO-C\equiv N\rightarrow O \;\; (2)
\xrightarrow{+CH\equiv CH}
\text{isoxazole-}CO\text{-isoxazole} \;\; (3)
\qquad (4)
$$

$$
2 \xrightarrow{dimer.} 4
$$

14 *Quilico, A., Freri, M.*: Gazz. Chim. Ital. **59**, 939 (1929).

15 *Quilico, A., Panizzi, L.*: Gazz. Chim. Ital. **72**, 458 (1942).

16 *Wieland, H.*: Ber. **40**, 418 (1907).

17 *Quilico, A., Freri, M.*: Gazz. Chim. Ital. **60**, 172 (1930).

18 *Quilico, A.*: Gazz. Chim. Ital. **61**, 265 (1931).

19 *Quilico, A., Freri, M.*: Gazz. Chim. Ital. **76**, 3 (1946).

Likewise, eulite (7),[20,21] a compound long before isolated in the reaction between citraconic acid and nitric acid, may be assumed to arise through an intermediate formation of nitrile oxide *5* (Eq. (5)).[22]

$$
\begin{array}{c}
CH_3-\underset{\underset{CH-COOH}{\|}}{C}-COOH \quad\xrightarrow{N_2O_3}\quad CH_3-\underset{\underset{NO}{|}}{C}-\underset{\underset{NO_2}{|}}{CH} \quad\xrightarrow{-CO_2}\quad CH_3-\underset{\underset{NOH}{\|}}{C}-CH_2NO_2
\end{array}
$$

$$
CH_3C(NO_2)_2C\equiv N\longrightarrow O \quad\xleftarrow{-HNO_2}\quad CH_3-\underset{\underset{NO_2}{|}}{C}-C\underset{NOH}{\overset{NO_2}{\diagdown}}
$$

5

$$ \text{(via } HNO_2 / HNO_3 \text{)} \qquad CH_3COCH_2NO_2 \quad \textit{6}$$

(5)

$$
5+6 \quad\xrightarrow{-H_2O}\quad CH_3-\underset{\underset{NO_2}{|}}{\overset{\overset{NO_2}{|}}{C}}-\text{(isoxazole ring)}-NO_2,\ CH_3
$$

7

In view of the close structural similarity between nitrile oxides, azides and aliphatic diazo compounds, this isoxazole synthesis has been extended by *Quilico* and his school to a wide variety of double and triple bond derivatives. Although postulated already by *Quilico*[4,13] and *Leandri*[23] as a 1,3-dipolar cycloaddition, only after *Huisgen's* extensive work[24-28] this type of reaction has been inserted into a scheme of much wider generality and its mechanism as well as its extension to hetero-dipolarophiles have been thoroughly investigated. Recent kinetic studies have confirmed the validity of *Huisgen's* scheme.

Mechanism. When a generalized unsaturated compound, here indicated as X=Y, enters reaction with a nitrile oxide, a cycloaddition takes place with formation of two new σ-bonds at the expense of two π-bonds. The following three pathways are *a priori* conceivable for such a reaction (Eq. (6)): (a) a one-step concerted mechanism through a simple transition

20 *Baup, S.:* Liebigs Ann. Chem. **81**, 102 (1852).

21 *Quilico, A., Fusco, R., Rosnati, V.:* Gazz. Chim. Ital. **76**, 30 (1946).

22 *Quilico, A., Fusco, R.:* Gazz. Chim. Ital. **76**, 195 (1946).

23 *Leandri, G., Pallotti, M.:* Ann. Chim. (Rome) **47**, 376 (1957).

24 *Huisgen, R.:* Festschrift der Zehnjahresfeier des Fonds der Chemischen Industrie, Düsseldorf 1960, p. 73.

25 *Huisgen, R.:* Proc. Chem. Soc. 357 (1961).

26 *Huisgen, R.:* Angew. Chem. **75**, 604 (1963); — Intern. Ed. Engl. **2**, 565 (1963).

27 *Huisgen, R.:* Angew. Chem. **75**, 741 (1963); — Intern. Ed. Engl. **2**, 633 (1963).

28 *Huisgen, R., Grashey, R., Sauer, J.:* The chemistry of Alkenes (ed. *S. Patai*), chap. 11: Cycloaddition reactions of alkenes. New York: Interscience 1964.

state (route A); (b) a two-step mechanism through a dipolar intermediate (route B); (c) a two-step mechanism through a diradical intermediate (route C).

$$
\begin{array}{c}
\text{(A)} \\
R-C{\equiv}N \to O \quad \text{(B)} \\
+ \\
X{=}Y \\
\text{(C)}
\end{array}
\qquad (6)
$$

The unlikelihood of a dipolar intermediate (route B) has already been evidenced by *Huisgen*,[27] who proposed for all 1,3-dipolar cyclo-addition reactions a concerted four-center mechanism (route A). The "diradical" mechanism has been recently supported,[29, 34a] but its draw-backs have been emphasized.[30]

In the light of our present knowledge, the cycloaddition reactions of nitrile oxides, especially those with double and triple C—C bond derivatives, where kinetic data are available, seem to follow route A on the ground of the following principal factors:

(a) *Stereospecificity.* As illustrated further on in Section V, B, a large number of *cis-trans* isomeric dipolarophiles have been reacted with nitrile oxides. In all cases the configuration of the reactants has been retained in the 2-isoxazoline thus obtained, showing without exception a strictly stereospecific *cis*-addition. This phenomenon would be con-ceivable in a two-step process only if the rotation about the single bond X—Y in the dipolar or diradical intermediate is considerably more slower than ring closure.

As a matter of fact, the *cis*-stereospecificity is a general pattern for all 1,3-dipolar cycloadditions so far reported; whatever the 1,3-dipole may be, and still represents the most valid argument against a two-step mechanism.

(b) *Rate dependence on the solvent.* As in most other types of 1,3-dipolar cycloadditions,[27] kinetic studies showed that the reaction rate is only moderately influenced by solvent polarity. Some pertinent data are presented in Table XV. This fact is hardly compatible with a highly polar intermediate.

29 *Firestone, R. A.:* J. Org. Chem. **33**, 2285 (1968).
30 *Huisgen, R.:* J. Org. Chem. **33**, 2291 (1968).

Table XV. Solvent effects on rate of cycloaddition

Nitrile oxide	Dipolarophile	Solvent	$10^2 \cdot K$ (at 25°) $(1 \cdot \text{mole}^{-1}\,\text{sec}^{-1})$	Ref.
$p\text{-ClC}_6\text{H}_4\text{CNO}$	Styrene	Carbon tetrachloride	1.66	[31]
	Styrene	1,2-Dichloroethane	1.22	[31]
	Styrene	Chloroform	0.96	[31]
	Styrene	Ethanol	1.55	[31]
	Styrene	Acetonitrile	1.12	[31]

Nitrile oxide	Dipolarophile	Solvent	$10^4 \cdot K$ (at 24.8°)	Ref.
(2,6-dichloro-3,4,5-trimethylphenyl) $\text{C}_6(\text{Cl})_2(\text{CH}_3)_3\text{CNO}$	$p\text{-NO}_2\text{C}_6\text{H}_4\text{C}\!\equiv\!\text{CH}$	Carbon tetrachloride	9.55	[32]
	$p\text{-NO}_2\text{C}_6\text{H}_4\text{C}\!\equiv\!\text{CH}$	Chloroform	3.19	[32]
	$p\text{-CH}_3\text{OC}_6\text{H}_4\text{C}\!\equiv\!\text{CH}$	Carbon tetrachloride	3.91	[32]
	$p\text{-CH}_3\text{OC}_6\text{H}_4\text{C}\!\equiv\!\text{CH}$	Chloroform	2.04	[32]

(c) *Activation parameters and substituent effects.* A concerted process predicts moderate activation enthalpies and strongly negative activation entropies. The values thus far obtained for cycloadditions of nitrile oxides with alkenes and alkynes (see Table XVI) are consistent with this general criterium, although it should be noted that entropy values are relatively high if compared with values from other 1,3-dipolar cyclo-additions[27] or Diels-Alder reactions,[33,34] which are usually around $-30\ \text{cal deg}^{-1}\,\text{mole}^{-1}$. This would imply a less rigid transition state in the cycloadditions involving nitrile oxides.

A recent re-evaluation of activation energies for the "diradical" mechanism of 1,3-dipolar cycloadditions, based on the use of Linnett structures, led to estimations in fairly good agreement with the experimental values.[34a]

The reaction of course follows a second order kinetic law, being first order with respect to both reagents. As far as the substituent effects on dipolarophiles of the hydrocarbon type, such as arylacetylenes and styrenes, are concerned, a mild rate acceleration by both electron-withdrawing and electron-releasing substituents, as shown by the available data summarized in Table XVII, has been reported.[31,32,35,36]

31 *Battaglia, A., Dondoni, A.:* Ric. Sci. **38**, 201 (1968).
32 *Beltrame, P., Veglio, C., Simonetta, M.:* J. Chem. Soc. (B) 867 (1967).
33 *Wassermann, A.:* Diels-Alder reactions. Amsterdam: Elsevier 1965.
34 *Sauer, J.:* Angew. Chem. **79**, 76; Intern. Ed. Engl. **6**, 16 (1967).
34a *Firestone, R. A.:* Communication at the Symposium on Orbital Symmetry Correlations, Cambridge, January 1969; J. Chem. Soc. (A) 1570 (1970).
35 *Dondoni, A.:* Tetrahedron Letters 2397 (1967) and personal communication.
36 *Beltrame, P., Veglio, C., Simonetta, M.:* Chem. Commun. 433 (1966).

Table XVI. Activation parameters

Nitrile oxide	Reactant	Solvent	$\Delta H^{\ddagger}$ (kcal/mole)	$\Delta S^{\ddagger}$ (e.u.)	Ref.
(2,6-dichloro-3,5-dimethylphenyl)CNO	p-NO$_2$C$_6$H$_4$C≡CH	CCl$_4$	15.1	−24	32
	p-CH$_3$OC$_6$H$_4$C≡CH	CCl$_4$	18.2	−15	32
p-ClC$_6$H$_4$CNO	C$_6$H$_5$C≡CH	CCl$_4$	14.4	−24	35
	C$_6$H$_5$CH=CH$_2$	CCl$_4$	12.4	−27	31
	C$_6$H$_5$CH=CH$_2$	CHCl$_3$	13.2	−26	31
(2,4,6-trimethylphenyl)CNO	C$_6$H$_5$CH=CH$_2$	CCl$_4$	14.3	−29	31

Table XVII. Rate constants for the reaction of nitrile oxides with arylacetylenes and styrenes

Nitrile oxide	Dipolarophile	Solvent	10^3 K at $25 \pm 0.1°$ [a] (l·mole^{-1} sec^{-1})	Ref.
C$_6$H$_5$CNO	p-NO$_2$C$_6$H$_4$C≡CH	CCl$_4$	4.29	35
	m-NO$_2$C$_6$H$_4$C≡CH	CCl$_4$	3.26	35
	p-ClC$_6$H$_4$C≡CH	CCl$_4$	2.23	35
	m-BrC$_6$H$_4$C≡CH	CCl$_4$	1.91	35
	C$_6$H$_5$C≡CH	CCl$_4$	1.78	35
	p-CH$_3$C$_6$H$_4$C≡CH	CCl$_4$	1.87	35
	p-CH$_3$OC$_6$H$_4$C≡CH	CCl$_4$	2.24	35

10^4 K at 24.8°

Nitrile oxide	Dipolarophile	Solvent	10^4 K at 24.8°	Ref.
(2,6-dichloro-3,5-dimethylphenyl)CNO	p-NO$_2$C$_6$H$_4$C≡CH	CCl$_4$	9.55 ±0.9	32, 36
	p-BrC$_6$H$_4$C≡CH	CCl$_4$	4.58 ±0.49	32
	p-ClC$_6$H$_4$C≡CH	CCl$_4$	4.52 ±0.36	32, 36
	C$_6$H$_5$C≡CH	CCl$_4$	3.25 ±0.22	32, 36
	p-CH$_3$C$_6$H$_4$C≡CH	CCl$_4$	3.73 ±0.23	32, 36
	p-CH$_3$OC$_6$H$_4$C≡CH	CCl$_4$	3.91 ±0.19	32, 36

10^4 K at $25 \pm 0.1°$ [a]

Nitrile oxide	Dipolarophile	Solvent	10^4 K at $25 \pm 0.1°$ [a]	Ref.
2,4,6-(CH$_3$)$_3$C$_6$H$_2$CNO	p-NO$_2$C$_6$H$_4$CH=CH$_2$	CCl$_4$	8.38	31
	p-ClC$_6$H$_4$CH=CH$_2$	CCl$_4$	3.97	31
	m-ClC$_6$H$_4$CH=CH$_2$	CCl$_4$	3.80	31
	C$_6$H$_5$CH=CH$_2$	CCl$_4$	2.92	31
	p-CH$_3$C$_6$H$_4$CH=CH$_2$	CCl$_4$	3.18	31
	p-CH$_3$OC$_6$H$_4$CH=CH$_2$	CCl$_4$	3.28	31

[a] Standard deviation 3–4%.

Correlation of these data according to Hammett equation gives rise to a V-shaped relationship, with slopes only slightly pronounced (ρ values between $+0.5$ and $+0.6$, resp. -0.2 and -0.3); no correlation, however, could be found with meta-substituents (Cl, Br, NO_2).

With substituted benzonitrile oxides (see Table XVIII) slightly positive ρ values ($+0.76$ for styrene[31] as dipolarophile and $+0.6$ for

Table XVIII. Rate constants for the reaction of substituted benzonitrile oxides with phenyl-acetylene[35] and styrene[31]

R	$RC_6H_4CNO + C_6H_5C{\equiv}CH$ $10^3 \cdot K$ at $25 \pm 0.1°$ in CCl_4 $(1 \cdot mole^{-1}\ sec^{-1})$	$RC_6H_4CNO + C_6H_5CH{=}CH_2$ $10^2 \cdot K$ at $25 \pm 0.1°$ in CCl_4[a] $(1 \cdot mole^{-1}\ sec^{-1})$
m-Cl	3.02 ± 0.05	2.20
p-Cl	2.63 ± 0.1	1.66
H	1.78 ± 0.08	1.10
p-CH$_3$	1.50 ± 0.02	0.855
p-CH$_3$O	1.21 ± 0.04	0.731

[a] Standard deviation 3–4%.

phenylacetylene[35]) were found, indicating the developing of small partial charges in the transition state.

A striking substituent effect is the strong promotion of dipolarophilic reactivity exerted by conjugation on the multiple bond. Since k_2 values with the most different dipolarophiles range over several powers of ten, the competition method was used to evaluate dipolarophilic activity. Table XIX lists the relative rate constants of the cycloaddition of benzonitrile oxide, generated *in situ* in ether solution, to various olefinic, acetylenic and hetero-bond dipolarophiles.[37] The values are related to the reaction with ethylene taken equal to unity.

As can be seen from the reported data, conjugation with a C=O bond increases the dipolarophilicity both of a double bond and of a triple bond: *e.g.* acrylic esters react about 25 times faster than propylene, and methyl propiolate 19 times faster than 1-hexyne. This conjugation effect has been ascribed to the π-electron polarizability of the multiple bond.[27] It is also noteworthy that conjugation is effective on both sides of the double or triple bond (acetylenedicarboxylate reacts 2.5 times faster than propiolate).

Steric factors seem also to affect somehow the rate of cycloaddition, since disubstituted ethylenes usually react slower than monosubstituted ones (*e.g.* acrylic ester reacts with benzonitrile oxide 100 times faster

---

37 *Huisgen, R.:* Personal communication. — *Christl, M.:* Ph. D. Thesis, Universität München, 1969.

Table XIX. Relative rate constants for the reaction of benzonitrile oxide with various dipolarophiles

Dipolarophile	k_2 (rel.)[a]	Dipolarophile	k_2 (rel.)[a]
β-Pyrrolidinostyrene	25.2	1,1-Diphenylethylene	(0.39)
Norbornene	15.3	Propylene	0.32
Diethyl mesoxalate	(11.1)	1-Hexene	0.31
Methyl acrylate	8.29	Cyclopentene	0.212
Ethyl acrylate	7.89	Dimethyl maleate	0.211
Dimethyl fumarate	6.06	Ethyl cyanoformate	(0.141)
Benzalmethylamine	(4.44)	Phenylacetylene	0.112
Methyl methacrylate	3.58	Methyl crotonate	0.082
Dimethyl acetylenedicarboxylate	3.06	Vinyl chloride	0.081
4-Nitrostyrene	(2.64)	Methyl cinnamate	0.071
Butyl vinyl ether	2.07	1-Hexyne	0.066
Methyl 3-pyrrolidinoacrylate	1.88	Methyl phenylpropiolate	0.064
4-Methoxystyrene	(1.88)	Methyl tetrolate	0.030
4-Chlorostyrene	(1.55)	*trans*-Stilbene	0.018
4-Methylstyrene	(1.33)	β-Isopropylstyrene	0.014
Methyl propiolate	1.24	Benzaldehyde	(0.0063)
Styrene	1.15	Methyl 3,3-dimethylacrylate	0.0062
Ethylene	1.00	Benzonitrile	(0.0062)
Cyclopentadiene	(0.44)	Cyclohexene	0.0025
Acetylene	0.404		

[a] Bracketted values are taken from unpublished work by *W. Mack* and *K. Bast*, Universität München, reported in Ref.[37].

than crotonic ester and 1300 times faster than β,β-dimethylacrylate). Angularly strained double bonds, such as that in norbornene, are highly dipolarophilic toward nitrile oxides and react even faster than conjugated double bonds, for instance, those in acrylic esters. This behaviour resembles that of phenylazide but differs from other 1,3-dipoles, such as nitrile imines or diazoalkanes: *e.g.* ethyl acrylate reacts with diphenylnitrilimine 15 times faster, and with diphenyldiazomethane 244 times faster than norbornene.[27] Possibly, this means that cycloadditions involving nitrile oxides and azides present a transition state having smaller partial negative charges on the dipolarophile than cycloadditions with other 1,3-dipoles.

The remarkably faster addition of *trans*-isomers as compared with *cis*-isomers (fumaric ester adds about 28 times faster than maleic ester) can also be accounted for by the concerted addition, which implies an increase of the *van der Waals* repulsion for *cis*-structures in the transition state.

Considering finally that alkenes add faster than the corresponding triple bond compounds (for cycloaddition of benzonitrile oxide

$$\frac{k_2(CH_2{=}CHC_6H_5)}{k_2(CH{\equiv}CC_6H_5)}=10.2 \text{ in ether at } 20°; =6.2 \text{ in } CCl_4 \text{ at } 25°),$$

and taking into account the *Woodward-Hoffmann* rules,[30, 38, 39] the cycloaddition of nitrile oxides with alkenes or alkynes seems to pass through a "two-planes" transition state like *8* (exemplified in the case of an alkene), where the non-synchronous formation of the two new σ-bonds, leading to small partial charges stabilized by the substituents, should be emphasized.

8

Owing to the higher π-electron polarizability of carbon and nitrogen atoms of the nitrile oxide[32] in contrast with the low polarizability of the oxygen atom, a concerted electrophilic attack from carbon and nucleophilic attack from oxygen may be expected to control the addition direction. As a matter of fact, 3,5-disubstituted isoxazoles or 2-isoxazolines are usually the predominant products of the cycloaddition of nitrile oxides with monosubstituted alkenes or alkynes, irrespectively of the electron-withdrawing or donating nature of the substituent. Nevertheless it should be noted that a recent reinvestigation[40] of some cycloadditions of this type has demonstrated that 3,4-disubstituted products are also present as by-products and sometimes might become the main product. For instance, acrylic esters react with various nitrile oxides to yield predominantly the isoxazoline-5-carboxylates and not more than 6—7% of the 4-isomers, which corresponds to a $\Delta\Delta G^{\ddagger}=1.5$ kcal/mole for the two energy profiles. On the contrary, methyl propiolate gave an isomer ratio of 72:28 in the cycloaddition with benzonitrile oxide and of 28:72 with mesitonitrile oxide. As steric effects should increase the ratio in the latter case, stabilization of partial charges in transition state may play a larger role and favour a passage through *10* instead of *9* when R is an electron-releasing group. A remarkable influence of the solvent polarity on the isomer ratio has also been observed in the latter case,[37] and seems to support the above conclusion.

9   *10*

38 *Woodward, R.B., Hoffmann, R.*: J. Am. Chem. Soc. **87**, 395, 2046, 2511, 4388, 4389 (1965). — Angew. Chem. **81**, 797 (1969); — Intern. Ed. Engl. **8**, 781 (1969).

39 *Hoffmann, R., Woodward, R.B.*: Accounts Chem. Res. **1**, 17 (1968).

40 *Christl, M., Huisgen, R.*: Tetrahedron Letters 5209 (1968).

Since also several other disubstituted α,β-unsaturated esters and ketones are known to give mixtures of the two predictable isomers (see Sections B and C), both steric and electronic effects of the substituents seem to control the orientation of the cycloaddition.[27, 30]

It should be added that the two-step diradical mechanism has been claimed to explain satisfactorily the orientation in most cases of 1,3-dipolar cycloadditions.[40a] However, in spite of the large amount of experimental data, orientation phenomena are still far from being thoroughly understood and represent perhaps the most fascinating problems to be solved in the field of nitrile oxide cycloaddition reactions.

Dipolarophiles containing hetero-atoms are usually less reactive (see Table XIX) in the cycloaddition than the corresponding C—C unsaturated dipolarophiles hitherto examined. The orientation in this case obeys the principle of maximum gain in σ-bond energy,[27] *viz.* the reactants join in the direction that allows the better compensation of the π-bond energy by the energy of the two new σ-bonds. This corresponds to saying that the electronegative end of the dipolarophile attacks the carbon atom of the nitrile oxide (Eq. (7), (8)).

$$R-C\equiv N\rightarrow O \;+\; R'-C\equiv N \;\longrightarrow\; \qquad (7)$$

$$R-C\equiv N\rightarrow O \;+\; R'-CH=O \;\longrightarrow\; \qquad (8)$$

It has been recently shown that the hetero-dipolarophile becomes more reactive if complexed with Lewis acids (*e.g.* boron trifluoride),[41–44] and the same catalytic effect has been observed subsequently for other 1,3-dipoles.[45]

Although several classes of hetero-dipolarophiles are known to react with nitrile oxides, the mechanism of their cycloaddition which might in principle be quite different from that with CC dipolarophiles, has not yet been studied, with one exception (nitrile oxides and sulfinylanilines, see Section V, G).

40a *Firestone, R. A.:* J. Chem. Soc. (A) 1570 (1970); Communication at the IUPAC-Symposium on Cycloaddition Reactions in Munich, September 1970.

41 *Morrocchi, S.: Ricca, A., Velo, L.:* Tetrahedron Letters 331 (1967).

42 *Morrocchi, S., Ricca, A.:* Chim. Ind. (Milan) **49**, 629 (1967).

43 *Morrocchi, S., Ricca, A., Zanarotti, A.:* Chim. Ind. (Milan) **50**, 352 (1968).

44 *Grundmann, C., Richter, R.:* Tetrahedron Letters 963 (1968).

45 *Hoberg, H.:* Liebigs Ann. Chem. **707**, 147 (1967).

B. Reactions with $C=C$ Derivatives

The 1,3-dipolar cycloaddition of nitrile oxides to alkenes is the most general method of preparation of 2-isoxazolines (Eq. (9))[1] and represents,

$$R-C\equiv N \longrightarrow O \; + \; \underset{/}{\overset{\backslash}{C}}=\underset{\backslash}{\overset{/}{C}} \longrightarrow \quad R \quad \longrightarrow \quad R \qquad (9)$$

through oxidation of the latter ones,[2] a convenient route to isoxazoles. The applicability of the reaction is so large that benzonitrile oxide has been suggested as a suitable reagent for the identification of olefins.[3]

The reaction is generally faster with less-substituted compounds, and conjugation greatly enhances the reactivity of the double bond, as already mentioned (see Section A). Since the cycloaddition reaction always competes with the dimerization reaction (see Chapter IV, C-2), compounds of low reactivity can be successfully added either adopting the technique of slowly generating the nitrile oxide *in situ* in the presence of an excess of dipolarophile, or applying the stable nitrile oxides, such as mesitonitrile oxide or its 3,5-dichloroderivative. It should be emphasized at this point that any comparison of reactivities presented in the following treatment is valid only for the experimental conditions reported in the individual case. As a rule (not devoid of exceptions), yields from most cycloadditions reported before 1962 are due to increase remarkably if the nitrile oxide is generated *in situ* from the corresponding hydroximic acid chloride (either by bases at or below room temperature or thermally, see Chapter III, D).

Attempts have been made to correlate the reactivity of the various types of double bond derivatives with physical properties, in order to predict if the cycloaddition to a particular ethylenic derivative will take place or not, at least with benzonitrile oxide under standard conditions, *i.e.* in boiling ether solution. To this end charge distribution[4] and bond lengths[5] have been evaluated in the cycloaddition to alkenes, inductive effects[6] and dipole moments[7] in the cycloaddition with α,β-unsaturated carbonyl compounds. In particular, a bond length not exceeding the critical value of 1.35 Å, corresponding to more than 80% double bond character, has been claimed to be necessary for an alkene to undergo 1,3-dipolar cycloaddition with benzonitrile oxide.[5]

1 *Quilico, A.:* The chemistry of heterocyclic compounds (ed. *A. Weissberger*), vol. XVII, chap.: Isoxazoles and related compounds, p. 99–115. New York: Interscience 1962.

2 *Bianchi, G., Grünanger, P.:* Tetrahedron **21**, 817 (1965).

3 *Quilico, A., Stagno d'Alcontres, G., Grünanger, P.:* Nature **166**, 226 (1950).

4 *Lo Vecchio, G., Monforte, P.:* Ann. Chim. (Rome) **46**, 76 (1956).

5 *Lo Vecchio, G.:* Gazz. Chim. Ital. **87**, 1413 (1957).

6 *Lo Vecchio, G.:* Ann. Chim. (Rome) **48**, 960 (1958).

7 *Lo Vecchio, G.:* Ann. Chim. (Rome) **48**, 969 (1958).

We will examine the reactivity of the several classes of ethylenic derivatives separately.

(a) Hydrocarbons. Ethylene and monoolefins with a terminal double bond react easily with nitrile oxides to give 2-isoxazolines;[8-15] use of carbethoxyfulmide allows an easy route to 2-amino-4-hydroxycarboxylic acids, following the sequence illustrated in Eq. (10):[16,17]

$$EtOOC-C\equiv N\rightarrow O$$
$$+$$
$$CH_2=CH-R$$

$$(10)$$

$$RCHCH_2CHCOOH$$

1,2-Disubstituted ethylenes react more sluggish, at least in ethereal solution. Tetramethylethylene does not react with benzonitrile oxide even when generated *in situ*,[18] but the reaction can be enforced with the stable 2,4,6-trimethylbenzonitrile oxide, which gave after refluxing for several days an 18% yield of 4,4,5,5-tetramethyl-3-mesityl-2-isoxazoline.[19]

Conjugation with an *aryl group* enhances the reactivity: styrene[20-29] and its ring-substituted derivatives[29-33] react in excellent yields with

8 *Stagno d'Alcontres, G.:* Gazz. Chim. Ital. **82**, 627 (1952).

9 *Andrisano, R., Pappalardo, G.:* Gazz. Chim. Ital. **88**, 174 (1958).

10 *Pappalardo, G.:* Gazz. Chim. Ital. **89**, 1736 (1959).

11 *Angeloni, A., Bellotti, A., Pappalardo, G.:* Gazz. Chim. Ital. **90**, 1616 (1960).

12 *Paul, R., Tchelitcheff, S.:* Bull. Soc. Chim. France 140 (1963).

13 *Aversa, M.C., Cum, G., Crisafulli, M.:* Gazz. Chim. Ital. **96**, 1046 (1966).

14 *Sokolov, S.D.:* Zh. Organ. Khim. **3**, 1532 (1967).

15 *Gaudiano, G., Ponti, P.P., Umani-Ronchi, A.:* Gazz. Chim. Ital. **98**, 48 (1968).

16 *Drefahl, G., Hörhold, H.-H.:* Ber. **97**, 159 (1964).

17 *Hörhold, H.-H.:* Ger. Pat. 37, 461 (1965); — Chem. Abstr. **63**, 10062 g (1965).

18 *Huisgen, R.:* Angew. Chem. **75**, 604 (1963); — Intern. Ed. Engl. **2**, 565 (1963).

19 *Grundmann, C., Frommeld, H.D., Flory, K., Datta, S.K.:* J. Org. Chem. **33**, 1464 (1968).

20 *Stagno d'Alcontres, G., Grünanger, P.:* Gazz. Chim. Ital. **80**, 831 (1950).

21 *Mukaiyama, T., Hoshino, T.:* J. Am. Chem. Soc. **82**, 5339 (1960).

22 *Vaughan, W.R., Spencer, J.L.:* J. Org. Chem. **25**, 1160 (1960).

23 *Arbasino, M., Grünanger, P.:* Ric. Sci. **34**, 561 (1964).

24 *Vita Finzi, P., Arbasino, M.:* Ric. Sci. **35**, 1484 (1965).

25 *Grundmann, C., Frommeld, H.D.:* J. Org. Chem. **31**, 4235 (1966).

26 *Sasaki, T., Yoshioka, T.:* Bull. Chem. Soc. Japan **40**, 2604 (1967).

27 *Sasaki, T., Yoshioka, T.:* Bull. Chem. Soc. Japan, **40**, 2608 (1967).

28 *Minami, S., Matsumoto, J.:* Chem. Pharm. Bull. **15**, 366 (1967).

29 *Dondoni, A., Taddei, F.:* Boll. Sci. Fac. Chim. Ind. Bologna **25**, 145 (1967).

30 *Grünanger, P.:* Gazz. Chim. Ital. **84**, 359 (1954).

31 *Monforte, P., Lo Vecchio, G.:* Atti Accad. Peloritana **49**, 183 (1950/65).

32 *Monforte, P., Lo Vecchio, G.:* Atti Accad. Peloritana **49**, 191 (1950/65).

33 *Vita Finzi, P., Grünanger, P.:* Chim. Ind. (Milan) **47**, 516 (1965).

nitrile oxides to give 3,5-disubstituted 2-isoxazolines; fulminic acid affords 5-phenyl-2-isoxazoline,[34] which can be ring-opened by bases to 3-phenyl-3-hydroxypropionitrile, a reaction typical for 2-isoxazolines and isoxazoles unsubstituted in position 3. Vinyl heterocyclics react as well,[35–40, 28] and cyanogen-bis-N-oxide and terephthalonitrile oxide yield with styrene a mixture of two stereoisomers.[41, 42]

The cycloaddition with styrene allowed to determine the course of the oximation of α,β-ethylenic ketones: contrary to previous claims,[43] attack on the carbonyl is demonstrated by obtention of the same isoxazoline by both routes A and B (Eq. (11)). This coincidence confirms simultaneously the orientation in the cycloaddition.[44]

$$
\begin{array}{ccc}
\text{Ar—C}{\equiv}\text{N}\longrightarrow\text{O} & & \text{Ar—CO—CH}{=}\text{CH—Ar}' \\
+ & & + \\
\text{CH}_2{=}\text{CH—Ar}' & & \text{NH}_2\text{OH}
\end{array}
\qquad (11)
$$

Cyclohexene does not react with equimolecular quantities of benzonitrile oxide in ethereal solution;[45] nevertheless the isoxazolines can be prepared with nitrile oxides generated *in situ*.[46] More strained double bonds, such as in cyclopentene,[45, 47] cyclobutene,[48] acenaphtylene[45, 47, 49] and norbornene,[34, 18] show higher dipolarophilic reactivity. It is also possible to distinguish between exocyclic double bonds (as in methylenecyclohexane) and endocyclic double bonds (as in 1-methylcyclohexene) by assaying their reactivity toward nitrile oxides: the former ones react far more easily to give 5-spiro-compounds, the latter ones either do not react or react more sluggishly.[50] While

34 *Huisgen, R., Christl, M.:* Ang. Chem. **79**, 471 (1967); — Intern. Ed. Engl. **6**, 456 (1967).

35 *Grünanger, P., Grasso, I.:* Gazz. Chim. Ital. **85**, 1271 (1955).

36 *Grünanger, P., Langella, M.R.:* Gazz. Chim. Ital. **91**, 1449 (1961).

37 *Grünanger, P., Vita Finzi, P.:* Rend. Accad. Naz. Lincei **31**, 277 (1961).

38 *Vita Finzi, P., Caramella, P., Grünanger, P.:* Ann. Chim. (Rome) **55**, 1233 (1965).

39 *Caramella, P.:* Ric. Sci. **36**, 986 (1966).

40 *Chang, M.S., Lowe, J.U., Jr.:* J. Org. Chem. **32**, 1577 (1967).

41 *Grundmann, C., Mini, V., Dean, J.M., Frommeld, H.D.:* Liebigs Ann. Chem. **687**, 191 (1965).

42 *Overberger, C.G., Fujimoto, S.:* Polymer Letters **3**, 735 (1965).

43 *Barnes, R.P., Dodson, L.B.:* J. Am. Chem. Soc. **67**, 132 (1945).

44 *Grünanger, P.:* Rend. Accad. Naz. Lincei **16**, 726 (1954).

45 *Barbulescu, N., Grünanger, P., Langella, M.R., Quilico, A.:* Tetrahedron Letters 89 (1961).

46 *Zinner, G., Günther, H.:* Ber. **98**, 1353 (1965).

47 *Barbulescu, N., Grünanger, P.:* Gazz. Chim. Ital. **92**, 138 (1962).

48 *Bianchi, G., Gandolfi, R.:* Personal communication.

49 *Bachmann, G.B., Strom. L.E.:* J. Org. Chem. **28**, 1150 (1963).

50 *Grünanger, P., Langella, M.R.:* Gazz. Chim. Ital. **91**, 1112 (1961).

β-pinene reacts promptly in ether solution,[45] the thermal generation *in situ* was found necessary in order to obtain the α-pinene derivative.[51]

The influence of conjugation on the orientation of the cycloaddition is illustrated by the reaction of nitrile oxides with indene: whereas benzonitrile oxide yields only *11*,[20,52] acetonitrile oxide yields both isomers *12* and *13* in a 94:6 ratio.[53] *Cis*- and *trans*-stilbene[20,54,55] yield the

11, R=C$_6$H$_5$
12, R=CH$_3$

13

diastereomeric 4,5-diphenyl-2-isoxazolines, the *trans* isomer giving as a rule better yields. Dicyan-di-N-oxide reacts with *trans*-stilbene to yield the two diastereomers *14* (meso-*trans*) and *15* (dl-*trans*).[41] Phenanthrene has been reported to be unreactive.[45]

14

15

Di- and poly-olefins will react, depending on the conditions, with one or more double bonds. *E.g.* 1,3-butadiene reacts with benzonitrile oxide to give both the 3-phenyl-5-vinyl-2-isoxazoline (*16*, R=C$_6$H$_5$) and the bis-adduct *17* (R=C$_6$H$_5$);[56] with other nitrile oxides only mono-adducts of type *16*[41,25,26,57] or bis-adducts of type *17*[58] have been reported.

16

17

18

51 *Sasaki, T., Yoshioka, T.:* Bull. Chem. Soc. Japan **41**, 2206 (1968).

52 *Perold, G. W., v. Reiche, F. V. K.:* J. African Chem. Inst. **10**, 5 (1957).

53 *Bianchi, G., Gandolfi, R., Grünanger, P., Perotti, A.:* J. Chem. Soc. (C) 1598 (1967).

54 *Grünanger, P., Gandini, C., Quilico, A.:* Rend. Ist. Lombardo Sci. Lettere **93**, 467 (1959).

55 *Huisgen, R.:* Personal communication. — *Christl, M.:* Ph. D. Thesis, Universität München, 1969.

56 *Quilico, A., Grünanger, P., Mazzini, R.:* Gazz. Chim. Ital. **82**, 349 (1952).

57 *Tartakovskii, V. A., Luk'yanov, O. A., Shlykova, N. I., Novikov, S. S.:* Zh. Organ. Khim. **3**, 980 (1967).

58 *Zinner, G., Günther, H.:* Angew. Chem. **76**, 440 (1964); — Intern. Ed. Engl. **3**, 383 (1964).

From the reaction between isoprene and benzonitrile oxide the two possible monoadducts and the bisadduct have been obtained.[59] On the contrary, 1-aryl-1,3-butadienes react only with their more reactive terminal bond to give *18*.[56, 60] Conjugated systems seem to be more reactive than heavily substituted isolated double bonds: *e.g.* myrcene reacts, although with low yield, with benzonitrile oxide to give exclusively 1:2 adducts.[61]

Mono- and bis-adducts are often easily separated because of the very low solubility of di-isoxazolines in ether.[8]

Cyclic dienes, such as cyclopentadiene,[45, 47] norbornadiene[34, 62] or 1,3-cyclohexadiene[63], react similarly; fulvenes add only on the two endocyclic double bonds,[56] whereas dicyclopentadiene yields only a mono-adduct.[56, 64] Benzonitrile oxide adds at low temperature cyclo-octatetraene to yield first the bicyclic adduct *19*, which at room temperature tautomerizes to the tricyclic isomer *20*, whose structure is assured by thermal ring opening (Eq. (12)).[65, 66] Fulminic acid seems to react with cyclic polyolefins more sluggishly.[34]

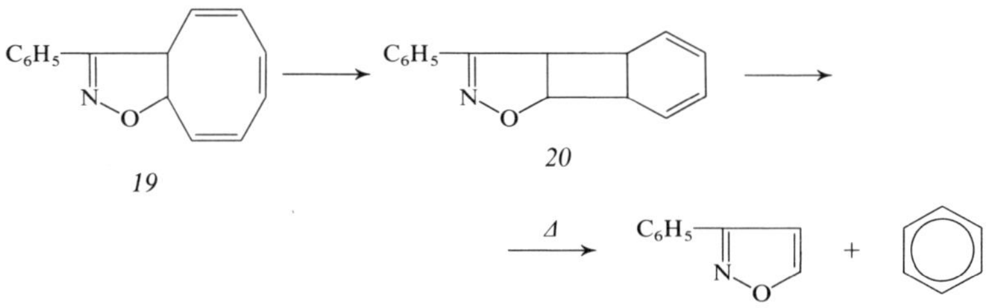

$$(12)$$

Allenes react with two molecules of nitrile oxides to yield spiro-compounds of type *21*;[67, 68] the elusive intermediate 5-methylene-2-iso-xazoline has never been isolated and is known to isomerize very rapidly to 5-methylisoxazole.[69, 70] The same spiro-compounds can be prepared

59 *Chistokletov, N. V., Troshchenko, A. T., Petrov, A. A.:* Zh. Obshchei Khim. **34**, 1891 (1964).

60 *Lo Vecchio, G., Monforte, P.:* Atti Soc. Peloritana **4**, 229 (1957/58).

61 *Sasaki, T., Eguchi, S., Ishii, T.:* Bull. Chem. Soc. Japan **42**, 558 (1969).

62 *Barbulescu, N., Lazar, R.:* Rev. Roumaine Chim. **11**, 1141 (1966).

63 *Bettinetti, G. F., Gamba, A.:* Gazz. Chim. Ital., in press.

64 *Perold, G. W., Steyn, A. P., v. Reiche, F. V. K.:* J. Am. Chem. Soc. **79**, 462 (1957).

65 *Bianchi, G., Gandolfi, R., Grünanger, P.:* Chim. Ind. (Milan) **49**, 757 (1967).

66 *Christl, M., Huisgen, R.:* Tetrahedron Letters 5209 (1968).

67 *Stagno d'Alcontres, G., Lo Vecchio, G.:* Gazz. Chim. Ital. **90**, 1239 (1960).

68 *Stagno d'Alcontres, G., Gattuso, M., Lo Vecchio, G., Crisafulli, M., Aversa, M. C.:* Gazz. Chim. Ital. **98**, 203 (1968).

69 *Sasaki, T., Yoshioka, T.:* Bull. Chem. Soc. Japan **42**, 258 (1969).

70 *Caramella, P., Gamba, A., Grünanger, P.:* Unpublished.

also by adding nitrile oxides to diketene (Eq. (13)).[71] Bifunctional nitrile oxides, such as cyanogen-di-N-oxide or phthalonitrile oxide, add allenes, cyclopentadiene and 1,3-butadiene to give, depending on the conditions, bis-adducts or 1:1 alternating copolymers containing 2-isoxazoline rings.[41, 42]

$$2\,R\!-\!C\!\equiv\!N\!\longrightarrow\!O \;+\; CH_2\!=\!C\!=\!CH_2 \;\longrightarrow\; \underset{21}{[\text{spiro isoxazoline}]} \;\xleftarrow{-CO_2}\; 2\,R\!-\!C\!\equiv\!N\!\longrightarrow\!O \;+\; CH_2\!=\!C(-O-CO-)CH_2 \tag{13}$$

Non-conjugated dienes react easily if the double bonds are terminal, as in diallyl,[56] but also one double bond in 1,4-cyclohexadiene[41] or 1,5-cyclooctadiene[51] can react.

Double bonds of a true aromatic system in benzenoid or heterocyclic rings do not react with nitrile oxides under standard conditions,[35] but also here the cycloaddition can be enforced with furans, pyrrole and thiophene by generating the nitrile oxide *in situ*.[72] Furan for instance yields as main product either the mono-adduct *22*, arising from attack of the electrophilic carbon atom of the nitrile oxide on the α-position of furan, or the bis-adduct *23*, where a second molecule of nitrile oxide has added to the vinyl ether double bond (Eq. (14)).

$$Ar\!-\!C(=\!N\!\rightarrow\!O) \;+\; [\text{furan}] \;\longrightarrow\; \underset{22}{[\text{mono-adduct}]} \;\longrightarrow\; \tag{14}$$

$$\xrightarrow{+\,ArCNO}\; \underset{23}{[\text{bis-adduct}]}$$

(b) Oxygen-Containing Compounds. Acyclic and cyclic vinyl ethers and esters react readily with nitrile oxides:[73–76, 21, 28] The initially formed

71 *Stagno d'Alcontres, G., Cum, G., Gattuso, M.*: Ric. Sci. **37**, 750 (1967).

72 *Corsico Coda, A., Grünanger, P., Veronesi, G.*: Tetrahedron Letters 2911 (1966).

73 *Stagno d'Alcontres, G., Grünanger, P.*: Gazz. Chim. Ital. **80**, 741 (1950).

74 *Just, G., Dahl, K.*: Tetrahedron Letters 2441 (1966).

75 *Desimoni, G., Grünanger, P.*: Gazz. Chim. Ital. **97**, 25 (1967).

76 *Just, G., Dahl, K.*: Tetrahedron **24**, 5251 (1968).

5-alkoxy- or 5-acyloxy-2-isoxazolines aromatize easily to isoxazoles either thermally or by treatment with acids or alkalies. This sequence represents the most convenient method for preparing 3-monosubstituted isoxazoles,[77-79] and is suitable for the synthesis of isoxazole itself,[80] if fulminic acid is used. When nitromethane and phenylisocyanate were used in an attempt to produce fulminic acid *in situ*, a 2-isoxazoline-3-carboxanilide was obtained,[12] as exemplified in Eq. (15); it could be demonstrated that nitromethane is first transformed by phenylisocyanate into the nitrile oxide *24*, which then enters cycloaddition:[55]

$$(15)$$

Dihydrofurans and -pyrans yield bicyclic isoxazolines,[12, 28, 77] the latter ones, as normal for six-membered rings, with lower yields. Ring scissions of these bicyclic compounds lead to interesting products: 3,4-Disubstituted isoxazoles are obtained by acidic or basic hydrolysis,[81] and substituted 1,4- or 1,5-diols are formed by hydrogenolysis (Eq. (16)).[77]

$$(16)$$

$n = 2$ or 3

77 *Paul, R., Tchelitcheff, S.*: Bull. Soc. Chim. France 2215 (1962).

78 *Langella, M.R., Vita Finzi, P.*: Chim. Ind. (Milan) **47**, 996 (1965).

79 *Bianchi, G., Cogoli, A., Grünanger, P.*: J. Organomet. Chem. **6**, 598 (1966).

80 *Stagno d'Alcontres, G., Mollica, G.*: Rend. Accad. Naz. Lincei **10**, 52 (1951).

81 *Adachi, I., Kano, H.*: Chem. Pharm. Bull. **16**, 117(1968).—*Kano, H., Adachi, I., Kido, R., Hirose, K.* (to Shionogi & Co.): Japan Pat. 29656 (1969); — Chem. Abstr. **72**, 43643 (1970).

The adducts *25* with 2,5-dihydrofuran provide an easy way to 3,4-disubstituted tetrahydrofurans (Eq. (17)).[77]

$$\text{(17)}$$

The cycloaddition of nitrile oxides to vinylidenecarbonate opens a convenient route to 4-hydroxyisoxazoles (Eq. (18)):[82]

$$\text{(18)}$$

α,β-Unsaturated carbonyl compounds are highly reactive dipolarophiles. Vinyl compounds of type $CH_2{=}CH{-}CO{-}R$ are reported to afford exclusively the 2-isoxazolines having the carbonyl group in 5-position: this is true *e.g.* for acrolein (and its diethylacetal)[83] and for vinyl alkyl ketones.[84–86, 13, 28] The 5-acyl-2-isoxazolines are characterized by their base-promoted fragmentation to nitrile and α-diketones.[86]

For cycloaddition reactions with nitrile oxides vinyl aryl ketones are preferably prepared *in situ* from Mannich bases. Direct reaction of these with hydroximic acid chlorides gave good yields of the expected 5-acyl-2-isoxazolines, formed most probably through nitrile oxides.[86–88]

α,β-Unsaturated ketones, such as benzalacetone[20,89] and benzalacetophenone,[89,90] add benzonitrile oxides to give variable amounts of the two possible structural isomers *27* and *28*; mesityl oxide, on the contrary, gives only 3-phenyl-4-acetyl-5,5-dimethyl-2-isoxazoline in low yield. Table XX illustrates the substituents effect on the regioselectivity of the cycloaddition.[89]

82 *Desimoni, G., Grünanger, P., Servi, S.:* Ann. Chim. (Rome) **58**, 1363 (1968).
83 *Stagno d'Alcontres, G., De Giacomo, G.:* Atti Soc. Peloritana **5**, 159 (1958/59).
84 *Quilico, A., Grünanger, P.:* Rend. Ist. Lombardo Sci. Lettere **88**, 990 (1955).
85 *Piozzi, F., Fuganti, C.:* Ann. Chim. (Rome) **57**, 486 (1967).
86 *Bianchi, G., Gandolfi, R., Grünanger, P.:* J. Heter. Chem. **5**, 49 (1968).
87 *Bianchi, G., Grünanger, P.:* Chim. Ind. (Milan) **46**, 1187 (1964).
88 *Bianchi, G., Galli, A., Gandolfi, R.:* Gazz. Chim. Ital. **98**, 331 (1968).
89 *Bianchi, G., Gandolfi, R., Vita Finzi, P.:* Personal communication.
90 *Stagno d'Alcontres, G., Fenech, G.:* Acta Scient. Venezolana **4**, 95 (1953).

Table XX. Cycloaddition of nitrile oxides to α,β-unsaturated ketones: isomers ratio[89]

R	R'	R"	27:28	R	R'	R"	27:28
C_6H_5	CH_3	C_6H_5	58:42	C_6H_5	α-C_4H_3S	C_6H_5	44:56
C_6H_5	C_2H_5	C_6H_5	48:52	C_6H_5	C_6H_5	α-C_4H_3S	63:37
p-$CH_3C_6H_4$	CH_3	C_6H_5	54:46	C_6H_5	α-C_4H_3S	α-C_4H_3O	76:24
2,4,6-$(CH_3)_3C_6H_2$	CH_3	C_6H_5	28:72	C_6H_5	α-C_4H_3O	α-C_4H_3S	87:13
p-BrC_6H_4	CH_3	C_6H_5	66:34	2,4,6-$(CH_3)_3C_6H_2$	C_6H_5	α-C_4H_3O	70:30
p-$CH_3OC_6H_4$	CH_3	C_6H_5	56:44	2,4,6-$(CH_3)_3C_6H_2$	α-C_4H_3O	C_6H_5	30:70
p-$NO_2C_6H_4$	CH_3	C_6H_5	67:33	2,4,6-$(CH_3)_3C_6H_2$	α-C_4H_3S	C_6H_5	14:86
C_6H_5	C_6H_5	C_6H_5	28:72	2,4,6-$(CH_3)_3C_6H_2$	α-C_4H_3O	α-C_4H_3O	87:13

$$R\text{—}C(=N)\text{—}C(\text{—COR}')\text{—}C(\text{—}R'')\text{—O}$$

27   *28*

Spiro-compounds of type *29* are obtained by use of cycloalkanones containing an α,β-semicyclic double bond such as alkylidene- (or arylidene) cyclo-hexanone (or -pentanone).[91-95] Starting from benzal-5-isoxazolones[96-98] or 5-pyrazolones[99,99a] spiro-heterocyclics of structure *30* are obtained.

29

30   X = O, NAr

The double bonds of p-quinones react easily with aromatic nitrile oxides[100,101,51] to give adducts, which in the absence of BF_3-etherate cannot be isolated, but oxidize to isoxazolo-quinones. In the presence

91 *Barbulescu, N., Quilico, A.:* Gazz. Chim. Ital. **91**, 326 (1961).

92 *Barbulescu, N., Lazar, R.:* Analele Univ. Bucaresti, Ser. Stiint. Nat. **12**, 115 (1963); — Chem. Abstr. **64**, 19587e (1966).

93 *Fritsch, W., Seidl, G., Ruschig, R.:* Liebigs Ann. Chem. **677**, 139 (1964).

94 *Awad, W.I., Moustafa, A.H., Raouf, A.R.A.:* J. Chem. U. A. R. **8**, 137 (1965).

95 *Fritsch, W., Stache, U.:* Ger. Pat. 1214224 (1966); — Chem. Abstr. **65**, 12266a (1966).

96 *Lo Vecchio, G., Cum, G., Stagno d'Alcontres, G.:* Tetrahedron Letters 3495 (1964).

97 *Lo Vecchio, G., Cum, G., Stagno d'Alcontres:* Gazz. Chim. Ital. **95**, 127 (1965).

98 *Lo Vecchio, G., Cum, G., Stagno d'Alcontres, G.:* Gazz. Chim. Ital. **95**, 206 (1965).

99 *Lo Vecchio, G., Gattuso, M., Stagno d'Alcontres, G.:* Gazz. Chim. Ital. **99**, 121 (1969); — Atti Soc. Peloritana **14**, 251 (1968).

99a *Lo Vecchio, G., Gattuso, M., Uccella, N.:* Atti Soc. Peloritana **14**, 291, 393 (1968).

100 *Quilico, A., Stagno d'Alcontres, G.:* Gazz. Chim. Ital. **80**, 140 (1950).

101 *Morrocchi, S., Ricca, A., Quilico, A., Selva, A.:* Gazz. Chim. Ital. **98**, 891 (1968).

of Lewis acids the primary adduct separates out, together with other compounds, formed by further addition of the nitrile oxide to the carbonyl group (see Section D).[101] n-Butyronitrile oxide, generated in situ from 1-nitrobutane (cf. Chapter III, C-2), seems to be unreactive toward quinones.[49]

1-Acetylcyclopentene is reported to give mainly the 5-acetyl bicyclic 2-isoxazoline;[102] several isoxazolino-steroids[93, 103–107] have been prepared through cycloaddition of nitrile oxides to the Δ^{16}-double bond, but the orientation of the cycloaddition does not seem to be uniform in all cases. Addition of the oxygen atom of nitrile oxide to C-17 and of the carbon atom to C-16 is supported in two papers[104, 105] by chemical proofs.

β-Dicarbonyl derivatives, such as β-diketones and β-ketoesters, as well as β-cyanoketones, react with nitrile oxides in the presence of small quantities of alkali hydroxides or alkoxides, giving good yields of isoxazole-4-ketones or -carboxylates or -nitriles.[108] Most probably the enolic form acts as dipolarophile in the cycloaddition (Eq. (19)) and the

$$
\begin{array}{c}
R-C\equiv N\rightarrow O \; + \; \underset{\underset{\text{OH}}{C}}{\overset{\text{CHCOR}'}{\underset{\text{R}''}{\parallel}}} \longrightarrow \left[\begin{array}{c} R\text{---}\\ N\diagdown_O \end{array} \right] \xrightarrow{-H_2O} \quad (19)
\end{array}
$$

31

R' = Alk or O-alk

intermediate 5-hydroxyisoxazoline *31* dehydrates immediately to the corresponding isoxazole.[109] Analogously, cyanoacetic esters can react in their keten-iminic tautomeric form to yield 5-aminoisoxazole-4-carboxylates.[108]

Indeed, the reaction between hydroximic acid chlorides and an alcoholic solution of the sodium salts of β-dicarbonyl or β-cyanocarbonyl derivatives, one of the most widely used method of synthesis of the isoxazole ring,[110] occurs very probably via the nitrile oxides, liberated *in situ* by the strong carbanion base. This view is supported by the formation of furoxans as by-products and by the isolation of the

102 *Corsico Coda, A., Grünanger, P.:* Rend. Accad. Naz. Lincei **40**, 586 (1966).
103 *Stache, U., Fritsch, W., Ruschig, H.:* Liebigs Ann. Chem. **685**, 228 (1965).
104 *Culbertson, T.P., Moersch, G.W., Neuklis, W.A.:* J. Heterocycl. Chem. **1**, 280 (1964).
105 *Moersch, G.W., Wittle, E.L., Neuklis, W.A.:* J. Org. Chem. **30**, 1272 (1965).
106 *Fritsch, W., Seidl, G.:* Ger. Pat. 1 210 821 (1966); — Chem. Abstr. **64**, 17 682 f (1966).
107 *Gerali, G., Parini, C., Sportoletti, G.C., Ius, A.:* Farmaco (Pavia) **24**, 231 (1969).
108 *Quilico, A., Speroni, G.:* Gazz. Chim. Ital. **76**, 148 (1946).
109 *Quilico, A., Stagno d'Alcontres, G., Grünanger, P.:* Gazz. Chim. Ital. **80**, 479 (1950).
110 Ref. 1, p. 19.

relatively stable nitrile oxide *33* from the corresponding hydroximic acid chloride *32* and sodium ethyl acetoacetate (Eq. (20)).[108]

$$C_6H_5 \text{—isoxazole—} C(Cl)=NOH \;(32) \;+\; Na^+[CH_3COCHCOOEt]^- \longrightarrow$$

$$\longrightarrow C_6H_5 \text{—isoxazole—} C\equiv N \rightarrow O \;(33) \xrightarrow{\Delta} \text{Furoxan}$$

$$C_6H_5 \text{—bis-isoxazole—} COOEt \tag{20}$$

The reaction of hydroximic acid chlorides with β-ketoesters and malonic derivatives in the presence of aqueous alkali to give isoxazolones probably follows a different mechanism, since intermediate open-chain sodium salts have been isolated.[111–115]

Free α,β-unsaturated acids can react either with their double bond to give isoxazoline derivatives or with their carboxylic group to give hydroxamic esters (see also Chapter VI, F). When the double bond is less substituted or activated by conjugation, as in acrylic acid, maleic acid or nitrocinnamic acids, the corresponding isoxazoline carboxylic acids are obtained, whereas cinnamic acid yields a mixture of the two products and crotonic and furylacrylic acids give only the hydroxamic acid derivatives.[54,116–118]

Acrylic esters yield predominantly the 5-carboxylate *35* (R=H), but more substituted esters, such as *cis*-[54,119] and *trans*-cinnamates[54,66,120]

111 *Stagno d'Alcontres, G., Lo Vecchio, G., Lamonica, G.:* Gazz. Chim. Ital. **91**, 1005 (1961).

112 *Lo Vecchio, G., Crisafulli, M.:* Atti Soc. Peloritana **7**, 223 (1961).

113 *Lo Vecchio, G., Cum, G., Lamonica, G.:* Atti Soc. Peloritana **8**, 351 (1962).

114 *Lo Vecchio, G., Lamonica, G., Cum, G.:* Gazz. Chim. Ital. **93**, 15 (1963).

115 *Stagno d'Alcontres, G., Lo Vecchio, G., Crisafulli, M., Gattuso, M.:* Gazz. Chim. Ital. **97**, 997 (1967).

116 *Grünanger, P., Vita Finzi, P.:* Rend. Accad. Naz. Lincei **26**, 386 (1959).

117 *Quilico, A., Grünanger, P.:* Gazz. Chim. Ital. **85**, 1449 (1955).

118 *Monforte, F., Lo Vecchio, G.:* Gazz. Chim. Ital. **83**, 416 (1953).

119 *Arbasino, M., Vita Finzi, P.:* Ric. Sci. **36**, 1339 (1966).

120 *Monforte, F.:* Gazz. Chim. Ital. **82**, 130 (1952).

and derivatives thereof[118] or *trans*-crotonate,[66] react with fulminic acid or nitrile oxides to give a mixture of the two possible isomers, *i.e.* 2-isoxazoline-4-carboxylate (*34*) and -5-carboxylate (*35*) (Eq. (21)).

$$
\begin{array}{l}
R-C\equiv N\rightarrow O \\
\qquad + \\
\end{array}
\quad\longrightarrow\quad
34 \quad + \quad 35 \tag{21}
$$

Table XXI. Ratio of isomers (Eq. (21))

R	R'=H *34:35*	R'=CH$_3$ *34:35*	R'=C$_6$H$_5$ *34:35*
H	0 :100	38:62	76:24
CH$_3$	5.1:94.9	64:36	70:30
(CH$_3$)$_3$C	0 :100	86:14	78:22
N≡C	2 :98	56:44	85:15
HON=CH		55:45	
C$_6$H$_5$	3.6:96.4	66:34	70:30
2,4,6-(CH$_3$)$_3$C$_6$H$_2$	6.6:93.4	73:27	64:36

Table XXI reports the ratio of the two isomers with several nitrile oxides.[66] Methyl β,β-dimethylacrylate yields with aliphatic as well as with aromatic nitrile oxides exclusively the isoxazoline-4-carboxylate. These results indicate that both steric and electronic effects from dipolarophile and nitrile oxide are at work during the cycloaddition.

Widely studied has been the cycloaddition of nitrile oxides with *cis-trans* isomers, which yields through a strict *cis*-addition the two diastereomers. *E.g.* dialkyl maleate and fumarate react with aromatic nitrile oxides[109, 121, 122] to give *cis*- (*36*) resp. *trans*- (*37*) 3-aryl-2-isoxazoline-4,5-dicarboxylate. In polar solvents the two diastereomers reach, possibly through enolization of the 4-carboxylate group, an equilibrium, where the lower melting *trans*-isomer largely prevails. The structure of the isomers are assured by the below reported interconversions (Eq. (22)). The insolubility of its potassium salt allows to isolate the *cis*-acid *38* in methanol from both isomeric esters, whereas the *trans*-acid *39* is the only product from aqueous hydrolysis.

Analogous reactions have been reported for the couple mesaconate -citraconate.[121]

Maleic anhydride is one of the most widely used dipolarophiles, since with nitrile oxides it forms often highly insoluble adducts.

121 *Quilico, A., Grünanger, P.:* Gazz. Chim. Ital. **82**, 140 (1952).
122 *Quilico, A., Grünanger, P.:* Gazz. Chim. Ital. **85**, 1250 (1955).

$$\text{(22)}$$

The carbonyl group can be replaced by a sulfonyl group: α,β-unsaturated sulfones[123] and sulfonic esters behave exactly as their carbonyl counterparts.

The cumulated double bond system of ketene and its homologs is very little reactive toward nitrile oxides, and usually the dimerization reactions of both types of compounds prevail. Only diphenylketene, more stable than other representatives, reacts with aliphatic nitrile oxide to yield 4,4-diphenyl-5-isoxazolones.[58, 124] The failure of free ketenes to undergo cycloadditions with nitrile oxides has been ascribed[125] to the similar prevailing electrophilic character of both classes of compounds.

On the contrary, ketene acetals,[125, 126] mercaptals[127] and aminals[128] add readily nitrile oxides: the acetals yield 5,5-dialkoxyisoxazolines, which aromatize in neutral medium to 5-alkoxyisoxazoles and are

123 *Corsico Coda, A., Limone, M.A.:* Ric. Sci. **36**, 345 (1966).
124 *Scarpati, R., Sorrentino, P.:* Gazz. Chim. Ital. **89**, 1525 (1959).
125 *Scarpati, R., Speroni, G.:* Gazz. Chim. Ital. **89**, 1511 (1959).
126 *Grünanger, P., Langella, M.R.:* Gazz. Chim. Ital. **89**, 1784 (1959).
127 *Scarpati, R., Santamaria, C., Sica, D.:* Gazz. Chim. Ital. **93**, 1706 (1963).
128 *Rajagopalan, P., Talaty, C.N.:* Tetrahedron Letters 4537 (1966).

further hydrolized by acids to 5-isoxazolones. With mercaptals and aminals the intermediate isoxazolines were not isolated and the corresponding 5-substituted isoxazoles have been directly obtained.

Finally, it should be mentioned that the cycloaddition of nitrile oxides to unsaturated compounds has been used in order to prepare isoxazoline and isoxazole C-glucosides.[128a]

(c) Nitrogen-containing Compounds. Enamines are highly reactive and versatile dipolarophiles and add nitrile oxides to yield as primary product the 2-isoxazolines, substituted in 5-position by the basic residue. Ready elimination of the base on acid treatment and formation of the isoxazole was generally observed (Eq. (23)).[129–132,27,28] 4,5-Trimethyl-

$$
\text{Ar–C}\!\!\underset{N\diagdown O}{\overset{\text{\tiny III}}{}} + \underset{\substack{C\text{–}R''\\|\\NR_2}}{\overset{CH\text{–}R'}{}} \longrightarrow \underset{N\diagdown O\diagdown NR_2}{Ar\text{–}\!\!\overset{R'}{\underset{R''}{}}} \xrightarrow[-R_2NH]{H^+} \underset{N\diagdown O\diagdown R''}{Ar\text{–}\!\!\overset{R'}{}} \qquad (23)
$$

ene-2-isoxazolines, prepared from cyclopentanone-enamines, are stable to acids, and elimination of the base is achieved through Hoffmann degradation.[131,133] Interesting applications of the cycloaddition of enamines derivatives are a convenient way to 3,4-disubstituted isoxazoles,[28,134] otherwise not easily accessible, a general synthesis of isoxazole-4-carboxylic acid derivatives,[129,134] an alternative to the above mentioned method utilizing β-ketoesters, and a preparation of indoxazenes and other polynuclear systems.[53,135] The existence of N-morpholino-ethylene in equilibrium with 1,1-di(N-morpholino)ethane could be demonstrated through capture with p-nitrobenzonitrile oxide.[136]

Both double bonds of dienamines are reactive and the cycloaddition to nitrile oxides affords an easy route to 4,5'-diisoxazoles.[137]

β-Iminoketones react both in their enamine and imino-enolic forms: *e.g.* iminoacetylacetone gives with benzonitrile oxide a mixture of a monoadduct *40* and a diadduct *42*.[138] With asymmetrical iminoketones (R ≠ R') the most reactive tautomeric form depends on the nature of the substituents. If the free carbonyl group is bound to a phenyl group, the

128a *Tronchet, J.M.J. et al.:* Helv. Chim. Acta **53**, 1484 (1970).
129 *Bianchetti, G., Pocar, D., Dalla Croce, P.:* Gazz. Chim. Ital. **93**, 1714 (1963).
130 *Bianchetti, G., Pocar, D., Dalla Croce, P.:* Gazz. Chim. Ital. **93**, 1726 (1963).
131 *Kuehne, M.E., Weaver, S.J., Franz, P.:* J. Org. Chem. **29**, 1582 (1964).
132 *Bianchetti, G., Pocar, D., Dalla Croce, P., Vigevani, A.:* Ber. **98**, 2715 (1965).
133 *Bianchi, G., Grünanger, P.:* Chim. Ind. (Milan) **46**, 425 (1964).
134 *Stork, G., Mc Murry, J.E.:* J. Am. Chem. Soc. **89**, 5461 (1967).
135 *Bianchi, G., Frati, E.:* Gazz. Chim. Ital. **96**, 559 (1966).
136 *Ferruti, P., Pocar, D., Bianchetti, G.:* Gazz. Chim. Ital. **97**, 109 (1967).
137 *Caramella, P., Bianchessi, P.:* Tetrahedron **26**, 5773 (1970).
138 *Morrocchi, S., Ricca, A., Velo, L.:* Chim. Ind. (Milan) **49**, 168 (1967). — *Ricca, A.:* Personal communication.

enaminic form reacts to give *41*, if it is bound to an alkyl group, the iminoenolic form yields the bis-adduct *43*. The higher dipolarophilic reactivity of the imino group in respect to the carbonyl group is well known (see Section E).

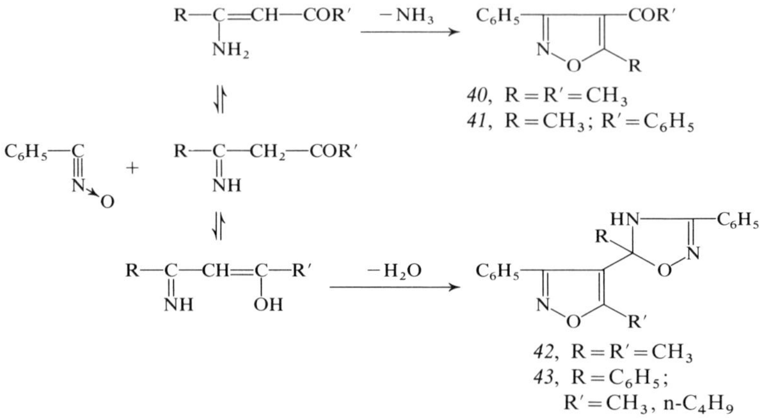

$$R-\underset{NH_2}{C}=CH-COR' \xrightarrow{-NH_3} \text{(isoxazole)} $$

40, $R = R' = CH_3$
41, $R = CH_3$; $R' = C_6H_5$

$$C_6H_5-C\equiv N\rightarrow O \;+\; R-\underset{NH}{C}-CH_2-COR'$$

$$R-\underset{NH}{C}-CH=\underset{OH}{C}-R' \xrightarrow{-H_2O}$$

42, $R = R' = CH_3$
43, $R = C_6H_5$;
$\qquad R' = CH_3$, n-C_4H_9

$$\tag{24}$$

Nevertheless, when imino-benzoylacetone reacts with benzonitrile oxide in the presence of a coordinating acidic agent, such as boron trifluoride, besides *41* the isomeric *44* is obtained, according to Eq. (25).[140]

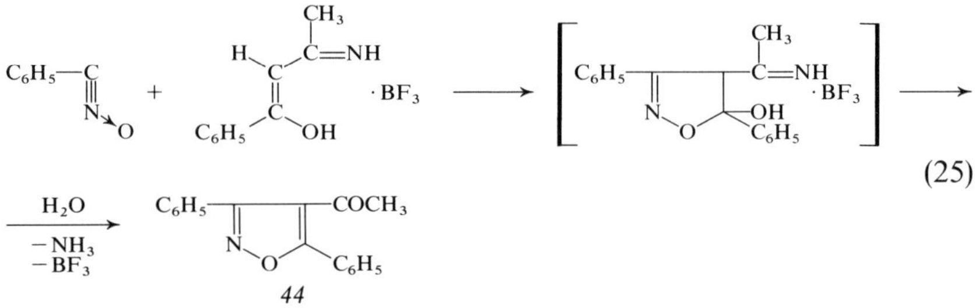

$$C_6H_5-C\equiv N\rightarrow O \;+\; \underset{C_6H_5}{\overset{CH_3}{\text{(enol)}}}\cdot BF_3 \longrightarrow \left[\cdots \cdot BF_3 \right] \longrightarrow$$

$$\xrightarrow[\substack{-NH_3\\-BF_3}]{H_2O} \text{(isoxazole)}$$

44

$$\tag{25}$$

Thus, the presence of boron trifluoride causes the inversion of the cycloaddition orientation.

(d) Miscellaneous Derivatives. Vinyl halides and β-nitrostyrene add nitrile oxides to give the corresponding 4- or 5-substituted 2-isoxazolines. Isoxazolines with a halogen atom in the 5-position are usually not isolable and lose spontaneously hydrochloric or hydrobromic acid to yield directly isoxazoles.[3,30,55,67] 4-Halogen- or 4-nitroisoxazolines aromatize thermally or by alkalies.[20,30,55,139]

139 *Monforte, P., Lo Vecchio, G.:* Ann. Chim. (Rome) **46**, 84 (1956).

Allyl halides yield easily 3,5-disubstituted isoxazolines;[16, 73] fulminic acid reacts in this case as its dimer.[140] β-Chlorovinyl-alkyl (or aryl) ketones yield the corresponding 4-chloro-5-acyl-2-isoxazolines, which could be isolated in pure form only in the case of the aryl derivatives; otherwise the adduct loses hydrogen chloride either thermally or better by treatment with alkali to give 5-acylisoxazoles.[141–143]

Vinyl boronate adds normally benzonitrile oxide to yield 3-phenyl-2-isoxazoline-5-boronic acid, which affords an easy route to 5-hydroxy-isoxazoline (Eq. (26)).[144]

$$
\begin{array}{c}
R-C\equiv N\rightarrow O \\
+ \\
CH_2=CHB(OC_4H_9)_2
\end{array}
\longrightarrow
\underset{\substack{}}{R\text{-isoxazoline}}B(OH)_2
\xrightarrow{H_2O_2}
\underset{\substack{}}{R\text{-isoxazoline}}OH
\qquad (26)
$$

All compounds obtained by cycloaddition of nitrile oxides to various types of olefinic compounds are compiled in Chapter IX (Tables XXVIII a–e).

140 *Stagno d'Alcontres, G., Fenech, G.:* Gazz. Chim. Ital. **82**, 175 (1952).
141 *Grünanger, P., Mangiapan, S.:* Gazz. Chim. Ital. **88**, 149 (1958).
142 *Grünanger, P., Fabbri, E.:* Gazz. Chim. Ital. **89**, 598 (1959).
143 *Vita Finzi, P., Arbasino, M.:* Ann. Chim. (Rome) **54**, 1165 (1964).
144 *Bianchi, G., Cogoli, A., Grünanger, P.:* Ric. Sci. **36**, 132 (1966).

C. Reactions with $C \equiv C$ Derivatives

It has already been mentioned (see Section A, Table XVIII) that triple bonds react with nitrile oxides more slowly than double bonds. However the triple bond $C \equiv C$ is by far more reactive than the triple bond $C \equiv N$: phenylacetylene is reported to be 20 times more reactive than benzonitrile and ethyl propiolate 10 times more than ethyl cyanoformate.[1]

Apart from the lower reactivity, the cycloaddition of nitrile oxides to acetylenic derivatives offers the same general picture as the cycloaddition to ethylenic compounds. Monosubstituted acetylenes react usually faster than disubstituted acetylenes, conjugation remarkably promotes the reaction, compounds of low reactivity can be successfully led to react by preparing the 1,3-dipole *in situ* in the presence of an excess of dipolarophile.

A recent study has pointed out that 1-alkynes can also enter a 1,3-addition reaction to nitrile oxides to give *anti*-aryl acetylenic oximes of type *1*, which are little stable, especially in basic medium, and cyclize readily to isoxazoles *2*. It could be demonstrated, however, that oximes *1* are not intermediates in the cycloaddition; therefore 1,3-addition and cycloaddition are here concurrent reactions (Eq. (1)).[2] Analogous results have been more recently achieved with other 1,3-dipoles, such as, *e.g.*, nitrilimines.[3]

$$
\begin{array}{l}
R-C\equiv N \rightarrow O \\
\quad + \\
CH \equiv C-R'
\end{array}
\quad
\begin{array}{l}
R-\underset{\underset{NOH}{\|}}{C}-C\equiv C-R' \\
\qquad\qquad 1
\end{array}
\qquad (1)
$$

The ratio of the two products *1* and *2* can be easily evaluated through deuterium incorporation,[4] and has been shown to depend mostly on the nature of R and R'. As Table XXII shows, electron-repelling groups on the arylacetylene and electron-attracting groups on the nitrile oxide side favour the formation of the acetylenic oxime.[4] Although the mechanism of the 1,3-addition reaction has not yet been fully explored, kinetic data so far available[5, 6] confirm that the formation of the acetylenic

1 *Huisgen, R.:* Angew. Chem. **75**, 742 (1963); — Intern. Ed. Engl. **2**, 633 (1963).

2 *Morrocchi, S., Ricca, A., Zanarotti, A., Bianchi, G., Gandolfi, R., Grünanger, P.:* Tetrahedron Letters 3329 (1969).

3 *Morrocchi, S., Ricca, A., Zanarotti, A.:* Tetrahedron Letters 3215 (1970).

4 *Morrocchi, S., Ricca, A., Selva, A., Zanarotti, A.:* Rend. Accad. Naz. Lincei **47**, 233 (1970).

5 *Battaglia, A., Dondoni, A.:* Tetrahedron Letters 1221 (1970).

6 *Beltrame, P., et al.:* Submitted to J. Chem. Soc.

oximes through 1,3-addition is concurrent with the cycloaddition to isoxazoles, without excluding a common intermediate. The absence of any primary isotope effect[6] in the reaction with DC≡CPh rules out a rather unlikely concerted mechanism for the 1,3-addition, and points out toward a two-step dipolar or preferably diradical mechanism.

Table XXII. Percentage yield of 1 in the reaction mixture

$$p\text{-}XC_6H_4C≡N→O \longrightarrow$$

$p\text{-}YC_6H_4C≡CH$ — Y↓ \ X→	NO_2	Cl	H	CH_3O
Cl	11	9	7	7
H	18	15	12	10
OCH_3	34	32	28	24
$N(CH_3)_2$	74	65	53	52

On the whole, the cycloaddition has a very wide applicability and represents one of the most important routes to isoxazole derivatives.[7] The reaction of fulminic acid, prepared *in situ*, with acetylene yielded isoxazole itself;[8,9] the yields are very poor and the main product is *3*, arising from the dimeric form of fulminic acid, *i.e.* oximinoacetonitrile oxide (see also Chapter IV, C-1) (Eq. (2)).[8,9]

$$HC≡N \longrightarrow O \quad \xrightarrow{+\,CH≡CH} \quad \text{(isoxazole)}$$

$$HON=CH-\overset{\underset{||}{N}}{\underset{O}{C}} \quad \xrightarrow{+\,CH≡CH} \quad HON=CH-\text{(isoxazole ring)} \quad \textit{3}$$

(2)

Interestingly enough, when the reaction was carried out in the presence of acetone (or other ketones) in order to increase the concentration of acetylene, compounds *5*, *7* and *9* were obtained.[10-12] A preliminary attack of fulminic acid on the ketone must therefore be inferred (Eq. (3)). Phenylacetylene behaves analogously, yielding *6* and *8* along with

7 *Quilico, A.:* The chemistry of heterocyclic compounds (ed. *A. Weissberger*), vol. XVII, chap.: Isoxazoles and related compounds, p. 19–32. New York: Interscience 1962.

8 *Quilico, A., Stagno d'Alcontres, G.:* Gazz. Chim. Ital. **79**, 703 (1949).

9 *Huisgen, R.:* Personal communication; — *Christl, M.:* Ph. D. Thesis, Universität München, 1969.

10 *Quilico, A., Speroni, G.:* Gazz. Chim. Ital. **69**, 508 (1939).

11 *Quilico, A., Speroni, G.:* Gazz. Chim. Ital. **70**, 779 (1940).

12 *Quilico, A., Stagno d'Alcontres, G.:* Gazz. Chim. Ital. **79**, 654 (1949).

5-phenylisoxazole (*4*).[8] The higher reactivity of fulminic acid with electrophiles, as compared with the homologous nitrile oxides, is well known and has been ascertained in other cases.[9]

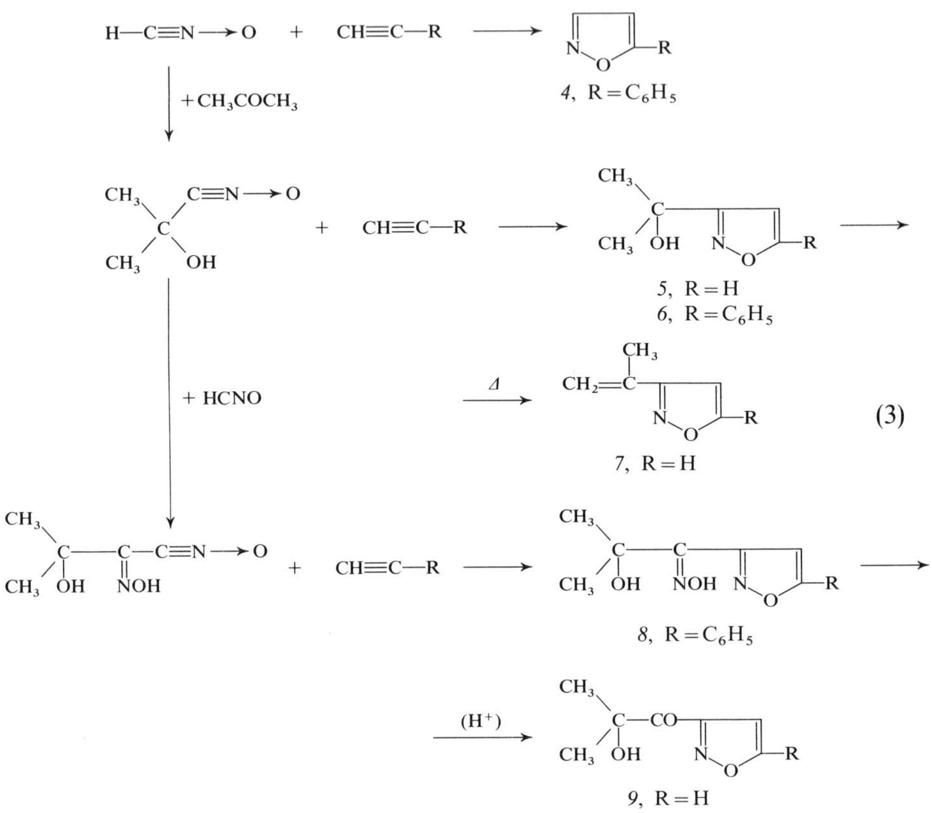

$$\tag{3}$$

Use of fulminic acid and monosubstituted acetylenes affords a convenient route to 5-alkyl- (or aryl-) isoxazoles. When fulminic acid is generated from the sodium salt in aqueous-alcoholic solution, yields are poor and the product is always accompanied by the corresponding isoxazole-3-aldoxime, which sometimes becomes the only product of the reaction.[13] The dimerization of fulminic acid can mostly be avoided by generating it *in situ* from formohydroximic acid iodide and triethylamine; yields of 3-unsubstituted isoxazoles are here remarkably higher.[14]

13 *Quilico, A., Panizzi, L.:* Gazz. Chim. Ital. **72**, 155 (1942).

14 *Huisgen, R., Christl, M.:* Angew. Chem. **79**, 471 (1967); — Intern. Ed. Engl. **6**, 456 (1967).

Nitrile oxides and acetylene or monosubstituted acetylenes yield 3-monosubstituted[8,15,16] resp. 3,5-disubstituted isoxazoles.[15,17–31] Both aliphatic and aromatic nitrile oxides are reactive, and no formation of 3,4-disubstituted isoxazoles has ever been recorded.

The same compounds are formed by directly thermal reaction of hydroximic acid chlorides with monosubstituted acetylenes.[30–32] The orientation of the cycloaddition can be assured by comparison of the isoxazole thus obtained with the compound prepared by oximation of β-diketones or α,β-acetylenic ketones.[33]

The long known synthesis of isoxazoles[34] from hydroximic acid chlorides and arylacetylides passes through an intermediate nitrile oxide (Eq. (4)), as already supposed by analogy with similar reactions[17] and recently demonstrated by trapping the nitrile oxide with norbornene.[35]

$$
\begin{array}{l}
\text{Ar—C—Cl} \\
\quad \| \\
\quad \text{NOH}
\end{array}
\;+\; \text{NaC} \equiv \text{C—Ar}' \;\longrightarrow
$$

$$\xrightarrow{-\text{NaCl}} \quad [\text{Ar—C} \equiv \text{N} \rightarrow \text{O} \;+\; \text{Ar}'\text{C} \equiv \text{CH}] \;\longrightarrow\; \underset{\text{(isoxazole)}}{\text{Ar–N=... –O–Ar}'}$$

$$\downarrow\; +\; \text{norbornene}$$

(4)

15 *Quilico, A., Simonetta, M.:* Gazz. Chim. Ital. **77**, 586 (1947).

16 *Lardicci, L., Botteghi, C., Salvadori, P.:* Gazz. Chim. Ital. **98**, 760 (1968).

17 *Quilico, A., Speroni, G.:* Gazz. Chim. Ital. **76**, 148 (1946).

18 *Quilico, A., Simonetta, M.:* Gazz. Chim. Ital. **76**, 200 (1946).

19 *Speroni, G., Bartoli, M.:* Sopra gli ossidi di nitrile, Nota IX, Stab. Tip. Marzocco, Firenze, 1952.

20 *Lo Vecchio, G., Monforte, P.:* Gazz. Chim. Ital. **86**, 399 (1956).

21 *Grundmann, C., Dean, J.D.:* Angew. Chem. **76**, 682 (1964); — Intern. Ed. Engl. **3**, 585 (1964).

22 *Vita Finzi, P., Grünanger, P.:* Chim. Ind. (Milan) **47**, 516 (1965).

23 *Grundmann, C., Dean, J.M.:* J. Org. Chem. **30**, 2809 (1965).

24 *Casnati, G., Quilico, A., Ricca, A., Vita Finzi, P.:* Gazz. Chim. Ital. **96**, 1064 (1966).

25 *Casnati, G., Quilico, A., Ricca, A., Vita Finzi, P.:* Gazz. Chim. Ital. **96**, 1073 (1966).

26 *Grundmann, C., Frommeld, H.D.:* J. Org. Chem. **31**, 4235 (1966).

27 *Beltrame, P., Veglio, C., Simonetta, M.:* J. Chem. Soc. (B) 867 (1967).

28 *Grundmann, C., Richter, R.:* J. Org. Chem. **32**, 2308 (1967).

29 *Casnati, G., Ricca, A.:* Tetrahedron Letters 327 (1967).

30 *Arbasino, M., Grünanger, P.:* Ric. Sci. **34**, 561 (1964).

31 *Vita Finzi, P., Arbasino, M.:* Ric. Sci. **35**, 1484 (1965).

32 *Sasaki, T., Yoshioka, T.:* Bull. Chem. Soc. Japan **40**, 2604 (1967).

33 *Johnston, K.M., Shotter, R.G.:* J. Chem. Soc. (C) 1774 (1968).

34 *Weygand, C., Bauer, E.:* Liebigs Ann. Chem. **459**, 123 (1927).

35 *Bettinetti, G.F., Fraschini, C.:* Gazz. Chim. Ital. **100**, 403 (1970).

Methyl propiolate reacts with fulminic acid and various nitrile oxides yielding a mixture of the two isomeric isoxazole esters *10* and *11*;[14, 36] the ratios are reported in Table XXIII and change with the nature of the 1,3-dipole but generally show a predominance of the 5-carboxylic ester, with the remarkable exception of mesitonitrile oxide. Tetrolate and phenylpropriate yield the 4-isomer (*12*) with a much higher predominancy as in the case of the corresponding double bond derivatives.[36]

10 *11* *12*

Table XXIII. Isomer ratios in the cycloaddition of nitrile oxides to methyl propiolate[9]

Nitrile oxide	Solvent	Total yield %	Isomer ratio *10* :*11*
H—CNO	Ether	50	84 :16
$(CH_3)_3C$—CNO	Ether	95	91 : 9
NC—CNO	Ether/Water	69	68.5:31.5
C_6H_5—CNO	Ether	98	72 :28
2,4,6-$(CH_3)_3C_6H_2$—CNO	Methanol	97	44 :56
2,4,6-$(CH_3)_3C_6H_2$—CNO	Ether	97	28 :72
2,4,6-$(CH_3)_3C_6H_2$—CNO	Cyclohexane	96.5	25 :75

Propargyl alcohols,[24, 29, 32, 37–42] halides[32, 43, 44] and amines[45, 46] react normally with high yields. 5-Isoxazolylcarbinols are starting materials for an interesting synthesis of 3-furanones *13* and 2,3-dihydro-γ-pyrones *14* (Eq. (5)),[29, 41, 47] whereas in the case of propargylamine the nucleophilic 1,3-addition reaction of the aminogroup (see Chapter VI, G) prevails over the cycloaddition by the triple bond and three products are obtained.[46] The higher reactivity of amino group as compared with that of triple bond is also confirmed by kinetic data.[48]

36 *Christl, M., Huisgen, R.:* Tetrahedron Letters 5209 (1968).
37 *Mugnaini, E., Grünanger, P.:* Rend. Accad. Naz. Lincei **14**, 95 (1953).
38 *Grünanger, P., Fabbri, E.:* Gazz. Chim. Ital. **89**, 598 (1959).
39 *Grünanger, P., Langella, M.R.:* Gazz. Chim. Ital. **91**, 1449 (1961).
40 *Vita Finzi, P., Arbasino, M.:* Ann. Chim. (Rome) **54**, 1165 (1964).
41 *Casnati, G., Quilico, A., Ricca, A., Vita Finzi, P.:* Gazz. Chim. Ital. **96**, 1073 (1966).
42 *Bertini, V., De Munno, A., Pelosi, P.: Pino, P.:* J. Heterocycl. Chem. **5**, 621 (1968).
43 *Stagno d'Alcontres, G., Cuzzocrea, G.:* Atti Soc. Peloritana **3**, 179 (1956/57).
44 *Stagno d'Alcontres, G., Lo Vecchio, G.:* Gazz. Chim. Ital. **90**, 1239 (1960).
45 *Kano, H., Adachi, I., Kido, R., Hirose, K.:* Fr. Pat. 1.380.177 (1964); — Chem. Abstr. **63**, 8367a (1965).
46 *Caramella, P., Vita Finzi, P.:* Chim. Ind. (Milan) **48**, 963 (1966).
47 *Casnati, G., Quilico, A., Ricca, A., Vita Finzi, P.:* Tetrahedron Letters 233 (1966).
48 *Dondoni, A.:* Personal communication.

$$C_6H_5\text{---}C\equiv N\longrightarrow O$$

$$+$$

$$CH\equiv C\text{---}(CH_2)_n\text{---}\underset{R'}{\overset{R}{\underset{|}{\overset{|}{C}}}}\text{---}OH \longrightarrow$$

(5)

A convenient synthesis of γ-substituted butyrolactams, starting from the *in situ* prepared methyl 3-fulmidopropionate and acetylenic ketones, should also be mentioned.[48a] Other isoxazoles, similarly prepared, are intermediates in a recently reported synthesis of semicorrins.[48b]

Alkoxyacetylenes and dialkylaminoacetylenes give high yields of 5-alkoxy-[49,50] resp. 5-dialkylamino-isoxazoles.[51-53] The former compounds are easily hydrolysed by mineral acids to 5-isoxazolones.

Dibutyl ethynylboronate and aromatic nitrile oxides in ether yielded directly the corresponding 5-isoxazoleboronic acids; oxidation of these latter with hydrogen peroxide afforded another route to 5-isoxazolones.[54,55]

Disubstituted acetylenes react with more difficulty: *e.g.* butindiol[37] and tolan[17] have been reported to be unreactive toward benzonitrile oxide in ether. Also here the conjugation enhances the reactivity: *e.g.* phenylpropiolic acid[17,18,20] and ester,[36] cyano-[56] and dicyanoacetylene[57], propynylphenyl ketone[58,59] are reported to react with moderate

---

48a *Traverso, G., Barco, A., Pollini, G.P., Anastasia, M., Sticchi, V., Pirillo, D.:* Farmaco (Pavia) **24**, 946 (1969); — *Traverso, G., Barco, A., Pollini, G.P.:* Farmaco (Pavia) **25**, 777 (1970).

48b *Stevens, R.V., Du Pree, L.E., Wentland, M.P.:* Chem. Commun. 821 (1970).

49 *Grünanger, P.:* Rend. Accad. Naz. Lincei **24**, 163 (1958).

50 *Grünanger, P., Langella, M.R.:* Gazz. Chim. Ital. **89**, 1784 (1959).

51 *Viehe, H.G., Fuks, R., Reinstein, M.:* Angew. Chem. **76**, 571 (1964); — Intern. Ed. Engl. **3**, 581 (1964).

52 Union Carbide Corp.: Neth. Pat. Appl. 6.507.886 (1965); — Chem. Abstr. **64**, 19630e (1966).

53 *Kuehne, M.E., Sheeran, P.J.:* J. Org. Chem. **33**, 4406 (1968).

54 *Bianchi, G., Cogoli, A., Grünanger, P.:* Ric. Sci. **36**, 132 (1966).

55 *Bianchi, G., Cogoli, A., Grünanger, P.:* J. Organomet. Chem. **6**, 598 (1966).

56 *Sasaki, T., Yoshioka, T.:* Bull. Chem. Soc. Japan **41**, 2212 (1968).

57 *Weis, C.D.:* J. Org. Chem. **28**, 74 (1963).

58 *Dal Piaz, V., Renzi, G.:* Chim. Ind. (Milan) **48**, 759 (1966).

59 *Renzi, G., Dal Piaz, V., Musante, C.:* Gazz. Chim. Ital. **98**, 656 (1968).

to good yields. Whereas the free acetylene dicarboxylic acid gives very low yields of the isoxazole derivative, the ethylester reacts normally to give *15*.[26, 32, 60, 61, 62] Also the disilyl derivative *16* has been obtained with moderate yield.[63]

$$R\text{—}\overset{\text{COOEt}}{\underset{\text{COOEt}}{\diagdown}}\quad\quad CH_3\text{—}\overset{\text{SiMe}_3}{\underset{\text{SiMe}_3}{\diagdown}}$$

15   *16*

Bis-nitrile oxides and phenylacetylene[23, 64–66] or substituted di-aminoacetylenes[52, 67] yield the corresponding bis-adducts; with di-ethynylbenzene polymers of type *17* are obtained.[66]

17

The highly reactive triple bond of benzyne affords a new route to indoxazenes (Eq. (6)).[68, 69]

$$C_6H_5\text{—}C\equiv\overset{+}{N}\text{—}O \;+\; \text{(benzenediazonium-2-carboxylate)} \xrightarrow[\substack{-N_2 \\ -CO_2}]{\Delta} \text{(indoxazene)} \tag{6}$$

Diynes with terminal triple bonds give both monoadducts and diadducts,[38, 39] depending on experimental conditions: *e.g.* butadiyne reacts with aromatic nitrile oxides to give either 3-aryl-5-ethynylisoxazole

$$
\begin{array}{ccc}
Ar\text{—}C\equiv N\text{—}O & & (ArCOCH_2CO)_2 \\
+ & & \\
CH\equiv C\text{—}C\equiv CH & & \downarrow\; NH_2OH\;(OH^{\ominus}) \\
\downarrow & & \\
\textit{18} & + & \textit{19}
\end{array}
\tag{7}
$$

$$+\; ArC\equiv N\text{—}O$$

60 *Erichomovitch, L., Chubb, L. F.*: Can. J. Chem. **44**, 2095 (1966).

61 *Peterson, L. I.*: Tetrahedron Letters 1727 (1966).

62 *Iwakura, Y., Uno, K., Shiraishi, S., Hongu, T.*: Bull. Chem. Soc. Japan **41**, 2954 (1968).

63 *Fritsch, C., Lefort, M.*: Fr. Pat. 1.371.325 (1964); — Chem. Abstr. **62**, 1689b (1965).

64 *Grundmann, C.*: Angew. Chem. **75**, 450 (1963); — Intern. Ed. Engl. **2**, 260 (1963).

65 *Grundmann, C., Mini, V., Dean, J. M., Frommeld, H. D.*: Liebigs Ann. Chem. **687**, 191 (1965).

66 *Overberger, C. G., Fujimoto, S.*: Polymer Letters **3**, 735 (1965).

67 *Fuks, R., Buijle, R., Viehe, H. G.*: Angew. Chem. **78**, 594 (1963); — Intern. Ed. Engl. **5**, 585 (1966).

68 *Minisci, F., Quilico, A.*: Chim. Ind. (Milan) **46**, 428 (1964).

69 *Sasaki, T., Yoshioka, T.*: Bull. Chem. Soc. Japan **42**, 826 (1969).

(*18*) or 3,3'-diaryl-5,5'-diisoxazole (*19*), the latter product being identical to the one obtained by alkaline oximation of the tetraketone (Eq. (7)).[38]

Monosubstituted butadiynes react only with their terminal triple bond.[70]

The cycloaddition of nitrile oxides to 1,4-pentadiyne[41] opens an interesting route to β-tetraketones (Eq. (8)).[71]

$$R-C\equiv N\rightarrow O$$
$$+$$
$$CH\equiv C-CH_2-C\equiv CH \longrightarrow$$

$$\xrightarrow{H_2} RCOCH_2COCH_2COCH_2COR \tag{8}$$

The behaviour of conjugated enynes toward fulminic acid and nitrile oxides falls within the usual pattern; vinylacetylene reacts preferably on its double bond[72-74] to give 5-ethynyl-2-isoxazolines; sometimes minor amounts of the isomeric 5-vinylisoxazole are present,[74] whereas in the presence of two equivalents of nitrile oxide the bis-adduct is formed. In isopropenylacetylene both unsaturations are reactive, and the ratio of obtained isoxazolines and isoxazole derivatives seems to depend on the nature of the nitrile oxide.[39, 74, 75] With monosubstituted enynes the terminal bond, whether double or triple, is always more reactive and often determines the nature of the only reaction product. For instance with 2-penten-4-yne,[74, 76] methyl 2-penten-4-ynecarboxylate[72] and 1-alkoxy-1-buten-3-yne[24] the isoxazoles *20a—c* were respectively obtained. Enyne ketones behave similarly.[77]

20a, R' = CH_3;   *b*, R' = COOCH_3;   *c*, R' = OAlk

All compounds obtained by cycloaddition reactions of nitrile oxides with various types of acetylenic compounds are compiled in Chapter IX (Tables XXIX a–c).

70 *Sokolov, L.B., Vagina, L.K., Chistokletov, V.N., Petrov, A.A.:* Zh. Organ. Khim. **2**, 615 (1966).

71 *Casnati, G., Quilico, A., Ricca, A., Vita Finzi, P.:* Chim. Ind. (Milan) **47**, 993 (1965).

72 *Quilico, A., Grünanger, P.:* Rend. Ist. Lombardo Sci. Lettere **88**, 990 (1955).

73 *Chistokletov, V.N., Troshchenko, A.T., Petrov, A.A.:* Doklady Akad. Nauk SSSR **135**, 631 (1960).

74 *Vagina, L.K., Chistokletov, V.N., Petrov, A.A.:* Zh. Organ. Khim. **2**, 417 (1966).

75 *Chistokletov, V.N., Troshchenko, A.T., Petrov, A.A.:* Zh. Obshchei Khim. **34**, 1891 (1964).

76 *Chistokletov, V.N., Troshchenko, A.T., Petrov, A.A.:* Zh. Obshchei Khim. **33**, 789 (1963).

77 *Bondarev, G.N., Ryzhov, V.A., Chistokletov, V.N., Petrov, A.A.:* Zh. Organ. Khim. **3**, 821 (1967).

D. Reactions with C=O and C=S Compounds

The carbonyl double bond of *aldehydes and ketones* reacts slowly with aromatic nitrile oxides, generated *in situ*, but only if activated by an adjacent electron-withdrawing group. The products are 1,3,4-dioxazoles (Eq. (9)),[1-3] a class of heterocyclics otherwise almost inaccessible. As

$$(9)$$

cyclic acetals, these compounds are readily cleaved by mineral acids to give the original carbonyl compound and the hydroxamic acid,[1] whilst the already known[4] thermal decomposition yields arylisocyanate along with the original carbonyl compound. Suitable addends are *e.g.* chloral, aromatic aldehydes, α-diketones, α-ketoesters.[1] It is worth noting that pyruvonitrile reacts first on its C≡N triple bond, and only afterwards the 5-acetyl-1,2,4-oxadiazole reacts on its C=O double bond.[2]

Unactivated aldehydes and ketones are reactive if the cycloaddition is catalysed by boron trifluoride etherate and give 1,3,4-dioxazoles, though with low to moderate yields.[5,6] Since it is known[7-9] that Lewis

1 *Huisgen, R., Mack, W.:* Tetrahedron Letters 583 (1961).
2 *Grünanger, P., Vita Finzi, P.:* Rend. Accad. Naz. Lincei **31**, 277 (1961).
3 *Simonyan, L. A., Zeifman, U. V., Gambaryan, N. P.:* Izv. Akad. Nauk SSSR 1916 (1968).
4 *Exner, O.:* Chem. Listy **50**, 779 (1956); — Chem. Abstr. **50**, 15477 f (1956).
5 *Morrocchi, S., Ricca, A., Velo, L.:* Tetrahedron Letters 331 (1967).
6 *Sasaki, T., Yoshioka, T.:* Bull. Chem. Soc. Japan **41**, 2206 (1968).
7 *Lombard, R., Stéphan, J. P.:* Bull. Soc. Chim. France 1369 (1957).
8 *Chalandon, P., Susz, B. P.:* Helv. Chim. Acta **41**, 697 (1958).
9 *Susz, B. P., Chalandon, P.:* Helv. Chim. Acta **41**, 1332 (1958).

acids give coordination compounds with aldehydes and ketones, a two-step cycloaddition might be conceivable (Eq. (10)).

$$(10)$$

The reaction of benzonitrile oxide and *amides* in the presence of BF_3-etherate provides diphenylurea in good yields, while the uncatalysed reaction hardly yields 10%[10].

The behaviour of *o- and p-quinones* is particularly interesting. As the reaction of benzonitrile oxide with o-naphthoquinone shows, both activated carbonyl groups of the o-quinone system react faster than the C=C double bond, giving rise to a mixture of the two isomeric mono-adducts. If the carbonyl α to the aromatic ring reacts first, an excess of nitrile oxide attacks further only the chalkone-like C=C double bond to yield a mixture of the bis-adduct and its oxidation product. If however the β-carbonyl group had reacted, both the C=O and the C=C bonds are still reactive and a 1:3 adduct can be obtained, as shown in Eq. (11).

$$(11)$$

---

10 *Morrocchi, S., Ricca, A., Zanarotti, A.:* Chim. Ind. (Milan) **50**, 352 (1968).

In the case of o-benzoquinone one carbonyl group and both C=C bonds are reactive.[11] When the quinone does not contain ethylenic bonds, like phenanthrene-9,10-quinone or chrysene-5,6-quinone, the reaction with nitrile oxide can lead to mono-[12,13] as well to bis-adducts[11] according to the solvent used.

The uncatalysed reaction of benzonitrile oxide with the double bond of *p-benzoquinone* yields only the bis-adduct *21*,[6,14] arising from the oxidation of the primary isoxazoline derivative *22* through air or a third molecule of quinone. The presence of boron trifluoride etherate allows to isolate the easily oxidable *22*, together with a second compound which contains 3 moles of nitrile oxide for one of quinone and whose structure *23* shows that one carbonyl group has also reacted.[15] Analogous results have been obtained with p-naphthoquinone.

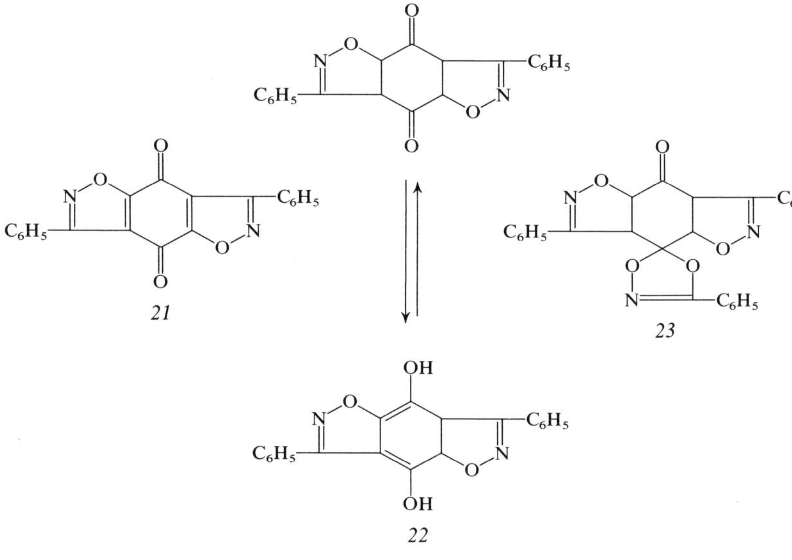

21

22

23

Several *thiocarbonyl derivatives* are reactive toward nitrile oxides prepared *in situ*, among them thioketones, thionocarbonates, dithiocarboxylates and trithiocarbonates.[16] This cycloaddition represents the most convenient synthesis of 1,4,2-oxathiazoles *24*, whose structure is

11 *Morrocchi, S., Ricca, A., Selva, A., Zanarotti, A.:* Gazz. Chim. Ital. **99**, 565 (1969).

12 *Awad, W.I., Omran, S.M.A.R., Sobhy, M.:* J. Org. Chem. **31**, 331 (1966).

13 *Awad, W.I., Sobhy, M.:* Can. J. Chem. **47**, 1473 (1969).

14 *Quilico, A., Stagno d'Alcontres, G.:* Gazz. Chim. Ital. **80**, 140 (1950).

15 *Morrocchi, S., Quilico, A., Ricca, A., Selva, A.:* Gazz. Chim. Ital. **98**, 891 (1968).

16 *Huisgen, R., Mack, W., Anneser, E.:* Angew. Chem. **73**, 656 (1961).

assured by their easy thermal decomposition into isothiocyanates and the oxygen analogue of the thiocarbonyl derivative employed (Eq. (12)).

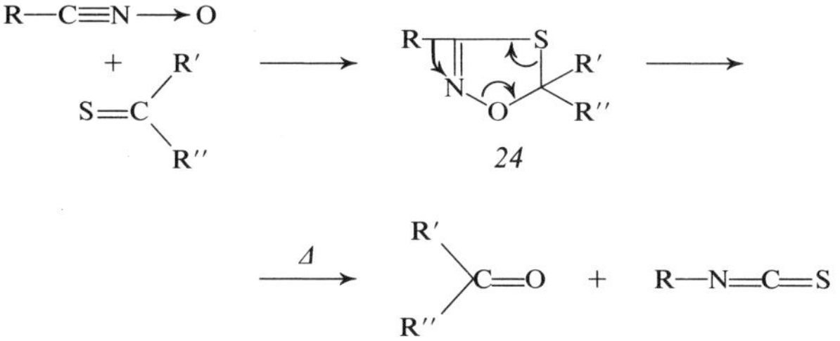

$$
\begin{array}{c}
R\text{—}C\equiv N\longrightarrow O \\[2pt]
+ \\[2pt]
S=C\!\!\begin{array}{l}R'\\ R''\end{array}
\end{array}
\quad\longrightarrow\quad
\underset{24}{\boxed{}}
\quad\longrightarrow\quad
\underset{25}{\left.\begin{array}{l}R'\\ R''\end{array}\right\}C=O}
\;+\;
\underset{26}{R\text{—}N=C=S}
\tag{12}
$$

Thioamides also react, but the 5-amino-1,4,2-oxathiazoles are unstable and the corresponding amides *25* ($R'=NH_2$) and isothiocyanates *26* are directly obtained.

Even 3-aryl-1,4,2-dithiazoline-5-thiones add nitrile oxides to give spiro-compounds of type *27*.[17]

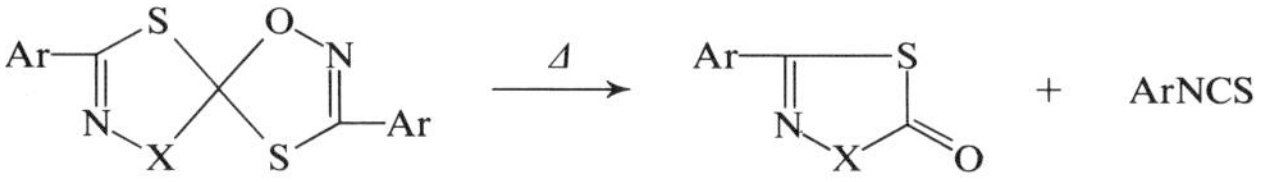

$$
\underset{\substack{27,\ X=S\\ 28,\ X=O}}{\text{(spiro ring)}}
\quad\overset{\Delta}{\longrightarrow}\quad
\underset{29}{\text{(ring)}}
\;+\;\text{ArNCS}
\tag{13}
$$

With *carbon disulfide* both C=S double bonds react with mesitonitrile oxide to yield a spiro-compound *28* which, contrary to the analogous compound obtained from nitrilimines,[18] decomposes spontaneously to 3-mesityl-1,4,2-oxathiazoline-5-one (*29*, X=O) and mesitylisothiocyanate (Eq. (13)).[19] Under the same conditions carbon oxysulfide does not react.

The C=S double bond seems on the whole to be more reactive than C=O double bond, at least when the reaction is uncatalysed. Consequently it is not surprising that thioketenes, prepared *in situ* as in Eq. (14),

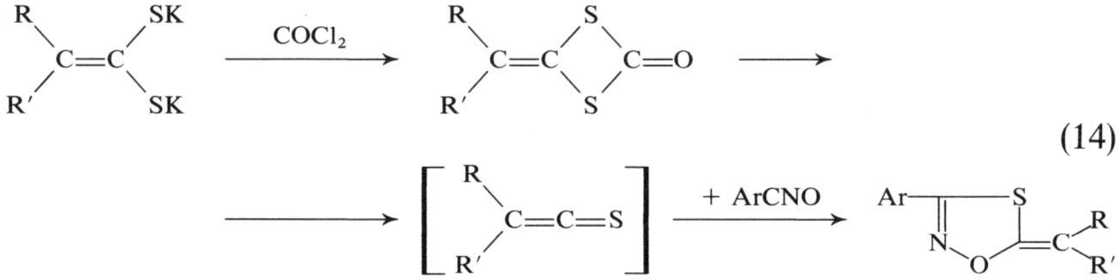

$$
\underset{R'}{\overset{R}{\diagdown}}C=C\underset{SK}{\overset{SK}{\diagup}}
\quad\xrightarrow{\;COCl_2\;}\quad
\underset{R'}{\overset{R}{\diagdown}}C=C\diagup\!\!\begin{array}{c}S\\ S\end{array}\!\!\diagdown C=O
\quad\longrightarrow
$$

$$
\longrightarrow\;
\left[\,\underset{R'}{\overset{R}{\diagdown}}C=C=S\,\right]
\;\xrightarrow{\;+\ \text{ArCNO}\;}\;
\text{(Ar ring)}
\tag{14}
$$

17 *Noël, D., Vialle, J.*: Bull. Soc. Chim. France 2239 (1967).

18 *Huisgen, R., Grashey, R., Seidel, M., Knupfer, H., Schmidt, R.*: Liebigs Ann. Chem. **658**, 169 (1962).

19 *Foye, W.O., Kauffman, J.M.*: J. Org. Chem. **31**, 2417 (1966).

add nitrile oxides on their thiocarbonyl bond,[20] whereas their oxygen analogues, *i.e.* ketene and derivatives, react, if at all, as already mentioned in Section B, with the C=C double bond.

Cycloaddition of a nitrile oxide to the CS double bond of methyl dithiocarbazate might be the key-step in the reaction of hydroximic acid chlorides with an excess of the above base to give N-substituted 2-amino-5-methylthio-1,3,4-thiadiazoles,[21] although some authors[22] considered such a mechanism less probable.

All compounds obtained by cycloaddition reactions of nitrile oxides with various types of carbonyl or thiocarbonyl compounds are compiled in Chapter IX (Tables XXXa and b).

---

20 *Dickoré, K., Wegler, R.:* Angew. Chem. **78**, 1023 (1966); — Intern. Ed. Engl. **5**, 970 (1966).

21. *Sasaki, T., Yoshioka, T.:* Bull. Chem. Soc. Japan **41**, 2211 (1968).

22 *Dornow, A., Fischer, K.:* Chem. Ber. **99**, 72 (1966).

E. Reactions with C = N Compounds

Imino-compounds, like aldimines (Schiff bases) or ketimines, either N-substituted or not, add easily to aromatic nitrile oxides, apparently with much less prerequisition than carbonyl compounds, yielding Δ^2-1,2,4-oxadiazolines.[1-4] The structure of the product is assured by hydrogenolysis (Eq. (14a)):

$$\tag{14a}$$

Yields are usually excellent, whereas the only known example of addition to an aliphatic nitrile oxide (by benzalaniline[5]) is reported to give a low yield. On the whole, the C=N double bond is more reactive than a comparable C=C double bond (see also the relative rate constant for benzalmethylamine in Table XVIII, Section A). Even N-substituted ketimines of cyclic ketones react with good yields to give spirocompounds,[3] in sharp contrast with alkylidenecyclanones. Furthermore, in a recent study, based on spectroscopic evidence, it has been reported that only the hetero-bond of conjugated imines, like cinnamylideneaniline (*30*), is reactive, as shown in Eq. (15).[6]

$$\tag{15}$$

Consequently, under comparable conditions the C=N group is a far better dipolarophile than the carbonyl group, as would be expected from the higher nucleophilicity of the former group. Indeed, reactions with phenanthrene- and chrysene-o-quinone monoimines seem to involve the $>$C=NH bond rather than the $>$C=O.[7,8]

1 *Fabbrini, L., De Sarlo, F.*: Chim. Ind. (Milan) **45**, 242 (1963).
2 *Huisgen, R.*: Angew. Chem. **75**, 604 (1963); — Intern. ed. Engl. **2**, 565 (1963).
3 *Lauria, F., Vecchietti, V., Tosolini, G.*: Gazz. Chim. Ital. **94**, 478 (1964).
4 *Morrocchi, S., Ricca, A., Velo, L.*: Chim. Ind. (Milan) **49**, 168 (1967).
5 *Mukaiyama, T., Hoshino, T.*: J. Am. Chem. Soc. **82**, 5339 (1960).
6 *Singh, N., Sandhu, J.S., Mohan, S.*: Tetrahedron Letters 4453 (1968).
7 *Awad, W.I., Omran, S.M.A.R., Sobhy, M.*: J. Org. Chem. **31**, 331 (1966).
8 *Awad, W.I., Sobhy, M.*: Can. J. Chem. **47**, 1473 (1969).

Most interesting is the behaviour of *monoimino-β-diketones* of type *31a—c*, for which three tautomeric forms are possible, with nitrile oxides. Depending on the nature of R and R', different adducts are

$$R-\underset{\underset{NH}{\|}}{C}-CH_2-CO-R' \quad \rightleftharpoons \quad R-\underset{\underset{NH_2}{|}}{C}=CH-CO-R' \quad \rightleftharpoons \quad R-\underset{\underset{NH}{\|}}{C}-CH=\underset{\underset{OH}{|}}{C}-R'$$

$$\textit{31a} \qquad\qquad\qquad \textit{31b} \qquad\qquad\qquad \textit{31c}$$

obtained.[4] When R is an alkyl group, both the keto-amino form (*31b*) and the enimino form (*31c*) can react: in the first case with evolution of ammonia a 4-acylisoxazole derivative *32* is obtained, in the second case cycloaddition to the enolic double bond furnishes with spontaneous dehydration the isoxazole ketimine *33*. This latter intermediate is not isolable and reacts further with a second molecule of nitrile oxide to yield bis-adducts of type *34* (Eq. (16)). Starting from iminobenzoylacetone

$$\tag{16}$$

(*31*, $R=C_6H_5$, $R'=CH_3$) only the bis-adduct is obtained, whereas iminobenzoylacetophenone (*31*, $R=R'=C_6H_5$) does not react. The isolation of *32* and the impossibility to isolate the hypothetical *33* shows that the $C=N$ bond is more reactive than the $C=O$ bond.

The rather strongly basic *alkyl imidates* react with nitrile oxides to give directly the 3,5-disubstituted 1,2,4-oxadiazoles (*36*) through the unstable intermediate dihydroderivative *35*.[4,9] This reaction provides an experimental basis to the hypothesis[10] that the synthesis of 1,2,4-oxadiazoles from hydroximic acid chlorides and alkylimidates[11,12] passes

9 *Rajagopalan, P.:* Tetrahedron Letters 311 (1969).

10 *Grundmann, G.:* Fortschr. Chem. Forsch. **7**, 62 (1966).

11 *Eloy, F., Lenaers, R.:* Bull. Soc. Chim. Belges **72**, 719 (1963).

12 *Sasaki, T., Yoshioka, T., Suzuki, Y.:* Bull. Soc. Chem. Japan **42**, 3335 (1969).

through the nitrile oxides, formed by action of the strongly basic reagent (Eq. (17)).

$$R-C\equiv N \rightarrow O \quad + \quad R-\underset{\underset{OR''}{|}}{C}=NH \longrightarrow \left[\text{35} \right] \longrightarrow$$

$$\xrightarrow{-R''OH} \quad \text{36}$$

(17)

However a nucleophilic substitution on the chlorine atom of hydroximic acid chloride, as proposed by the original authors,[11] cannot be excluded *a priori*.

The same mechanistic ambiguity exists for the addition of hydroximic acid chlorides to benzamidine to yield 3,5-disubstituted 1,2,4-oxadiazoles.[12, 13]

The C=N double bond of the *oximes* is reactive toward nitrile oxide only in the presence of an acidic catalyst, *e.g.* boron trifluoride etherate.[14] Both aldoximes and ketoximes react, the latter ones with lower yields. The hitherto unknown 4-hydroxy-1,2,4-oxadiazolines *37* are characterized by their behaviour toward acids, illustrated in Eq. (18).

$$C_6H_5-C + \quad \underset{}{\overset{BF_3}{N}}-OH \quad \longrightarrow \quad 37 \quad + \quad BF_3$$

(18)

$$(R=H) \quad \xrightarrow{-H_2O} \quad (R\neq H)$$

$$C_6H_5-CN + \underset{R'}{\overset{R}{\diagdown}}CO \quad \left(+\tfrac{1}{2}(HNO)_2\right)$$

An acid-catalysed condensation of benzonitrile oxide to benzaldoxime has also been invoked to explain the formation of 3,5-diphenyl-1,2,4-oxadiazole during the lead tetraacetate oxidation of aromatic *anti*-aldoximes.[15, 16]

13 *Musante, C.:* Gazz. Chim. Ital. **68**, 331 (1938).
14 *Morrocchi, S., Ricca, A.:* Chim. Ind. (Milan) **49**, 629 (1967).
15 *Just, G., Dahl, K.:* Tetrahedron **24**, 5251 (1968).
16 *Williams, W.M., Dolbier, W.R.:* J. Org. Chem. **34**, 155 (1969).

The cumulated C=N bond system of *aromatic carbodiimides* reacts with nitrile oxides to give first 5-imino-1,2,4-oxadiazoles (*38*) and successively, only in the presence of BF$_3$-etherate, the spiro-compounds of type *39* (Eq. (19)).[1, 17] *Aliphatic carbodiimides* do not seem to be reactive,

$$R—C\equiv N\rightarrow O \;+\; Ar—N=C=N—Ar \longrightarrow \quad \textbf{38} \quad \xrightarrow[\text{(BF}_3)]{+RCNO} \quad \textbf{39} \qquad (19)$$

at least with isolated nitrile oxides; the thermal generation of the nitrile oxide by dissociation of the hydroximic chloride enforces the reaction, but in the presence of hydrogen chloride the rearranged 1,2,4-triazoline-5-one *41* instead of the expected *40* is obtained (Eq. (20)).[18]

$$ (20) $$

The C=N bond of *cyclic amidines*, such as 1,5-diazabicyclo[4.3.0]non-5-ene and 1,5-diazabicyclo[5.4.0]undec-5-ene, is a good dipolarophile, yielding the corresponding condensed oxadiazolines.[19] Even the oxime-like C=N bond of *2-isoxazolines* add in some cases nitrile oxides, yielding new heterocycles. E.g., the monoadduct *42* from benzonitrile oxide and 1,3-cyclohexadiene adds a second molecule of 1,3-dipole to give both the bis-adducts *43* and the bis-adduct *44*. The structure of the latter compound derives from its cleavage to 3,5-diphenyl-1,2,4-oxadiazole (*45*). In this case the reactivity of the C=N bond is competitive with that of the little reactive C=C bond in the cyclohexene system (Eq. (21)).[20]

17 *Grundmann, C., Richter, R.:* Tetrahedron Letters 963 (1968).
18 *Sasaki, T., Yoshioka, T.:* Bull. Chem. Soc. Japan **42**, 258 (1969).
19 *Caramella, P., Gamba, A., Grünanger, P.:* Unpublished results.
20 *Bettinetti, G. F., Gamba, A.:* Gazz. Chim. Ital. (in press).

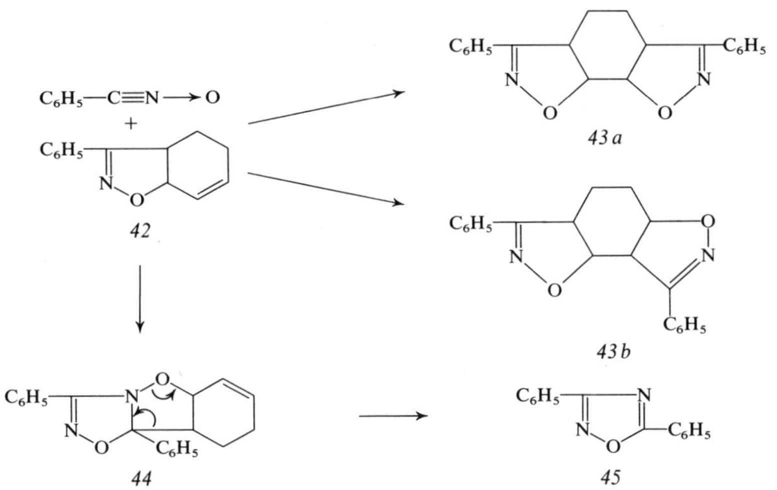

$$C_6H_5-C\equiv N\rightarrow O$$

$$\tag{21}$$

An analogous bis-adduct has been isolated as a by-product in the cycloaddition of benzonitrile oxide to cyclohexene.[21]

It is likely that the formation of 3,5-diphenyl-1,2,4-oxadiazole, sometimes observed as a by-product in the cycloaddition of nitrile oxides to double bonds, could be ascribed to an analogous side reaction, leading to spontaneously decomposing bis-adducts.

All compounds obtained by cycloaddition reactions of nitrile oxides to various types of C=N-compounds are compiled in Chapter IX (Table XXXI).

---

21 *Bianchi, G., Gandolfi, R.:* Personal communication.

F. Reactions with Nitriles

The triple bond $-C\equiv N$ adds nitrile oxides and 3,5-disubstituted 1,2,4-oxadiazoles (*46*) are invariably obtained (Eq. (22)), owing to the polarisation of the nitrile bond; no isomer has ever been recorded.

$$\tag{22}$$

As expected, nitriles as dipolarophiles possess a lower order of reactivity than the structurally analogous alkenes or alkynes, as the cycloaddition of nitrile oxides with acrylonitrile[1] or propiolonitrile,[2] which afford exclusively cyano-isoxazolines or -isoxazoles, and kinetic comparative data[3] support. On the other side, the $C\equiv N$ group is more reactive than the $C=O$ group, as illustrated for instance by the reaction of benzonitrile oxide with pyruvonitrile, where a mixture of mono-adduct *47* and the bis-adduct *48* was obtained (Eq. (23)).[4,5] This reactivity

$$C_6H_5CNO \quad + \quad CH_3COCN \quad \longrightarrow$$

$$\tag{23}$$

sequence can be explained by the principle of maximum gain in σ-bond energy,[3] cited by *Huisgen* as one proof of his concerted mechanism (see also Section A).

Nevertheless, it should be noted that recently two steroidal α,β-ethylenic nitriles have been shown to add nitrile oxides to their $C\equiv N$ bond in preference to their (exocyclic or endocyclic) $C=C$ double bond.[5a]

1 *Stagno d'Alcontres, G., Grünanger, P.:* Gazz. Chim. Ital. **80**, 741 (1950).
2 *Sasaki, T., Yoshioka, T.:* Bull. Chem. Soc. Japan **41**, 2212 (1968).
3 *Huisgen, R.:* Angew. Chem. **75**, 742 (1963); — Intern. Ed. Engl. **2**, 633 (1963).
4 *Grünanger, P., Vita Finzi, P.:* Rend. Accad. Naz. Lincei **31**, 277 (1961).
5 *Huisgen, R., Mack, W., Anneser, E.:* Tetrahedron Letters 587 (1961).
5a *Ferrara, L.:* Private communication.

Aromatic[6,7] and heterocyclic[8,9] nitriles as well as aliphatic nitriles activated by electron-withdrawing substituents[5] react without catalyst both with aromatic and aliphatic nitrile oxides. The highly reactive cyanic esters, ArOCN, afforded easily the 5-aryloxy-1,2,4-oxadiazoles.[10,11] Difunctional nitrile oxides and difunctional nitriles yield polymeric 1,2,4-oxadiazoles,[12,13] p-cyanobenzonitrile oxide polymerizes in solid state.[13,14] Because of the sluggishness of the reaction, it is convenient to carry out the cycloaddition by generating the nitrile oxide *in situ* by one of the several techniques described in Chapter III, D. Particularly useful is the thermal reaction between aromatic hydroximic acid chlorides and nitriles.[15,16]

Non-activated *aliphatic nitriles* can add nitrile oxides if the reaction is catalyzed by BF_3 etherate;[17] in this case, as already mentioned in Section D, the cycloaddition could present, instead of a concerted one, a two-step mechanism.

In the reaction with *aromatic nitriles*, activation by electron withdrawing substituents and desactivation by electron-donating para-substituents could be noticed.[7] The order of reactivity is different from that of the cycloaddition of alkenes and alkynes; a kinetic study of the reaction seems therefore desirable.

In *α-N-fluoroimidonitriles* the fluorine atom decreases the reactivity of the C=N bond to such an extent that only the cyano group enters cycloaddition.[18]

It is worth noting that, while cyanamide and its alkyl- or aryl-derivatives react with their nucleophilic amine group (see Chapter VI, G), cyanoguanidine reacts with nitrile oxides[19-21] or directly with hydroximyl chlorides[15] to yield 5-guanidino-1,2,4-oxadiazoles. The reaction,

6 *Leandri, G.:* Boll. Sci. Fac. Chim. Ind. Bologna **14**, 80 (1956).

7 *Leandri, G., Pallotti, M.:* Ann. Chim. (Rome) **47**, 376 (1957).

8 *Chang, M.S., Lowe, J.U., Jr.:* J. Org. Chem. **32**, 1577 (1967).

9 *Eloy, F.:* Bull. Soc. Chim. Belges **73**, 793 (1964).

10 *Martin, D., Herrmann, H.J., Rackow, S., Nadolsky, K.:* Angew. Chem. **77**, 96 (1965); — Intern. Ed. Engl. **4**, 73 (1965).

11 *Martin, D., Weise, A.:* Chem. Ber. **99**, 317 (1966).

12 *Overberger, C.G., Fujimoto, S.:* Polymer Letters **3**, 735 (1965); — J. Polymer Sci. C **16**, 4161 (1968).

13 *Klein, D.A., Fouty, R.A.:* Macromolecules **1**, 318 (1968).

14 *Akiyama, M., Iwakura, Y., Shiraishi, S., Imai, Y.:* Polymer Letters **4**, 305 (1966).

15 *Lenaers, R., Eloy, F.:* Helv. Chim. Acta **46**, 1067 (1963).

16 *Sasaki, T., Yoshioka, T.:* Bull. Chem. Soc. Japan **41**, 2206 (1968).

17 *Morrocchi, S., Ricca, A., Velo, L.:* Tetrahedron Letters 331 (1967).

18 *Stevens, T.O.:* J. Org. Chem. **33**, 2660 (1968).

19 *Fabbrini, L., Speroni, G.:* Chim. Ind. (Milan) **43**, 807 (1961).

20 *Plenkiewicz, J., Eckstein, Z.:* Bull. Acad. Polon. Sci. **15**, 99 (1967).

21 *Sasaki, T., Yoshioka, T.:* Bull. Chem. Soc. Japan **40**, 2608 (1967).

which represents the only method of practical value to prepare these latter compounds, has been extended to disubstituted cyanoguanidines[22] and to benzoylcyanamide;[23] the latter compound yielded 5-benzoyl-amino-1,2,4-oxadiazole. It is also interesting that cyanoacetic acid esters which react in basic medium with the C—C double bond of the tautomeric imino form (see p. 105) add to the C—N triple bond in the presence of an acidic catalyst.[17]

The chemistry of 1,2,4-oxadiazoles has been the subject of a recent exhaustive review,[24] where the methods of preparation of this nucleus starting from nitrile oxides or from hydroximic acid chlorides are evaluated and compared with other methods.

All compounds obtained by the cycloaddition reactions of nitrile oxides to nitriles are compiled in Chapter IX (Table XXXII).

22 *Rembarz, G., Brandner, H., Bebenroth, E. M.:* J. Prakt. Chem. [4] **31**, 221 (1966).
23 *D'Alò, G., Invernizzi, L., Grünanger, P.:* Boll. Chim. Farm. **108**, 792 (1969).
24 *Eloy, F.:* Fortschr. Chem. Forsch. **4**, 807 (1965).

G. Reactions with Other Unsatured Systems

While azobenzene does not react with benzonitrile oxide prepared *in situ*[1] or with the stable mesitonitrile oxide, even on prolonged refluxing at 78°,[2] the more activated *dimethyl or diethyl azodicarboxylate* enters at 15° 1,3-dipolar cycloaddition to aromatic nitrile oxides.[3,4] The primary 4-aryl-1,2,3,5-oxatriazoline dicarboxylate (*49*) is unstable and sometimes not isolable even at low temperature: ring cleavage yields, probably through the dipole *50*, the open-chain compound *52*. This latter one can eventually add a second molecule of nitrile oxide to yield *51* as by-product (Eq. (24)).

The $N{=}S$ double bond of both aliphatic

$$(24)$$

and aromatic *N-sulfinylamines* is a good dipolarophile for aliphatic and aromatic nitrile oxides, yielding 2-oxo-1,2,3,5-oxathiadiazoles (*53*) (Eq. (25), p. 134).[5–10, 10a]

Since heating readily and quantitatively cleaves this labile heterocyclic ring with evolution of sulfur dioxide, the cycloaddition provides

1 *Grünanger, P.:* Unpublished results.

2 *Grundmann, C., Richter, R.H.:* Unpublished results.

3 *Rajagopalan, P.:* Tetrahedron Letters 887 (1964).

4 *Huisgen, R., Blaschke, H., Brunn, E.:* Tetrahedron Letters 405 (1966).

5 *Rajagopalan, P., Daeniker, H.U.:* Angew. Chem. **75**, 91 (1963); — Intern. Ed. Engl. **2**, 46 (1963).

6 *Schmitt, E.E.:* U.S. Pat. 3.118.902 (1964); — Chem. Abstr. **60**, 10691 h (1964).

7 *Schmitt, E.E.:* U.S. Pat. 3.118.903 (1964); — Chem. Abstr. **60**, 10692 a (1964).

8 *Rajagopalan, P., Advani, B.G.:* J. Org. Chem. **30**, 3369 (1965).

9 *Eloy, F., Lenaers, R.:* Bull. Soc. Chim. Belges **74**, 129 (1965).

10 *Eloy, F.:* Helv. Chim. Acta **48**, 380 (1965).

10a *Levchenko, E.S., Ugarov, B.N.:* Zh. Organ. Khim. **5**, 148 (1969).

a method of general applicability for the preparation of unsymmetrical carbodiimides *54*.[8] When R is an aliphatic group and R′ an aryl group, pyrolysis of *53* led, along with the corresponding carbodiimide, to up to 40% of the rearranged compound *55*.[10]

$$\text{(25)}$$

Recent kinetic studies[11] have established that the mechanism of this cycloaddition cannot be visualized either as a pure two-step reaction with a dipolar intermediate, formed through a nucleophilic attack of the oxygen atom on the sulfur atom from the sulfinylamine, or as a perfectly concerted one-step reaction. A non-synchronous bond formation in the multicenter mechanism, leading to a polarized transition state, explains better the experimental results.

Sulfur dioxide reacts analogously, the formed 1,3,2,4-dioxathiazole S-oxides, decompose thermally to isocyanates and sulfur dioxide.[11a]

While neither dimeric nor trimeric borimide are reactive, the recently isolated *monomeric borimides*[12] *56* react quantitatively with benzonitrile oxide already at $-25°$ to yield 1,3,5,2-oxadiazaboroles *57* (Eq. (26)).[13, 14]

$$\text{(26)}$$

As frequently encountered in boron-nitrogen chemistry, the N—B multiple bond, notwithstanding the formal charges of mesomeric form

11 *Beltrame, P., Comotti, A., Veglio, C.*: Chem. Commun. 996 (1967); — *Beltrame, P., Vintani, C.*: J. Chem. Soc. (C) 873 (1970).

11a *Burk, E.H., Carlos, D.D.*: J. Heterocycl. Chem. **7**, 177 (1970).

12 *Paetzold, P.I., Simson, W.M.*: Angew. Chem. **78**, 825 (1966); — Intern. Ed. Engl. **5**, 842 (1966).

13 *Paetzold, P.I.*: Angew. Chem. **79**, 583 (1967); — Intern. Ed. Engl. **6**, 572 (1967).

14 *Paetzold, P.I., Stohr, G.*: Chem. Ber. **101**, 2874 (1968).

56b, leaves a higher electron density on the more electronegative nitrogen atom, which attacks the positively charged carbon atom of nitrile oxide, thus controlling the direction of the cycloaddition. According to the 6π-electron availability of the heterocyclic ring, a remarkable stability of *57* to solvolysis was observed.

In contrast to the definite nucleophilic attack of sulfur- and arsen-ylides (see Chapter VII), *phosphor-ylides* and *phosphor-imines* seem to react with nitrile oxides through a true cycloaddition reaction. In some cases the primarily formed 5-membered cycloadduct could be isolated, but more often its formation could only be deduced through isolation of various types of compounds, arising from ring cleavage and possibly

$$(27)$$

further reactions of some intermediates with a second molecule of nitrile oxide or phosphorane. Eq. (27) (p. 135) illustrates the various cases possible for phosphor-ylides (*58*).

At the present time there is not sufficient evidence to decide whether the reaction follows a concerted one-step mechanism (pathway A) or a two-step mechanism through a dipolar intermediate *60* (pathway B).

The stability of the 1,2,5-oxazaphosphol-2-ines (*59*) depends on the nature of the substituents R, R′ and R″:

(a) when R′ and R″ are electron-attracting, even at room temperature C—P cleavage takes place to give, through a stabilized dipolar intermediate *62* and rearrangement of R, the ketenimines *64*. These latter compounds are not very stable and sometimes can react with a second molecule of *58* to give a new phosphor-ylide *65*;

(b) when R′ and/or R″ are electron-repelling, the cycloadduct *59* is isolable and only at higher temperature it decomposes with cleavage of the C—P bond to give the azirines *63*. The nature of the nitrile oxide may also influence the fragmentation of the cycloadduct: whereas starting from *66a* the ketene N-phenylimine *67* is isolated, *66b* cleaves to the azirine *68* (Eq. (28)), owing to the impossibility of the carbethoxy group to migrate;

$$
\begin{array}{ccc}
\underset{\text{EtOOC}}{\overset{\text{CH}_3}{\diagup}}\!\!\text{C}{=}\text{C}{=}\text{N}{-}\text{C}_6\text{H}_5 & \xleftarrow{\;-\ \text{Ph}_3\text{PO}\;} & \text{R-cycloadduct} \\
\textit{67} & \textit{66a},\ \text{R}=\text{C}_6\text{H}_5 & \\
 & \textit{66b},\ \text{R}=\text{COOEt} &
\end{array}
\qquad (28)
$$

$$
\xrightarrow{\;-\ \text{Ph}_3\text{PO}\;}\quad \underset{\textit{68}}{\text{EtOOC}-\text{C}{=}\text{N azirine (CH}_3,\ \text{COOEt)}}
$$

(c) when R′ and R″ are both electron-repelling and R is electron-attracting, O—P cleavage previals and thermal decomposition of the isolable cycloadduct leads to the α,β-unsaturated oxime *61*.[15]

A few examples should illustrate the above general conclusions. The simple *methylene-phosphorane* (*58*, R′=R″=H) yields a relatively stable cycloadduct *59* (R′=R″=H),[16,17] easily cleaved in the presence of hydrobromic acid to the oxime of triphenylphenacylphosphonium

15 *Bestmann, H.J., Kunstmann, R.:* Chem. Ber. **102**, 1816 (1969).
16 *Huisgen, R., Wulff, J.:* Tetrahedron Letters 917 (1967).
17 *Huisgen, R., Wulff, J.:* Chem. Ber. **102**, 1833 (1969).

bromide (*69*), which in turn cyclizes to *59* in alkaline solution.[18] A method of wider applicability for preparing *2-oximino-phosphonium halides*, such as *69*, and through them the oxazaphospholines *59*, employs as starting products α-haloketoximes and triphenylphosphine (Eq. (29)).[19, 20] Ther-

$$R-\underset{\underset{NOH}{\parallel}}{C}-\underset{\underset{R''}{|}}{\overset{\overset{R'}{|}}{C}}-Br \quad + \quad Ph_3P \quad \longrightarrow$$

$$R-\underset{\underset{NOH}{\parallel}}{C}-\underset{\underset{R''}{|}}{\overset{\overset{R'}{|}}{C}}-\overset{\oplus}{PPh_3} \quad Br^{\ominus} \quad \underset{HBr}{\overset{base}{\rightleftharpoons}} \quad R-\underset{N-O-PPh_3}{\overset{R'}{\underset{R''}{\bigtriangleup}}} \tag{29}$$

$$\qquad\qquad 69 \qquad\qquad\qquad\qquad\qquad\qquad 59$$

mal cleavage of the cycloadduct *59* (R=C_6H_5; R'=R''=H) at 140° led to a mixture of 38% of 2-phenylazirine (*63*, R=C_6H_5; R'=R''=H) and 34% of ketenimine *65* (R=C_6H_5; R'=R''=H), isolated as N,N'-diphenylacetamidin.[17] When the cycloaddition was carried out in DMSO, 3-phenyl-5-anilinoisoxazole was isolated as by-product,[18] which clearly arose by reaction of a second molecule of benzonitrile oxide on the intermediate ketene N-phenylimine (*64*, R=C_6H_5; R'=R''=H).

The course of the cycloaddition of *benzylidene phosphorane* (*58*, R''=H, R'=C_6H_5) with benzonitrile oxide is different; the primary cycloadduct is not isolable and spontaneously cleaves at room temperature to phenylketene N-phenylimine (*64*, R=R'=C_6H_5; R''=H). This compound, for electronic and steric reasons, is less reactive toward nitrile oxides and reacts therefore electrophilically with a second molecule of phosphorane to yield, as the only reaction product, the yellow phosphorane *65* (R=R'=C_6H_5; R''=H).[16, 18] Methoxycarbonylmethylene triphenylphosphorane behaves analogously.[16, 21]

Carbethoxy-fulmide reacted with isopropyl or *sec*-butyl triphenylphosphonium bromide to give fairly stable cycloadducts *59* (R=COOEt). These latter compounds, either through an intramolecular β-elimination or an intermediate formation of a nitroso-alkene, yielded the α,β-un-

18 *Umani-Ronchi, A., Acampora, M., Gaudiano, G., Selva, A.:* Chim. Ind. (Milano) **49**, 388 (1967).

19 *Gaudiano, G., Mondelli, R.: Ponti, P. P., Ticozzi, C., Umani-Ronchi, A.:* J. Org. Chem. **33**, 4431 (1968).

20 *Masaki, M., Fukui, K., Ohta, M.:* J. Org. Chem. **32**, 3564 (1967).

21 *Bestmann, H. J., Kunstmann, R.:* Angew. Chem. **78**, 1059 (1966); — Intern. Ed. Engl. **5**, 1039 (1966).

saturated oxime *61* (R=COOEt), resp. a mixture of the oximes *70* and *71* (Eq. (30)).[15]

$$EtOOC-C=N-O^{\ominus}$$
$$Ph_3\overset{\oplus}{P}-C-CH_2-H$$
$$CH-H$$
$$CH_3$$

(30)

$$\xrightarrow{-Ph_3P}$$

$$EtOOC-\underset{\underset{NOH}{\|}}{C}-\overset{\overset{CH_3}{|}}{C}=CHCH_3 \quad + \quad EtOOC-\underset{\underset{NOH}{\|}}{C}-\overset{\overset{CH_2CH_3}{|}}{C}=CH_2$$

70   *71*

Table XXIV. Reaction of nitrile oxides with phosphor-ylides (Eq. (27))

Nitrile oxide	Phosphor-ylide *58*		Solvent	% Yield of:				Ref.
	R'	R''		*59*	*63*	*64*	*61*	
EtOOCCNO	H	COOMe	Benzene		40[b]			[15]
	CH_3	CH_3	Benzene	51			75	[15]
	CH_3	C_2H_5	Benzene	36			71	[15]
	CH_3	COOEt	Benzene			27		[15, 21]
	CH_3	C_6H_5	Benzene			18		[15]
C_6H_5CNO	H	H	DMSO	40[a]				[18, 19]
	H	H	Benzene	66	38	34[d]		[16, 17]
	H	CH_3	DMSO	65[a]				[18, 19]
	H	CH_3	Benzene	75	80	20		[15]
	H	$n-C_3H_7$	DMSO	45[a]				[19]
	H	COOEt	Benzene			73[c]		[15, 21]
	H	COOMe	Benzene			98[c]		[16, 17]
	H	C_6H_5	DMSO			32[c]		[18]
	H	C_6H_5	Benzene			75[c]		[16, 17]
	$-(CH_2)_2-$		Benzene	61	84			[15, 22]
	CH_3	CH_3	Benzene	54	88			[15, 21]
	CH_3	C_2H_5	Benzene	55	79			[15]
	CH_3	COOEt	Benzene			68		[15, 21]
	CH_3	C_6H_5	Benzene			69		[15]
$p-ClC_6H_4CNO$	H	CH_3	Benzene	85	92	8		[15]
$2,4,6-(CH_3)_3C_6H_2CNO$	H	H	Ether	76				[16, 17]
	H	C_6H_5	Benzene			87[c]		[16, 17]

[a] Isolated as open-chain 2-oximinophosphonium bromide.
[b] Adduct 1:1 from azirine and phosphorane.
[c] Isolated as phosphorane *65*.
[d] Isolated as N,N'-diphenylacetamidine.

22 *Bestmann, H.J., Denzel, T., Kunstmann, R., Lengyel, J.:* Tetrahedron Letters 2895 (1968).

Experimentally, the nitrile oxides are as a rule generated *in situ* from the hydroximic acid chlorides, using as dehydrohalogenating agent a second molecule of the ylide itself.

Table XXIV summarizes the available data on the cycloaddition of nitrile oxides to phosphor-ylides.

The reaction of aromatic nitrile oxides with *N-aryliminophosphoranes* (72) did not allow to isolate the cycloaddition product 73, which decomposed into triphenylphosphine oxide and the corresponding carbodiimide 74,[23, 24] isolated as the disubstituted urea. The carbodiimide can react with a further molecule of iminophosphorane to yield a different carbodiimide (75) (Eq. (31), route b). With a symmetrical carbodiimide

$$(31)$$

(a) $Ar = Ar' = C_6H_5$
(b) $Ar = p\text{-}NO_2C_6H_4$
$Ar' = p\text{-}CH_3OC_6H_4$

the 4-membered cycloadduct, carbodiimide-iminophosphorane, can add a further molecule of nitrile oxide to yield as by-product a 1,2,4-oxadiazolin-5-one anil (76).

23 *Huisgen, R., Wulff, J.:* Tetrahedron Letters 921 (1967).
24 *Huisgen, R., Wulff, J.:* Chem. Ber. **102**, 1848 (1969).

VI. Addition Reactions Leading to Open-Chain Structures; 1,3-Additions

A. General

In the last two decades the main interest of the investigators remained centered on the 1,3-dipolar cycloaddition reactions and this overshadowed the ability of nitrile oxides to react promptly with a very large number of nucleophiles to yield open-chain 1,3-addition products.

Notwithstanding *Wieland's* erroneous statement that benzonitrile oxide "is unreactive both with ammonia, aniline, phenylhydrazine and with bromide, iodine, hydrogen chloride and phosphorus pentachloride"[1] (see also Chapter VI, G, p. 166ff.), *Ponzio*[2-3] and later *Speroni*[4] pointed out the reactivity of nitrile oxides towards ammonia, amines and other nucleophiles as well as toward hydrogen halides and carboxylic acids. Several reasons contributed to leave these findings unnoticed by the chemical world, and only recently *Grundmann,*[5-7] and subsequently several other investigators, extended and generalized the study of 1,3-addition reactions of nitrile oxides.

Owing to the electrophilicity of the carbon atom of nitrile oxides, the reaction with a generalized nucleophile B—H leads to a product where the nucleophilic part of the reactant has joined the carbon atom of the nitrile oxide and the oxygen bears the remaining proton, as shown by Eq. (1). Of course, the primary product *1* will occasionally rearrange or transform further in the reaction medium.

$$R-C\equiv N \rightarrow O \quad + \quad B-H \longrightarrow \tag{1}$$

$$\longrightarrow \underset{\underset{\textstyle 1}{}}{R-\underset{\underset{\textstyle NOH}{\|}}{C}-B} \quad \xleftarrow{\; -\,BH_2^{\oplus}Cl^{\ominus}\;} \quad R-\underset{\underset{\textstyle NOH}{\|}}{C}-Cl \quad + \quad 2\,B-H$$

---

1 *Wieland, H.:* Ber. **40**, 1667 (1907).

2 *Ponzio, G.:* Gazz. Chim. Ital. **53**, 379 (1923).

3 *Ponzio, G.:* Gazz. Chim. Ital. **66**, 117, 127 (1936).

4 *Speroni, G., Bartoli, M.:* Sopra gli ossidi di nitrile, VIII, Stab. Tip. Marzocco, Firenze, 1952.

5 *Grundmann, C.:* Angew. Chem. **75**, 450 (1963); — Intern. Ed. Engl. **2**, 260 (1963).

6 *Grundmann, C., Mini, V., Dean, J.M., Frommeld, H.D.:* Liebigs Ann. Chem. **687**, 191 (1965).

7 *Grundmann, C., Frommeld, H.D.:* J. Org. Chem. **31**, 157 (1966).

The same addition product *1* is usually obtained directly from hydroximic acid chlorides, this latter reaction being as a rule much longer known and experimentally more convenient than the nitrile oxide route, at least whenever the chlorides are easily available. Since hydroximic acid chlorides dehydrohalogenate very rapidly under the action of bases (see Chapter III, C-1, pp. 47–51), there seems to be little doubt that most, if not all, of their reactions infer as first step a dehydrohalogenation to nitrile oxide, which then reacts further. However, a direct substitution reaction cannot *a priori* be discarded, since O-alkylated and O-acylated hydroximic acid chlorides are also known to react with nucleophiles, *e. g.* with alkoxides[8, 9] or ammonia[10, 11] to yield alkyl O-alkylhydroxamates resp. O-substituted amidoximes. Direct evidence of an intermediate nitrile oxide is with certain exceptions still lacking, but would now not be difficult to reach for instance, through interception with a dipolarophile.

Threfore, the following treatment will chiefly illustrate the behaviour of isolated nitrile oxides in 1,3-addition reactions. The reactivity of the corresponding hydroximic acid halides toward the same nucleophiles will briefly be mentioned, only when some evidence or strong suspicion exists for the assumption of intermediate formation of nitrile oxides.

8 *Tiemann, F., Krüger, P.:* Ber. **18**, 727 (1885).
9 *Lossen, W., Jacobson, R.:* Liebigs Ann. Chem. **281**, 216 (1894).
10 *Houben, J., Kauffmann, H.:* Ber. **46**, 2827 (1913).
11 *Ponzio, G., Durio, E.:* Gazz. Chim. Ital. **60**, 436 (1930).

B. Addition of Hydrogen

As it has been already mentioned in Chapter IV, B, acidic reduction of nitrile oxides invariably leads by deoxygenation to nitriles.

Sodium borohydride, on the contrary, reduced O-methylpodocarponitrile oxide quantitatively to the *syn*-aldoxime.[1]

Polarographic reduction in aqueous solution has also been reported to yield *syn*-aldoximes.[2, 3] It is interesting to mention here that the other N-oxides, *e.g.* nitrones,[4] are, on the contrary, deoxygenated under the same conditions.

Both reductions can be visualized as a nucleophilic attack on the positively charged carbon atom of the nitrile oxide, from the hydride ion in the first case and from the electrons in the second case.

---

1 *Just, G., Dahl, K.:* Tetrahedron **24**, 5251 (1968).
2 *Armand, J.:* Bull. Soc. Chim. France 882 (1966).
3 *Armand, J., Souchay, P., Valentini, F.:* Bull. Soc. Chim. France 4585 (1968).
4 *Zuman, P., Exner, O.:* Collection Czech. Chem. Commun. **30**, 1832 (1965).

C. Addition of Water (Hydrolysis)

Nitrile oxides are stable in water, at least as long as the dimerization process allows it. *E.g.*, benzonitrile oxide can be steam-distilled,[1] although under extensive formation of furoxan. Acidic catalysis, however, promotes the addition of water: *e.g.* mesitonitrile oxide yields the corresponding hydroxamic acid by treatment with dilute sulfuric acid at 35°.[2] The reaction probably involves the protonation of the nitrile oxide to the conjugate acid, as illustrated by Eq. (2).

$$R-C{\equiv}N{\longrightarrow}O \;\xrightleftharpoons{H^+}\; R-\overset{\oplus}{C}{=}NOH \;\longrightarrow$$

$$\xrightarrow{H_2O} \; R-\underset{\overset{|}{\underset{\oplus}{O}H_2}}{C}{=}NOH \;\xrightleftharpoons{-H^+}\; R-\underset{\overset{\|}{N}OH}{C}-OH \tag{2}$$

A protonated nitrile oxide was also invoked to explain the formation of hydroxamic acids by the action of concentrated sulfuric acid on nitronic esters or nitro alkanes.[2a]

Terephthalo-bis-N-oxide was hydrolyzed to the bis-hydroxamic acid by treatment with formic acid and afterwards with water.[3] The hydroxamic acid *2* has analogously been prepared by action of acetic acid on oximinophenylacetonitrile oxide;[4] the same product arose also by treatment of the hydroximic acid iodide with sodium acetate and dilute acetic acid.[5,6] Interestingly, with propionic acid 3-phenyl-1,2,4-oxadiazolin-5-one (*4*) has instead been obtained. As the hydroxamic acid *2* itself yields under the same conditions the oxadiazolone, a Lossen type rearrangement to oximinobenzoyl isocyanate (*3*) seems reasonable to be invoked; this last compound, in attempts through different routes, could never be isolated because of its ready cyclisation to the oxadiazolone.[7-9]

$$C_6H_5-\underset{\overset{|}{CO-NHOH}}{C}{=}NOH \;\longrightarrow\; C_6H_5-\underset{\overset{\|}{N}OH}{C}-NCO \;\longrightarrow\;$$

$$\qquad\qquad 2 \q\qquad\qquad\qquad\qquad 3 \qquad\qquad\qquad\qquad 4$$

1 *Wieland, H.:* Ber. **40**, 1667 (1907).

2 *Grundmann, C., Frommeld, H.-D.:* J. Org. Chem. **31**, 157 (1966).

2a *Kornblum, N., Brown, R.A.:* J. Am. Chem. Soc. **87**, 1742 (1965).

3 *Overberger, C.G., Fujimoto, S.:* Polymer Letters **3**, 735 (1965); J. Polymer Sci. (C) **16**, 4161 (1968).

4 *Ponzio, G.:* Gazz. Chim. Ital. **66**, 117, 127 (1936).

5 *Carbone, G.:* Gazz. Chim. Ital. **62**, 428 (1932).

6 *Ponzio, G., Baldracco, F.:* Gazz. Chim. Ital. **60**, 415 (1930).

7 *Musante, C.:* Gazz. Chim. Ital. **68**, 331 (1938).

8 *Vaughan, W.R., Spencer, J.L.:* J. Org. Chem. **25**, 1077 (1960).

9 *Golankiewicz, K.:* Bull. Acad. Polon. Sci. **10**, 417 (1962).

Under more drastic conditions, the hydrolysis proceeds further, and the aromatic carboxylic acid together with hydroxylamine are produced: *e.g.* treatment of the ether solution of benzonitrile oxide with concentrated hydrochloric acid leads to benzoic acid and hydroxylamine hydrochloride.[1, 10] Analogous results have been obtained by treatment of oximinophenylacetonitrile oxide with boiling conc. HCl.[4, 11] Under different conditions, addition of hydrogen chloride takes place (see Section F).

The same type of hydrolyzed products can be obtained starting from the oligomers of nitrile oxides, including the parent compound, *Wieland's* so-called "trifulmin". This polymeric material arises by mild alkaline treatment of nitrolic acids or from benzonitrile oxide and NEt_3 or from cyanoformohydroximic acid chloride with water (see Chapter IV, C-2, pp. 83–84); treatment with aqueous acids leads to the hydroxamic acid and further to hydroxylamine and the corresponding carboxylic acid.[2, 12–14] Starting from poly-benzonitrile oxide the main product is the dimer 3,5-diphenyl-1,2,4-oxadiazole-4-oxide.[12, 14]

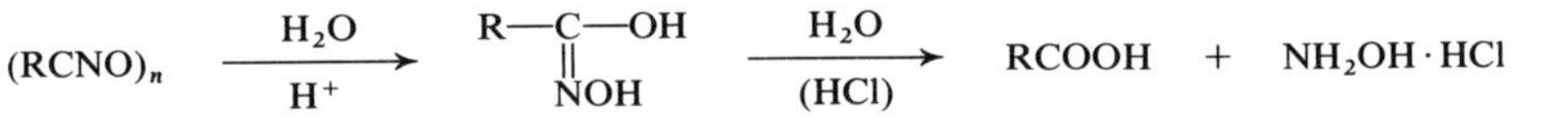

$$(RCNO)_n \xrightarrow[\text{H}^+]{\text{H}_2\text{O}} \underset{\overset{||}{\text{NOH}}}{\text{R—C—OH}} \xrightarrow[\text{(HCl)}]{\text{H}_2\text{O}} RCOOH + NH_2OH \cdot HCl \qquad (3)$$

Alkaline hydrolysis is not so straightforward, and several products have been obtained by treatment of benzonitrile oxide with an aqueous-ethanolic solution of sodium hydroxide.[14] When the reaction takes place at low temperature, an insoluble unstable plastic mass separates out, possibly similar to *Wieland's* polymeric nitrile oxide, for which structures 5 or 6 have been proposed. When the reaction is carried out at room

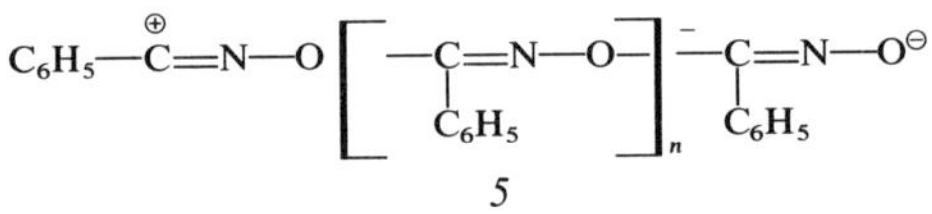

$$C_6H_5-\overset{\oplus}{C}=N-O\left[-\underset{C_6H_5}{C}=N-O-\right]_n-\underset{C_6H_5}{C}=N-O^{\ominus}$$

5

$$C_6H_5-\overset{\oplus}{C}=\overset{\overset{O}{\uparrow}}{N}\left[-\underset{C_6H_5}{C}=\overset{\overset{O}{\uparrow}}{N}-\right]_n-\underset{C_6H_5}{C}=N-O^{\ominus}$$

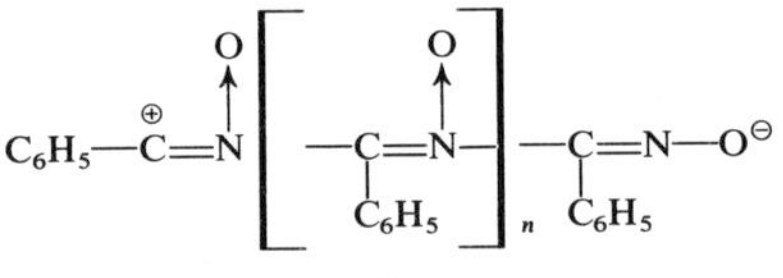

6

10 *Werner, A., Buss, H.:* Ber. **27**, 2193 (1894).

11 *Ponzio, G.:* Gazz. Chim. Ital. **53**, 379 (1923).

12 *Wieland, H.:* Ber. **42**, 803 (1909).

13 *Wieland, H.:* Ber. **42**, 816 (1909).

14 *Speroni, G., Bartoli, M.:* Sopra gli ossidi di nitrile, VIII, Stab. Tip. Marzocco, Firenze, 1952.

temperature without cooling, the following products could be isolated: benzoic acid, benzhydroxamic acid, dibenzhydroxamic acid, ethyl benzoate, diphenylurea, 3,5-diphenyl-1,2,4-oxadiazole, possibly along with benzonitrile, phenylisocyanate and N_2O. All these products can be generated by further action of alkali and nitrile oxide on the primary hydrolysis (or alkoholysis) products, benzhydroxamic acid (or its ethyl ester). When the reaction is uncontrolled, only benzoic acid and traces of diphenyloxadiazole are obtained, all other products having been destroyed by the alkali.

Cyanogen-bis-N-oxide reacts with NaOH to furnish yellow, very sensitive, water-soluble compounds; in absolute ethanolic solution an adduct with one mole each of NaOH and NaOEt was isolated and analyzed. Barium and copper salts were also obtained, but all attempts to isolate one of the well-known oxaldihydroxamic acids by cautious acidification failed, as *e.g.* only NaCl and NH_3OHCl could be recovered by treatment of the disodium salt with conc. HCl. A structure 7 was therefore proposed for the disodium salt, according to a mechanism involving an initial attack of OH^- to a positive nitrogen atom, subsequent prototropic rearrangement and finally a "normal" attack of the alkoxide anion to the positive carbon atom. (Eq. (4)).[15]

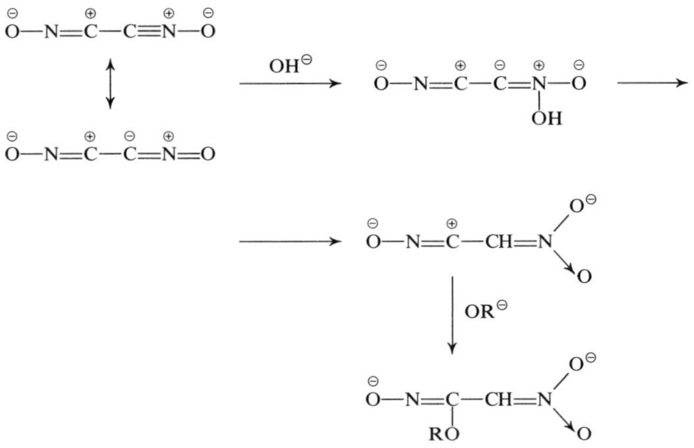

$$\overset{\ominus}{O}-N=\overset{\oplus}{C}-C\equiv\overset{\oplus}{N}-\overset{\ominus}{O} \quad\updownarrow\quad \overset{\ominus}{O}-N=\overset{\oplus}{C}-\overset{\ominus}{C}=\overset{\oplus}{N}=O$$

$$\xrightarrow{OH^{\ominus}}\quad \overset{\ominus}{O}-N=\overset{\oplus}{C}-\overset{\ominus}{C}=\overset{\oplus}{N}-\overset{\ominus}{O} \;(\text{OH}) \longrightarrow$$

$$\longrightarrow\quad \overset{\ominus}{O}-N=\overset{\oplus}{C}-CH=N\overset{O^{\ominus}}{\diagdown O} \tag{4}$$

$$\Big\downarrow OR^{\ominus}$$

$$\overset{\ominus}{O}-N=\underset{RO}{C}-CH=N\overset{O^{\ominus}}{\diagdown O}$$

7

The rather astonishing claim of the formation of a mixture of 3-phenyl-1,2,4-oxadiazolin-5-one, 5-phenyl-1,2,4-oxadiazolin-3-one and phenylmethazonic acid by treatment of oximinophenylacetonitrile oxide with cold 5 % NaOH[16] needs further confirmation because of the still contro-

---

15 *Grundmann, C., Mini, V., Dean, J. M., Frommeld, H.-D.:* Liebigs Ann. Chem. **687**, 191 (1965).

16 *Ponzio, G.:* Gazz. Chim. Ital. **56**, 490 (1926).

versial structure of the starting material (see Chapter III, C-2, p. 52). For the formation of the former product an initial rearrangement of the nitrile oxide to the isocyanate has been invoked,[4] but this seems unlikely considering the alkaline medium and the mild conditions employed.

Treatment of aromatic hydroximic acid chlorides with strong aqueous bases sometimes yielded the corresponding hydroxamic acids as by-products;[10, 17] the formation of furoxan as main products does not exclude a passage through the nitrile oxides, although on the whole the action of bases on the latter ones needs additional studies.

---

17 *Werner, A.:* Ber. **27**, 2846 (1894).

D. Addition of Alcohols and Phenols

Nitrile oxides do not react with alcohols, at least under 100°, unless the addition reaction is promoted by an acidic catalyst, such as sulfuric acid; it is reasonable to deduce that the conjugate acid of the nitrile oxide is here involved (Eq. (5)).[1]

$$\text{Mes—C}\equiv\text{N}\longrightarrow\text{O} \xrightleftharpoons{+H^+} \text{Mes—}\overset{\oplus}{\text{C}}\text{=N—OH} \longrightarrow \tag{5}$$

$$\xrightarrow{\text{ROH}} \underset{\underset{\text{NOH}}{\|}}{\text{Mes—}\overset{\oplus}{\text{C}}\text{—}\underset{\text{H}}{\text{OR}}} \xrightleftharpoons{-H^+} \underset{\underset{\text{NOH}}{\|}}{\text{Mes—C—OR}}$$

Mes = 2,4,6-trimethylphenyl; R = CH_3, C_2H_5

High yields of the alkyl mesitohydroximic acids (8) have also been obtained using a basic catalyst, such as sodium alkoxide; in this case a nucleophilic attack takes place first (Eq. (6)). The structure of the alkyl hydroximic acids, a class of compounds now readily accessible by this route, could be demonstrated by comparison with the isomeric alkyl hydroximates (9), prepared from alkylhydroxylamine and aroyl chloride; with diazomethane both isomers led to the same product, methyl alkyl-hydroximate (10).[1]

$$\text{Mes—C}\equiv\text{N}\longrightarrow\text{O} \xrightarrow{\text{RO}^{\ominus}} \underset{\underset{\text{NO}^{\ominus}}{\|}}{\text{Mes—C—OR}} \xrightarrow{\text{H}^+} \underset{\underset{\underset{8}{\text{NOH}}}{\|}}{\text{Mes—C—OR}}$$

$$(\text{R}=\text{CH}_3)\ \Big\downarrow\ \text{CH}_2\text{N}_2 \tag{6}$$

$$\text{CH}_3\text{ONH}_2\ +\ \text{MesCOCl} \longrightarrow \underset{9}{\text{MesCONHOCH}_3} \xrightarrow{\text{CH}_2\text{N}_2} \underset{\underset{\underset{10}{\text{NOCH}_3}}{\|}}{\text{Mes—C—OCH}_3}$$

An intramolecular displacement followed the addition of methanol under basic conditions to the interesting bridgehead nitrile oxide, 1-fulmido-2-bromo-7,7-dimethylnorcamphane.[2]

Treatment of benzonitrile oxide with an ethanolic solution of hydrogen chloride led to quantitative isolation of hydroxylamine hydrochloride and ethyl benzoate.[3] Clearly, under the conditions employed the primarily formed ethyl benzhydroximic acid is not stable and hydrolyses to the reaction products. Phenol, being more acidic than alcohols, adds benzonitrile oxide without the aid of a catalyst, to give a 5% yield of phenyl hydroximic acid at room temperature, whereas at −15° no reaction occurs.[4] As expected, use of phenoxide anion instead of phenol remarkably increases the yields.[5]

1 *Grundmann, C., Frommeld, H.-D.:* J. Org. Chem. **31**, 157 (1966).

2 *Ranganathan, S., Singh, B.B., Panda, C.S.:* Tetrahedron Letters 1225 (1970).

3 *Speroni, G., Bartoli, M.:* Sopra gli ossidi di nitrile, VIII, Stab. Tip. Marzocco, Firenze, 1952.

4 *Alexandrou, N.E., Nicolaides, D.N.:* Chim. Chronika **30**, 49 (1965).

5 *Alexandrou, N.E., Nicolaides, D.N.:* Tetrahedron Letters 2497 (1966).

E. Addition of Sulfides and Thiols

Free hydrogen sulfide reacts slowly with mesitonitrile oxide to yield di-mesityl-oximinosulfide (*13a*, Ar=Ar′). Better yields were obtained with sodium hydrogen sulfide at pH 7–8. At pH 10–12, the reaction yielded upon acidification only the thiohydroxamic acid *14*, which itself can add a second molecule of nitrile oxide (see Section F) to give *13*. Other nitrile oxides behave analogously. This pH-dependent behaviour may be explained by assuming that at high pH (>10) the formed species is the thiohydroximate dianion (*11*), which would be incapable to react further with the nitrile oxide. At a lower pH (7–8) the primary product could be the thiohydroximate monoanion (*12*), which adds further to the nitrile oxide to form *13*. The symmetrical nature of the

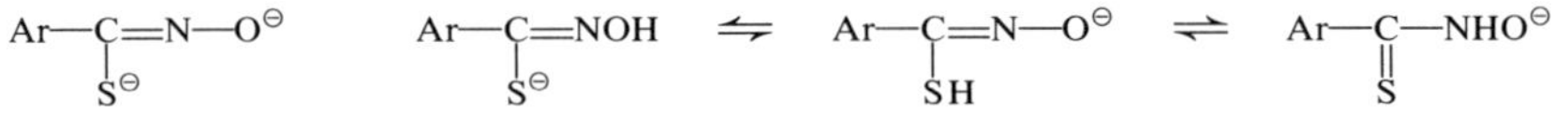

$$\underset{\textit{11}}{Ar-\underset{S^{\ominus}}{\overset{|}{C}}=N-O^{\ominus}} \qquad Ar-\underset{S^{\ominus}}{\overset{|}{C}}=NOH \;\rightleftharpoons\; Ar-\underset{SH}{\overset{|}{C}}=N-O^{\ominus} \;\rightleftharpoons\; \underset{\textit{12}}{Ar-\underset{S}{\overset{\parallel}{C}}-NHO^{\ominus}}$$

diadduct (*13a*, Ar=Ar′) is supported by the NMR spectrum and, indirectly, by the synthesis of the same diadduct in which the two groups Ar and Ar′ are different *via* two synthetic routes starting from *14a* resp. from *14c* and the corresponding nitrile oxides. (Eq. (7)).[1]

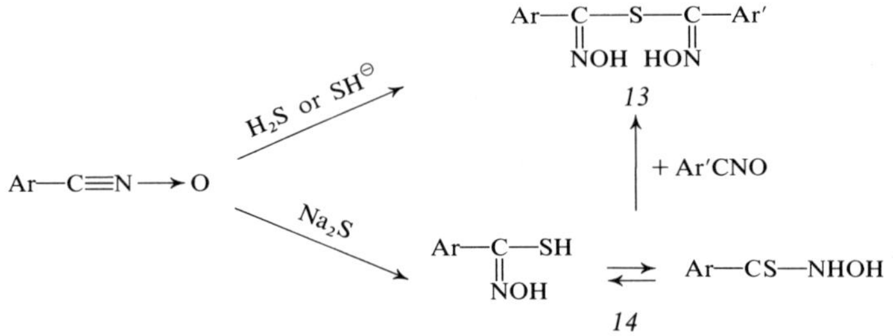

$$Ar-C\equiv N\rightarrow O \quad\begin{array}{c} \xrightarrow{H_2S \text{ or } SH^{\ominus}} \\ \\ \xrightarrow{Na_2S} \end{array}\quad \begin{array}{c} Ar-\underset{NOH}{\overset{\parallel}{C}}-S-\underset{HON}{\overset{\parallel}{C}}-Ar' \\ \textit{13} \\ \uparrow\; +Ar'CNO \\ \\ Ar-\underset{NOH}{\overset{\parallel}{C}}-SH \;\rightleftharpoons\; Ar-CS-NHOH \\ \textit{14} \end{array} \tag{7}$$

(a: 2,4,6-Me$_3$C$_6$H$_2$, b: 2,3,5,6-Me$_4$C$_6$H, c: 2,6-Cl$_2$C$_6$H$_3$)

The thermal decomposition of *13a* (Ar=Ar′) at 200° gives an almost quantitative yield of mesityl isothiocyanate (*17*) and 42 % of dimesitylurea (*16*).[1] Two mechanism are conceivable here (Eq. (8)): (route A) an acyl migration takes place to *15*, analogous to those observed in the case of acyl benzhydroximates, followed by an α-elimination of thiohydroxamic acid (*14a*), which undergoes a Lossen rearrangement to the isothio-

---

1 *Grundmann, C., Frommeld, H.-D.:* J. Org. Chem. **31**, 157 (1966).

cyanate (*17*), and a nitrene intermediate which leads to the urea *16* through the isocyanate, or (route B) a dissociation takes place into the starting materials, the thiohydroxamic acid (*14a*) and the nitrile oxide, which immediately isomerizes to the isocyanate. This dissociation is analogous to the one proposed for benzyl thiobenzhydroximate.[2] Both reaction schemes, A and B, are supported by the fact that *14a* itself decomposes $> 110°$ with the formation of at least 70% of *17*[3]; however, other aromatic thiohydroxamic acids are reported to yield under similar conditions nitriles and sulfur.[2, 4, 5]

2,4,6-Trimethoxybenzonitrile oxide does not add sulfide ion under the same conditions, but is readily reduced to 2,4,6-trimethoxybenzonitrile. The electron-repelling power of the methoxy groups obviously increases the oxidation potential of the nitrile group.

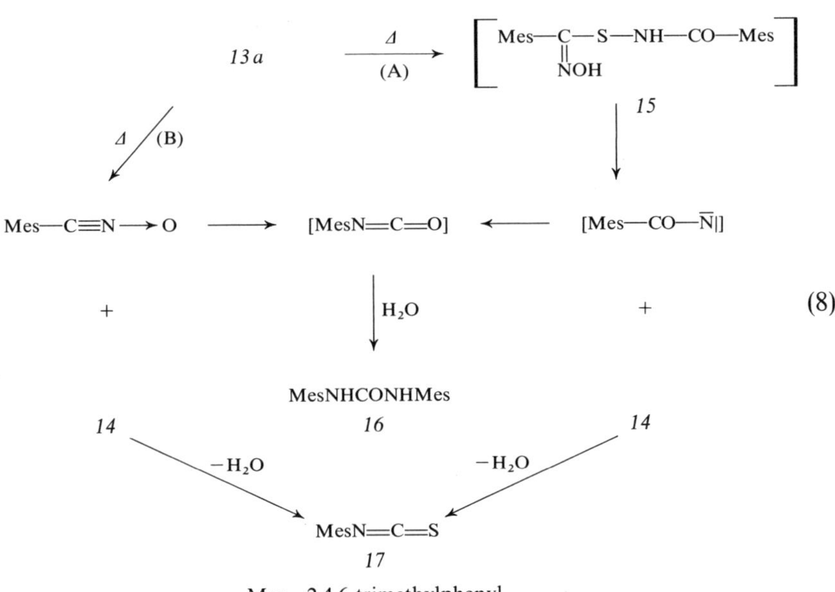

Oxalonitrile-bis-N-oxide does not react with free hydrogen sulfide, but adds sodium sulfide to give with low yields a mixture of the oxalo-bis-thiohydroximic acid (*18*) and—by further addition of a second mole

2 *Bacchetti, T., Alemagna, A.:* Rend. Ist. Lombardo Sci. Lettere **91**, 574 (1957).

3 *Grundmann, C.:* Unpublished.

4 *Bacchetti, T., Alemagna, A.:* Rend. Ist. Lombardo Sci. Lettere **91**, 30 (1957).

5 *Cambi, L.:* Rend. Accad. Reale Lincei [5] **12**, 687 (1909).

of nitrile oxide—of the cyclic tetraoximino-1,4-dithian (*19*), along with a high percentage of polymeric material (Eq. (9)).[6]

$$O \leftarrow N \equiv C - C \equiv N \rightarrow O \quad + \quad 2 S^{--} \quad \longrightarrow \quad ^{\ominus}S-C-C-S^{\ominus}$$

(9)

18

19

According to their greater degree of nucleophilicity, mercaptans add easily to nitrile oxides, usually generated *in situ*, to yield the thiohydroximic acids (*20*). The reaction seems to have a wide range of applicability, both aliphatic and aromatic nitrile oxides and thiols having been used.[1, 4, 7, 8] One example of a selenomercaptan has also entered the addition.[4] The available data are collected in Table XXV.

Since a slight excess of triethylamine appears to increase the yields, a mechanism involving a *trans* addition of the thiolate ion has been proposed (Eq. (10)).[4]

$$R-C \equiv N \rightarrow O \quad \longrightarrow \quad \left[\begin{array}{c} R \\ C=N \\ R'S \quad O^{\ominus} \end{array} \right] \quad \xrightarrow{H^+} \quad R \quad SR'$$

(10)

20

Sterically hindered thiols, such as *t*-butyl mercaptan, give lower yields, the majority of the nitrile oxide being dimerized to furoxan.[8] A secondary reaction, observed when the mixture of benzonitrile oxide and benzyl mercaptan was not adequately cooled, is the formation of disulfide and benzonitrile as by-products. This can be regarded as an oxidation-reduction process.[2]

When difunctional nitrile oxides, such as oxalonitrile-bis-N-oxide, react with dithiols, *e.g.* o-phenylenedithiol, in dilute solution, moderate yields of 2,3-bishydroxyimino-1,4-benzodithians are obtained.[19]

---

6 *Grundmann, C., Mini, V., Dean, J.M., Frommeld, H.D.:* Liebigs Ann. Chem. **687**, 191 (1965).

7 *Zinner, G., Günther, M.:* Angew. Chem. **76**, 440 (1964); — Intern. Ed. Engl. **3**, 383 (1964).

8 *Benn, M.H.:* Can. J. Chem. **42**, 2393 (1964).

Table XXV. Addition of thiols to nitrile oxides

Thiol	Nitrile oxide	Thiohydroximic acid % yield	Ref.
Methylmercaptan	C_6H_5CNO	50	[8]
	$C_6H_5CH_2CNO$	65	[8]
Ethylmercaptan	C_6H_5CNO	87	[8]
	$(CH_3)_3C_6H_2CNO$	93	[1]
Mercaptoacetone	C_6H_5CNO	85	[9]
3-Thiolpropionic acid	C_6H_5CNO	50	[8]
n-Butylmercaptan	CH_3CNO	b	[7]
t-Butylmercaptan	C_6H_5CNO	12	[8]
Thiophenol	C_6H_5CNO	91	[8]
Benzylmercaptan	HCNO	b	[10]
	$(CH_3)_3CCNO$	b	[7]
	$(C_2H_5)_2CHCNO$	b	[7]
	C_6H_5CNO	91	[2, 4, 8]
p-Methylthiophenol	$(CH_3)_2CHCNO$	66	[8]
	C_6H_5CNO	84	[8]
	$C_6H_5CH_2CNO$	70	[8]
2,3,4,6-Tetra-O-acetyl-1-mercapto-β-D-glucopyranose	HCNO	b	[10]
	CH_3CNO	86	[11]
	$CH_2{=}CHCH_2CNO$	b	[12]
	$(CH_3)_2CHCNO$	60	[13]
	$CH_2{=}CHCH_2CH_2CNO$	b	[14]
	$CH_3CH_2CH(CH_3)CNO$	63	[15]
	$C_6H_5CH_2CNO$	b	[16]
	$C_6H_5CH_2CH_2CNO$	86	[17]
	$p\text{-}CH_3OC_6H_4CH_2CNO$	54	[18]
	$p\text{-}CH_3COOC_6H_4CH_2CNO$	60	[18]
Ethylene dithiol	$(C{\equiv}N \rightarrow O)_2$	b	[19]

9 *Bacchetti, T., Alemagna, A.:* Rend. Ist. Lombardo Sci. Lettere **91**, 481 (1957).
10 *Kjersgaard, D., Kjaer, A.:* Acta Chem. Scand. **24**, 1367 (1970).
11 *Benn, M.H.:* Can. J. Chem. **42**, 163 (1964).
12 *Benn, M.H., Ettlinger, M.G.:* Chem. Commun. 445 (1965).
13 *Benn, M.H., Meakin, D.:* Can. J. Chem. **43**, 1874 (1965).
14 *Kjaer, A., Jensen, S.R.:* Acta Chem. Scand. **22**, 3324 (1968).
15 *Benn, M.H., Yelland, L.:* Can. J. Chem. **45**, 1595 (1967).
16 *Benn, M.H.:* Can. J. Chem. **41**, 2836 (1963).
17 *Benn, M.H.:* J. Chem. Soc. 4072 (1964).
18 *Benn, M.H.:* Can. J. Chem. **43**, 1 (1965).
19 *Alexandrou, N.E., Nikolaides, D.N.:* J. Chem. Soc. (C) 2319 (1969).

Table XXV (continued)

Thiol	Nitrile oxide	Thiohydroximic acid % yield	Ref.
Toluene-3,4-dithiol	$(C\equiv N \rightarrow O)_2$	[b]	[19]
Selenophenol	C_6H_5CNO	35[a]	[8]

[a] Phenyl-benzo-seleno-hydroximic acid.   [b] Not reported.

The addition of thiols to nitrile oxides has been the key step for the synthesis of several naturally occuring mustard oil glucosides by the route illustrated in Eq. (11). Using this method the following glucosides

$$\text{(11)}$$

$$CH_2=CH-(CH_2)_n-C\equiv N \rightarrow O$$
$$\textit{21}$$
$$n = 1, 2$$

were synthesized: glucoapparin,[11] glucoaubrietin,[18] glucocochlearin,[15] gluconapin,[14] gluconasturtiin,[17] glucoputranjivin,[13] glucosinalbin,[18] glucotropaeolin[16] and sinigrin.[12] Interestingly enough, two cases are known, in which unsaturated nitrile oxides, such as *21*, were required and therefore prepared *in situ* from the corresponding hydroximic acid chlorides. Notwithstanding the presence of the chain-terminal double bond, a potentially highly reactive dipolarophilic group, good yields of the required unsaturated thiohydroximic acids were reported.[12, 14]

Finally, it should be noted that the same thiohydroximates can be prepared directly from hydroximic acid chlorides and sodium thiolates; a convenient procedure in aqueous medium has been developed.[20]

20 *Davies, J. H., Davis, R. H., Kirby, P.:* J. Chem. Soc. (C) 431 (1968); see also Neth. Pat. Appl. 8, 122 (1967); Chem. Abstr. **69**, 51610 (1968).

F. Addition of Inorganic and Organic Acids and their Derivatives

The addition of *hydrogen chloride* to fulminate salts has been known since 1894,[1,2] but has been considered for a long time a strong evidence for the divalent carbon structure *22* of the initially formed fulminic acid (see i.a.[3]). In view of the marked additive properties of the higher nitrile oxides toward hydrogen halides, this reaction perfectly supports the formonitrile oxide structure (*23*). The adduct, formhydroximic acid chloride (*24*, X=Cl) is very unstable and looses hydrogen chloride to yield the oligomers of fulminic acid. With dilute acid hydrolysis takes place to hydroxylamine hydrochloride and formic acid[4] or carbon monoxide (Eq. (12)).

$$(CNO)M \xrightarrow{HX} \left\{ \begin{array}{c} C{=}NOH \\ 22 \\ HC{\equiv}N \longrightarrow O \end{array} \right\} \underset{-HX}{\overset{+HX}{\rightleftharpoons}} \underset{24}{\overset{HCX}{\underset{NOH}{\parallel}}} \xrightarrow[(HX)]{H_2O} \begin{array}{c} HCOOH \\ + \\ NH_2OH \cdot HCl \end{array} \qquad (12)$$

M = Na, Ag   *23*

The formhydroximic bromide (*24*, X = Br) and iodide (*24*, X = I), similarly prepared from sodium fulminate and hydrogen halides, are relatively more stable;[5,6] the iodide, more conveniently obtained directly from mercuric fulminate, according to Eq. (13),[7] represents the most suitable starting material to generate *in situ* fulminic acid.[8]

$$Hg(CNO)_2 \ + \ 4\,HI \ + \ 2\,KI \ \longrightarrow \ 2\,IHC{=}NOH \ + \ K_2HgI_4 \qquad (13)$$

Contrary to an early statement,[9] the 1,3-addition of hydrogen chloride, bromide or iodide to both aliphatic and aromatic nitrile oxides is a quite general reaction (Eq. (14)). Hydrogen fluoride however does not react because of the less nucleophilic nature of the fluoride anion.

$$R{-}C{\equiv}N \longrightarrow O \ + \ HX \underset{(2)}{\overset{(1)}{\rightleftharpoons}} \underset{25}{\overset{R{-}C{-}X}{\underset{NOH}{\parallel}}} \qquad (14)$$

1 *Nef, J. U.:* Liebigs Ann. Chem. **280**, 291 (1894).

2 *Scholl, R.:* Ber. **27**, 2816 (1894).

3 *Kurtz, P.:* Methoden der organischen Chemie, 4th ed. (ed. *E. Müller*), vol. 8, chapt.: Methoden zur Herstellung und Umwandlung von Knallsäure, p. 355-358. Stuttgart: G. Thieme, 1952.

4 *Carstanjen, E., Ehrenberg, A.:* J. Prakt. Chem. (2) **25**, 232 (1882).

5 *Palazzo, F. C.:* Rend. Accad. Reale Lincei (5) **16**, 545 (1907).

6 *Quilico, A., Panizzi, L.:* Gazz. Chim. Ital. **72**, 155 (1942).

7 *Huisgen, R.:* Private communication; — *Christl, M.:* Ph. D. Thesis, Universität München, 1969.

8 *Huisgen, R., Christl, M.:* Angew. Chem. **79**, 471 (1967); — Intern. Ed. Engl. **6**, 456 (1967).

9 *Wieland, H.:* Chem. Ber. **40**, 1667 (1907).

The nitrile oxides Nos. 6,[10,11] 9,[12] 23,[12,13] 25,[12,13] 28,[12] 29,[14] 31,[15] 32,[16] 33,[17] 34,[18] 47,[14] 48,[12] of Table III (p. 16) have thus been converted to the corresponding hydroximic acid chlorides (or bromides or iodides). The addition has little preparative importance because of the ready availability of hydroximic chlorides (*25*, X = Cl) by direct chlorination of aldoximes, so that usually nitrile oxides are prepared from the former compounds by action of bases (see Chapter III, C-1, pp. 47–51). It is however noteworthy to mention that benzhydroximic bromide and iodide have been obtained for the first time by hydrohalogenation of benzonitrile oxide[19,20] and that basic, stable nitrile oxides add hydrogen chloride preferentially at the CNO group and not at the amine nitrogen.[14]

At higher temperatures in an inert solvent (*e.g.* toluene), Eq. (14) can be reversed, especially in the presence of a dipolarophile trapping the slowly forming nitrile oxide (see Chapter III, D, p. 56).

The equilibrium hydroximic acid chloride-nitrile oxide in aqueous dilute solutions (10^{-3}–10^{-4} M) has been extensively studied both with polarographic and UV spectroscopic methods.[19–23] Of course, Eq.(14) is shifted far to the right even in dilute solution, but the reaction is reversible already in acidic medium, so that at as low as pH = 2 dissociation becomes measurable. The configuration of the hydroximic acid chlorides in equilibrium with several nitrile oxides has been demonstrated *syn*-Cl by NMR spectroscopy[24] and therefore during the addition the lone pair of electrons develops on the nitrogen atom *trans* to the attacking chloride anion. The dissociation constants for hydroximic acid chlorides (Eq. (14), direction 2), measured at 21° are collected in Table XXVI.

Kinetic data support the following two-step mechanism (Eq. (15)), where the rate-controlling step is the nucleophilic attack of the chloride ion to the positively charged carbon atom of nitrile oxide. The stability of

10 *Grundmann, C.:* Angew. Chem. **75**, 450 (1963); — Intern. Ed. Engl. **2**, 260 (1963).

11 *Grundmann, C., Mini, V., Dean, J.M., Frommeld, H.D.:* Liebigs Ann. Chem. **687**, 191 (1965).

12 *Grundmann, C., Dean, J.M.:* J. Org. Chem. **30**, 2809 (1965).

13 *Grundmann, C., Dean, J.M.:* Angew. Chem. **76**, 682 (1964); — Intern. Ed. Engl. **3**, 585 (1964).

14 *Grundmann, C., Richter, R.:* J. Org. Chem. **32**, 2308 (1967).

15 *Wieland, H., Rosenfeld, B.:* Liebigs Ann. Chem. **484**, 236 (1930).

16 *Ponzio, G.:* Gazz. Chim. Ital. **66**, 114, 127 (1936).

17 *Ponzio, G.:* Gazz. Chim. Ital. **66**, 134 (1936).

18 *Ponzio, G.:* Gazz. Chim. Ital. **71**, 693 (1941).

19 *Souchay, P., Armand, J.:* Compt. Rend. **256**, 4907 (1963).

20 *Armand, J.:* Bull. Soc. Chim. France 882 (1966).

21 *Souchay, P., Armand, J., Valentini, F.:* Compt. Rend. **262**, 985 (1966).

22 *Armand, J., Guetté, J.P., Valentini, F.:* Compt. Rend. **263**, 1388 (1963).

23 *Armand, J., Souchay, P., Valentini, F.:* Bull. Soc. Chim. France 4585 (1968).

24 *Guetté, J.P., Armand, J., Lacombe, L.:* Compt. Rend. **264**, 1509 (1967).

Table XXVI. Dissociation constants for hydroximic acid chlorides, $R—C(=NOH)Cl$

R	$10^5 \cdot K$	Ref.
CH_3	130	[20]
C_6H_5	90	[20]
C_6H_5CO	23.5	[23]
$C_6H_5C(NOH)$	6.5	[23]
$CH_3C(NOH)$	2.75	[23]
$HC(NOH)(anti)$	1.2	[20]
$EtOOC$	0.095	[23]

the hydroximic acid chloride anion seems to depend on the electron-attracting power of the substituent R. This mechanism parallels the well-known classical mechanism of the cyanohydrin formation.[25]

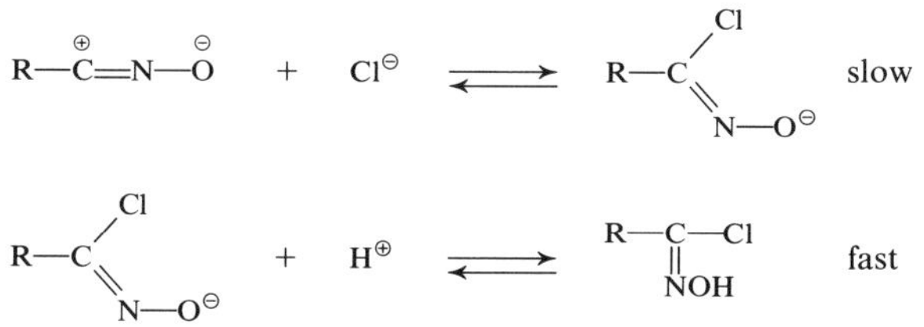

$$(15)$$

During a study toward a possible reactivation of "aged" *phosphonylated acetylcholinesterase*, it has been reported that derivatives of *phosphinic acid* of structure *26* react with p-nitrobenzonitrile oxide to yield a more or less stable adduct *27*.[26] When $R'=OEt$, easy hydrolysis to *28* was observed (Eq. (16)).

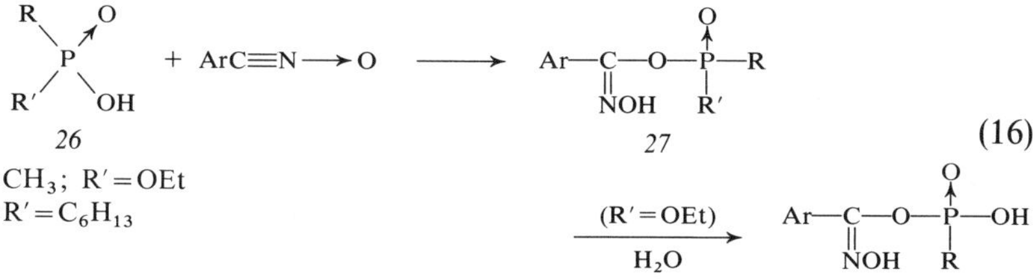

$$R = CH_3; \ R' = OEt$$
$$R = R' = C_6H_{13}$$

$$(16)$$

Organic Acids. Already *Wieland*[9] postulated an addition reaction of benzonitrile oxide to benzoic acid to explain the formation of benzoyl

25 *Lapworth, A.*: J. Chem. Soc. **83**, 995 (1903); **85**, 1206 (1904).
26 *Cadogan, J.I.G., Maynard, J.A.*: Chem. Commun. 854 (1966).

benzhydroxamate (dibenzhydroxamic acid) as a by-product observed during the acidic decomposition of benzhydroximic acid chloride. On the other side, the reaction between benzhydroximic acid chloride and silver benzoate was already known to give benzoyl benzhydroxamate.[27] Although the alleged isolation of primarily formed benzoyl benzhydroximic acid has been questioned later on,[28] the isolation of diphenylfuroxan as a side or main product[29] demonstrates that the reaction intermediate is the nitrile oxide.

It was only much later, however, that the addition of *carboxylic acids* to nitrile oxides was recognized to have fairly general applicability. Usually the aroyl hydroxamates *30* are isolated, most probably because under the reaction conditions the initially formed benzoylhydroximic acid *29* isomerizes promptly (Eq. (17)).

$$\begin{array}{ccc}
\text{Ar—C}\equiv\text{N}\longrightarrow\text{O} & & \text{Ar—C(OH)}{=}\text{NOCOR} \\
+ \quad\longrightarrow\quad \left[\begin{array}{c}\text{Ar—C—OCOR}\\ \parallel\\ \text{NOH}\end{array}\right] \longrightarrow & & \Updownarrow \qquad\qquad (17)\\
\text{RCOOH} & & \text{ArCONHOCOR}\\
& 29 & 30
\end{array}$$

Benzonitrile oxide has added in this manner acetic acid,[30,31] benzoic acid[30] and diphenylacetic acid;[32] the reaction has been extended to other aromatic and aliphatic nitrile oxides[31,33,34]. The structure of the benzhydroxamic acid esters can always be assured by the well-known synthesis from benzhydroxamic acid and the corresponding acyl chloride.

The addition, of course, takes easily place also when carboxylate anions are used; the pH of the reaction mixture, however, should not exceed 8 in order to avoid hydrolysis of the product to hydroxamic acid and the starting carboxylic acid.

Most probably, both in acidic and in alkaline medium, a 5-membered ring is involved as intermediate during the 1,4-acyl migration from the primarily formed acyl hydroximic acid (*29*) to the acyl hydroxamate (*30*) (Eq. (18)).[31,35] In the case of Ar = 2,4,6-trimethylphenyl, *29* was actually isolated, but rearranged quickly to the stable isomer *30*. Compounds of type *29* can be distinguished from *30* by the red color reaction with $FeCl_3$, due to the free =NOH group.[33,27]

27 *Werner, A., Buss, H.:* Ber. **27**, 2193 (1894).

28 *Sutherland, J.K., Widdowson, D.A.:* J. Chem. Soc. 4651 (1964).

29 *Werner, A., Skiba, W.:* Ber. **32**, 1654 (1899).

30 *Speroni, G., Bartoli, M.:* Sopra gli ossidi di nitrili, Nota VIII, Stab. Tip. Marzocco, Firenze, 1952.

31 *Just, G., Dahl, K.:* Tetrahedron **24**, 5251 (1968); Can. J. Chem. **48**, 966 (1970).

32 *Scarpati, R., Sorrentino, P.:* Gazz. Chim. Ital. **89**, 1525 (1959).

33 *Grundmann, C., Frommeld, H.D.:* J. Org. Chem. **31**, 157 (1966).

34 *Govindachari, T.R., Nagarajan, K., Rajappa, S., Akerkar, A.S., Iyer, V.S.:* Tetrahedron **22**, 3367 (1966).

35 *Alexandrou, N.E., Nicolaides, D.N.:* Tetrahedron Letters 2497 (1966).

The additive activity of the carboxylic group toward nitrile oxides seems to be lower than that of thiols and competitive with the dipolarophilic activity of the double bond. Indeed β-thiolpropionic acid adds to nitrile oxides to yield the carboxyethyl thiohydroximic acid, the SH group being the more reactive.[36]

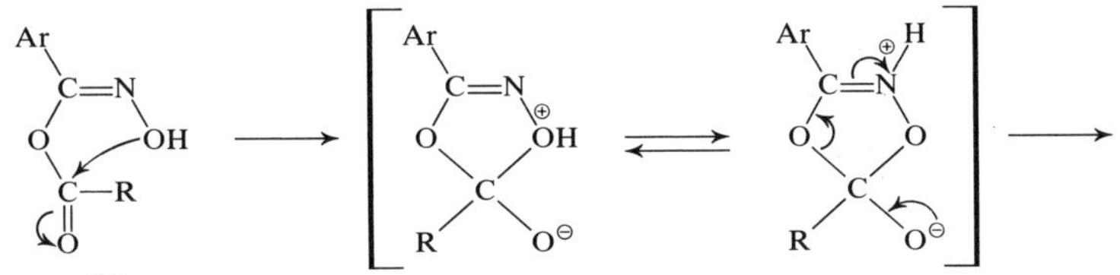

$$\longrightarrow ArCONHOCOR \qquad (18)$$

29

30

α,β-Unsaturated carboxylic acids can react either on the double bond to give 2-isoxazoline carboxylic acids through a 1,3-dipolar cycloaddition (Eq. (19), route A) (see also Chapter V, B, p. 96), or on the free car-

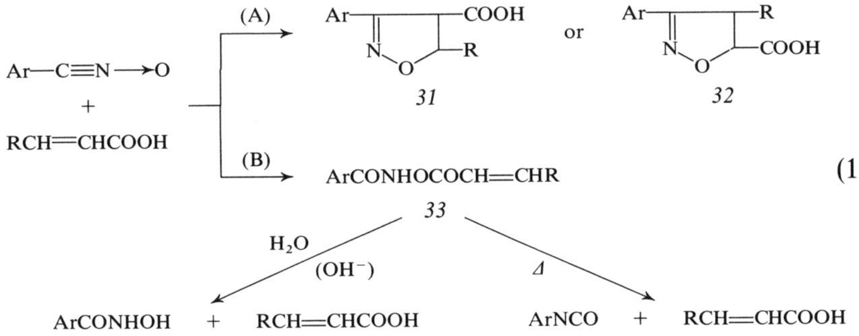

$$Ar-C\equiv N \rightarrow O \; + \; RCH=CHCOOH \qquad (19)$$

(A) → 31   or   32

(B) → ArCONHOCOCH=CHR

33

H_2O (OH⁻) → ArCONHOH + RCH=CHCOOH

Δ → ArNCO + RCH=CHCOOH

boxylic group to give the open-chain 1,3-adduct (Eq. (19), route B).[37] The prevailing of one or the other way of addition depends on the structure of the carboxylic acid: acrylic acid,[37] o-hydroxy-[38] and o-, m-, and p-nitrocinnamic acids,[39] isoxazole- and isoxazoline-5-β-acrylic acids[40] give more or less exclusively the isoxazoline-carboxylic acids, crotonic acid,[37] *cis*-cinnamic acid,[41] thienyl-, furyl-[37] and 3-indolyl-β-

36 *Benn, M.H.:* Can. J. Chem. **42**, 2393 (1964).
37 *Grünanger, P., Vita Finzi, P.:* Rend. Accad. Naz. Lincei **26**, 386 (1959).
38 *Monforte, P., Lo Vecchio, G.:* Atti Accad. Peloritana **49**, 183, 191 (1950/65).
39 *Monforte, F., Lo Vecchio, G.:* Gazz. Chim. Ital. **83**, 416 (1953).
40 *Stagno d'Alcontres, G., De Giacomo, G.:* Atti Soc. Peloritana **5**, 169 (1958/59).
41 *Grünanger, P., Gandini, C., Quilico, A.:* Rend. Ist. Lombardo Sci. Lettere **93**, 467 (1959).

acrylic acids[42] are reported to afford, although in low yields, only the hydroxamates, whereas *trans*-cinnamic acid[41] and its p-chloroderivative[37] yield a mixture of the two products.

Contrary to earlier statements,[43,44] the 2-isoxazoline acids bear always the carboxylic group in 4-position (structure *31*), except in the case of acrylic acid, which yields as expected the 2-isoxazoline-5-carboxylic acid *32* (R' = H). The thermal decomposition to isocyanates and starting carboxylic acid, claimed[45] to support the structure of 3,4-diphenyl-2-isoxazoline-5-carboxylic acid (*32*, Ar = R' = C_6H_5) for a product obtained from benzonitrile oxide and *trans*-cinnamic acid,[43] subsequently identified as cinnamoyl benzhydroxamate (*33*, Ar = R = C_6H_5),[41] is a characteristic reaction of aroyl hydroxamates.[46,47] The latter compounds are also identified by their hydrolysis to the hydroxamic acids and the starting carboxylic acids.

Equimolar quantities of *maleic acid* and benzonitrile oxide react to give the *cis*-3-phenyl-2-isoxazoline-4,5-dicarboxylic acid, but a second mole of benzonitrile oxide adds further to yield a mono-hydroxamate.[48] In the case of benzonitrile oxide and 5-phenyl (or o-, m- and p-nitrophenyl)-2,4-pentadienoic acid, a low yield of the isoxazoline-5-carboxylic acid was claimed on the basis of the oxidation products.[49]

An interesting application of the 1,3-addition of carboxylic acids to nitrile oxides is the preparation of activated esters from *N-protected amino acids* and nitrile oxides, especially p-chlorobenzonitrile oxide[34] and pivalonitrile oxide.[50] The resulting acyl hydroxamates couple with amino acids esters to give di- and tri-peptides in reasonable yields with no impairment of optical activity. Unfortunately, application of this method to larger peptides is hindered by the occurence of *Lossen* rearrangement.[51]

Although not extensively studied, the addition of *acid anhydrides* to nitrile oxides seems to give diacylation products. Oximinophenylacetonitrile oxide[16] and its benzoylderivative[17] add benzoic anhydride to yield *34*. Mesitonitrile oxide adds acetic anhydride at room temperature to form diacetylmesitohydroximic acid (*35*); in the presence of conc. sulfuric acid, however, *Lossen* rearrangement takes place and the only

42 *Piozzi, F., Fuganti, C.:* Ann. Chim. (Rome) **57**, 486 (1967).

43 *Monforte, F.:* Gazz. Chim. Ital. **82**, 130 (1952).

44 *Grünanger, P., Grasso, I.:* Gazz. Chim. Ital. **85**, 1271 (1955).

45 *Lo Vecchio, G., Monforte, P.:* Atti Soc. Peloritana **4**, 245 (1957/58).

46 *Pietschel, E.:* Liebigs Ann. Chem. **175**, 305 (1875).

47 *Lossen, W.:* Liebigs Ann. Chem. **175**, 313, 320 (1875).

48 *Quilico, A., Grünanger, P.:* Gazz. Chim. Ital. **85**, 1449 (1955).

49 *Lo Vecchio, G., Monforte, P.:* Atti Soc. Peloritana **4**, 229 (1957/58).

50 *Rajappa, S., Nagarajan, K., Iyer, V.S.:* Tetrahedron **23**, 4805 (1967).

51 *Govindachari, T.R., Rajappa, S., Akerkar, A.S., Iyer, V.S.:* Tetrahedron **23**, 4811 (1967).

identifiable product was N-acetylmesidine (*36*)[33]

$$C_6H_5-\underset{\underset{NOR}{\|}}{C}-\underset{\underset{NOCOC_6H_5}{\|}}{COCOC_6H_5} \qquad Mes-\underset{\underset{NOCOCH_3}{\|}}{COCOCH_3} \qquad MesNHCOCH_3$$

$$\textit{34} \qquad\qquad\qquad \textit{35} \qquad\qquad \textit{36}$$

$$R = H, COC_6H_5 \qquad\qquad Mes = 2,4,6\text{-trimethylphenyl}$$

Thiobenzoic acid reacts with benzonitrile oxide to give a mixture of benzonitrile, benzoic acid, sulfur and dibenzoyl sulfide. The presence of the first three products may be best explained by assuming a high instability of the initially formed benzoyl thiohydroximic acid (*37*), which decomposes presumably through a cyclic intermediate (Eq. (20)). The formation of low yields of dibenzoyl sulfide clearly depends on a redox side process.[52]

$$\begin{array}{c} C_6H_5-C\equiv N\longrightarrow O \\ + \\ C_6H_5COSH \end{array} \longrightarrow \left[\begin{array}{c} C_6H_5-\underset{\underset{SCOC_6H_5}{|}}{C}=NOH \\ \textit{37} \end{array} \right] \longrightarrow$$

$$\longrightarrow \left[\begin{array}{c} C_6H_5 \\ C=N \\ S \quad\quad O \\ C \\ HO \quad\quad C_6H_5 \end{array} \right] \longrightarrow \begin{array}{c} C_6H_5CN \\ + \\ C_6H_5COOH \\ + \\ S \end{array} \tag{20}$$

Both *carboxylic and sulfonic acid chlorides* add to aromatic nitrile oxides in the presence of triethylamine as catalyst, yielding arylhydroxamoyl chloride esters *38* resp. *39* (Eq. (21)).[53, 54]

$$Ar-C\equiv N\longrightarrow O \left\{ \begin{array}{l} + \quad RCOCl \longrightarrow Ar-\underset{\underset{NOCOR}{\|}}{C}-Cl \quad \textit{38} \\ \\ + \quad RSO_2Cl \longrightarrow Ar-\underset{\underset{NOSO_2R}{\|}}{C}-Cl \quad \textit{39} \end{array} \right. \tag{21}$$

Several mechanisms have been proposed for the reaction, but the alleged formation of an intermediate "sulfene"[55, 56] does not appear

52 *Alemagna, A., Bacchetti, T.:* Rend. Ist. Lombardo Sci. Lettere **97**, 182 (1963).
53 *Rajagopalan, P., Talaty, C.N.:* Tetrahedron Letters 2161 (1966).
54 *Eloy, F., Overstraeten, A. van:* Bull. Soc. Chim. Belges **76**, 63 (1967).
55 *Truce, W.E., Naik, A.R.:* Can. J. Chem. **44**, 297 (1966).
56 *King, J.F., Durst, T.:* Can. J. Chem. **44**, 409 (1966).

necessary, and is, moreover, impossible when R in Eq. (21) is a phenyl group. It is perhaps more reasonable to assume that triethylamine is apt to enhance either the nucleophilic power of the oxygen atom of the nitrile oxide (through binding to the carbon atom) or the electrophilic activity of the acid chloride (through complex formation).[53] It should be added that a recent report shows that the addition of benzoyl chloride to benzonitrile oxide occurs also in the absence of a basic catalyst, though presumably with lower yields.[57] The long known reaction of primary nitro-compounds with benzoyl chloride and pyridine[58] might also proceed through a nitrile oxide intermediate.

Nitrile oxides behave in their reaction with sulfonic acid chlorides and triethylamine quite differently than nitrones, a class of structurally related 1,3-dipoles, which react with the same reagents to give cyclic seven-membered azasultones, resulting from the 1,3-dipolar cycloaddition of nitrones to sulfenes and subsequent ring enlargement.[54, 59]

57 *Awad, W.I., Sobhy, M.:* Can. J. Chem. **47**, 1473 (1969).

58 *Wieland, H., Kitasato, Z.:* Ber. **62**, 1250 (1929).

59 *Truce, W.E., Norrell, J.R., Campbell, R.W., Brady, D.G., Fieldhouse, J.W.:* Chem. Ind. 1870 (1965).

G. Addition of Ammonia and its Derivatives

The reactions of benzhydroximic acid chloride with ammonia, hydrazine or hydroxylamine to give benzamidoxime,[1] benzhydrazoxime[2] resp. benzoxyamidoxime,[3] were discovered at the very beginning of the chemistry of hydroximic acid derivatives and have general applicability. Owing to the very fast dehydroalogenation of hydroximic acid halides by action of any kind of base, it is highly probable that the first step of the reaction leads to the nitrile oxide, which subsequently adds the nucleophile.

As a matter of fact, despite an early erroneous statement,[4] the addition of *ammonia, hydrazine* and *hydroxylamine* to nitrile oxides is a quite general reaction and occurs under mild conditions, yielding the expected adducts *40* to *42* (Eq. (22)). Yields are moderate to good in ethanolic

$$
R\text{---}C\equiv N\longrightarrow O \quad
\begin{cases}
+\ NH_3 \longrightarrow \underset{\displaystyle NOH}{\overset{\displaystyle R\text{---}C\text{---}NH_2}{\|}} \quad 40 \\[2ex]
+\ NH_2NH_2 \longrightarrow \underset{\displaystyle NOH}{\overset{\displaystyle R\text{---}C\text{---}NHNH_2}{\|}} \quad 41 \\[2ex]
+\ NH_2OH \longrightarrow \underset{\displaystyle NOH}{\overset{\displaystyle R\text{---}C\text{---}NHOH}{\|}} \quad 42
\end{cases}
\qquad (22)
$$

solution; an exception is oxalo-bis-N-oxide, which gave the bis-amidoxime with 10% yield, owing to the conceivably easy formation of polymeric material.[5]

The reaction of hydroximic acid chlorides with 2 moles of *amines* to yield N-substituted amidoximes also was discovered very early in the nineties. It was first applied to the simplest homolog of the series, formhydroximic acid chloride, by *Nef,*[6] who isolated with aniline the so-called "phenylisouretin"; soon afterwards it was extended to aromatic hydroximic acid chlorides and to secondary amines.[7] The reaction found soon wide application and still now represents the best method of preparation of N-mono- or di-substituted amidoximes.[8]

1 *Werner, A., Buss, H.:* Ber. **27**, 2193 (1894).

2 *Wieland, H.:* Ber. **42**, 4199 (1909).

3 *Ley, H.:* Ber. **31**, 2126 (1898).

4 *Wieland, H.:* Ber. **40**, 1667 (1907).

5 *Grundmann, C., Mini, V., Dean, J.M., Frommeld, H.G.:* Liebigs Ann. Chem. **687**, 191 (1965).

6 *Nef, J.U.:* Liebigs Ann. **280**, 291 (1894).

7 *Werner, A.:* Ber. **27**, 2846 (1894).

8 *Eloy, F., Lenaers, R.:* Chem. Rev. **62**, 155 (1962).

The nucleophilic 1,3-addition of a previously isolated nitrile oxide to amines was first discovered by *Ponzio* with oximinophenylacetonitrile[9] and oximinomalonomononitrile oxides.[10] Because of the controversial structures of the starting material (see Chapter III, C-2, p. 52) these reports remained almost unnoticed. More recently, however, various investigators have confirmed that the addition is quite a general one, both with aliphatic and aromatic nitrile oxides, as well as with primary, or secondary aliphatic or aromatic amines.

The wide applicability of the reaction is illustrated in Table XXVII which summarizes the data hitherto reported on the addition of isolated nitrile oxides to ammonia and its inorganic or organic derivatives.

Table XXVII. Addition reaction of nitrile oxides to ammonia and its derivatives

$$R-C\equiv N\rightarrow O \quad + \quad HN\big\langle{}^{R'}_{R''} \quad \longrightarrow \quad R-\underset{\underset{NOH}{\|}}{C}-N\big\langle{}^{R'}_{R''}$$

R	R'	R''	Product	Ref.
CH_3		$-(CH_2)_5-$	a	11
$HOOCC(=NOH)$	H	C_6H_5	a	10
$(CH_3)_3C$		$-(CH_2)_5-$	a	11
$(C_2H_5)_2CH$		$-(CH_2)_5-$	a	11
$C\equiv NO$	H	H	b	5
	H	$CH_2CH_2NH_2$	c	5
	C_2H_5	C_2H_5	d	5
	H	$2\text{-}HO\text{-}5\text{-}ClC_6H_3$	c	12
	H	C_6H_5	d	5, 13
	H	$2\text{-}HOC_6H_4$	c	12
	H	$2\text{-}H_2NC_6H_4$	c	5, 12
	H	NHC_6H_5	d, e	5
	H	C_6H_{11}	d	5
	H	$2\text{-}HO\text{-}4\text{-}CH_3C_6H_3$	c	12
	H	$2\text{-}HO\text{-}5\text{-}CH_3C_6H_3$	c	12
	CH_3	$2\text{-}HOC_6H_4$	c	12
	H	$2\text{-}H_2N\text{-}3\text{-}CH_3C_6H_3$	c	12
	CH_3	$2\text{-}CH_3NHC_6H_4$	c	12
	H	$2\text{-}HOC_{10}H_6$	c	12
	H	$2\text{-}H_2NC_{10}H_6$	c	12

9 *Ponzio, G.:* Gazz. Chim. Ital. **66**, 127 (1936).

10 *Ponzio, G., Paolini, I. de:* Gazz. Chim. Ital. **56**, 247 (1926).

11 *Zinner, G., Günther, H.:* Angew. Chem. **76**, 440 (1964); — Intern. Ed. Engl. **3**, 383 (1964).

12 *Alexandrou, N.E., Nicolaides, D.N.:* J. Chem. Soc. (C) 2319 (1969).

13 *Grundmann, C.:* Angew. Chem. **75**, 450 (1963); — Intern. Ed. Engl. **2**, 260 (1963).

Table XXVII (continued)

R	R'	R''	Product	Ref.
2-bromo-7,7-dimethyl-1-norbornyl	H	H	a	13a
C_6H_5	H	H	a	14, 15
	H	NH_2	a	15
	H	OH	a	15
	H	CH_3	a	16
	H	CN	f	17
		$-CH_2CH_2-$	a	18
		$-CH(CH_3)CH_2-$	a	18
	H	$CH_2C\equiv CH$	a	19
	H	C_6H_5	a	14, 15, 20
	H	NHC_6H_5	g	14
	CH_3	C_6H_5	a	20
	CN	C_6H_5	a	17
	H	3-phenyl-5-isoxazolylmethyl	a	19
$o\text{-}ClC_6H_4$	H	H	a	21
$m\text{-}ClC_6H_4$	H	H	a	21
$p\text{-}ClC_6H_4$	H	H	a	21
		$-CH_2CH_2-$	a	18
	H	CH_2CH_2Br	a	18
	H	C_6H_5	a	21, 22
	H	$p\text{-}ClC_6H_4$	a	21
$2,6\text{-}Cl_2C_6H_3$	H	H	a	21
	H	C_6H_5	a	21
$3,4\text{-}Cl_2C_6H_3$		$-CH_2CH_2-$	a	18
$o\text{-}NO_2C_6H_4$	H	H	a	23
	H	C_6H_5	a	23

13a *Ranganathan, S., Singh, B.B., Panda, C.S.:* Tetrahedron Letters 1225 (1970).

14 *Grundmann, C., Frommeld, H.-D.:* J. Org. Chem. **31**, 157 (1966).

15 *Speroni, G., Bartoli, M.:* Sopra gli ossidi di nitrili, VIII, Stab. Tip. Marzocco, Firenze, 1952.

16 *De Sarlo, F., Fabbrini, L., Ruggini, U.:* Personal communication.

17 *Fabbrini, L., Speroni, G.:* Chim. Ind. (Milan) **43**, 807 (1961).

18 *Rajagopalan, P., Talaty, C.N.:* J. Am. Chem. Soc. **88**, 5048 (1966).

19 *Caramella, P., Vita Finzi, P.:* Chim. Ind. (Milan) **48**, 963 (1966).

20 *Alexandrou, N.E., Nicolaides, D.N.:* Chim. Chronika **30**, 49 (1965).

21 *Speroni, G., Bartoli, M.:* Sopra gli ossidi di nitrili, IX, Stab. Tip. Marzocco, Firenze, 1952.

22 *Barbaro, G., Battaglia, A., Dondoni, A.:* Boll. Sci. Fac. Chim. Ind. Bologna **27**, 149 (1969).

23 *Speroni, G., Bartoli, M.:* Sopra gli ossidi di nitrili, X, Stab. Tip. Marzocco, Firenze, 1952.

Table XXVII (continued)

R	R'	R''	Product	Ref.
$m\text{-}NO_2C_6H_4$	H	H	a	23
	H	CN	f	17
		$-CH_2CH_2-$	a	18
		$-(CH_2)_4-$	a	24
	H	C_6H_5	a	23
$p\text{-}NO_2C_6H_4$	H	H	a	23
	H	C_6H_5	a	23
$2,4,6\text{-}(CH_3)_3C_6H_2$	CH_3	CH_3	a	14
	H	C_6H_5	a	25, 26
	H	C_6H_{11}	a	14
	H	$2,4,6\text{-}(CH_3)_3C_6H_2$	a	27
$2,4,6\text{-}(CH_3)_3\text{-}3,5\text{-}Cl_2C_6$	H	C_6H_5	a	28, 22
$2,3,5,6\text{-}(CH_3)_4C_6H$	H	C_6H_5	a	25, 26
$2,4,6\text{-}(CH_3O)_3C_6H_2$	H	C_6H_5	a	25, 26
$2,6\text{-}(CH_3)_2\text{-}4\text{-}(CH_3)_2NC_6H_2$	H	C_6H_5	a	29
$C_6H_5C(=NOH)$	H	H	a	9
	H	C_6H_5	a	9
$p\text{-}CH_3C_6H_4C(=NOH)$	H	H	a	30
(structure: $(CH_3)_2N$-substituted benzene ring with CH_3, CH_3, CH_3, CNO substituents)	H	C_6H_5	d	29
(structure: ONC-substituted benzene ring with four CH_3 substituents)	H	C_6H_5	d	25, 26

24 *Rajagopalan, P., Talaty, C. N.:* Tetrahedron Letters 2101 (1966).
25 *Grundmann, C., Dean, J. M.:* Angew. Chem. **76**, 682; Intern. Ed. Engl. **3**, 585 (1964).
26 *Grundmann, C., Dean, J. M.:* J. Org. Chem. **30**, 2809 (1965).
27 *Grundmann, C., Frommeld, H. D., Flory, K., Datta, S. K.:* J. Org. Chem. **33**, 1464 (1968).
28 *Speroni, G., Scarpati, R.:* Personal communication.
29 *Grundmann, C., Richter, R.:* J. Org. Chem. **32**, 2308 (1967).
30 *Ponzio, G.:* Gazz. Chim. Ital. **61**, 561 (1931).

Table XXVII (continued)

R	R′	R″	Product	Ref.
5-NO$_2$-2-furyl		—CH$_2$CH$_2$—	a	31
(CH$_3$)$_2$N—[pyrimidine ring, 4,6-dimethyl]	H	C$_6$H$_5$	a	29

[a] Amidoxime.
[b] Bis-adduct, besides polymers.
[c] 1,4-Diazine or 1,4-Oxazine.
[d] Bis-adduct.
[e] Furazan derivative.
[f] 1,2,4-Oxadiazolonimine.
[g] Azoderivative.

Recent kinetic studies[22] of the reaction of 3,5-dichloro-2,4,6-trimethylbenzonitrile oxide with *aniline* in CCl$_4$ have ascertained that the addition does not follow a simple second-order kinetic law, but rather is the result of three processes:

(i) a bimolecular reaction of aniline and nitrile oxide;

(ii) a ternary reaction involving an additional molecule of aniline;

(iii) a ternary reaction involving a molecule of the product,

this latter catalytic action being more significant than the other two processes. Inhibition by triethylamine suggests that the catalytic effect of the product depends on its acidity.

The reaction of nitrile oxides with n-butylamine is so fast and complete that it has been chosen as basis for a volumetric analysis of nitrile oxides in kinetic experiments.[32, 33]

Sometimes the formation of the amidoxime is followed by ring closure with loss, *e.g.*, of hydroxylamine, as in the case of the reaction with o-aminophenol (or -thiophenol), which led to benzoxazoles resp. benzothiazoles (Eq. (23)).[34]

$$\text{(23)}$$

31 *Sasaki, T., Yoshioka, T.:* Bull. Chem. Soc. Japan **42**, 556 (1969).
32 *Battaglia, A., Dondoni, A.:* Ric. Sci. **38**, 201 (1968).
33 *Barbaro, G., Battaglia, A., Dondoni, A.:* J. Chem. Soc. (B) 588 (1970).
34 *Sasaki, T., Yoshioka, T., Susuki, Y.:* Bull. Chem. Soc. Japan **42**, 3335 (1969).

Bifunctional nitrile oxides react with *bifunctional amines* to yield polyamidoximes of various degree of polymerisation; these possess an interesting chelating affinity for various transition metals,[35] *e.g.* the polymer *43* from oxalonitrile-bis-N-oxide and p-phenylene diamine. The same bis-nitrile oxide undergoes ring closure with 1,2-diamines to yield 1,4-diazines: *e.g.* with o-phenylene diamine, 2,3-dioximino-1,2,3,4-tetra-hydro-quinoxaline (*44*) is obtained.[5,12] Analogously o-aminophenols give 1,4-oxazines and dithiols 1,4-dithianes (see above, Section E), the following reactivity order being observed: $NH_2 > SH > OH$ (Eq. (24)).[12]

$$
\begin{array}{c}
H_2N-\left[-\bigcirc-NH-\underset{\underset{HON}{\|}}{C}-\underset{\underset{NOH}{\|}}{C}-NH-\right]_n-\bigcirc-NH_2 \\
43
\end{array}
$$

$$
\begin{array}{c}
C\equiv N\longrightarrow O \\
| \\
C\equiv N\longrightarrow O
\end{array}
\left\{
\begin{array}{l}
+ \; H_2N-\bigcirc-NH_2 \\[2em]
+ \; \bigcirc\genfrac{}{}{0pt}{}{NH_2}{NH_2}
\end{array}
\right.
\longrightarrow \;\; 44
$$

(24)

The nucleophilic additive activity of the primary amine group is competitive with the dipolarophile activity of the unsaturated bond: *e.g.*, propargyl amine reacts with benzonitrile oxide to yield a mixture of the three possible adducts *45*, *46* and *47*, where the products derived from a 1,3-addition are prevailing on those from the cycloaddition to isoxazole derivatives.[19]

$$
\underset{45}{C_6H_5-\underset{\underset{NOH}{\|}}{C}-NHCH_2C\equiv CH}
\qquad
\underset{46}{C_6H_5-\!\!\!\overset{\displaystyle\diagup\!\!\diagdown}{\underset{N\diagdown_O\diagup}{}}\!\!\!-CH_2NH_2}
$$

Yields %:   36−43   5−6

$$
\underset{47}{C_6H_5-\!\!\!\overset{\displaystyle\diagup\!\!\diagdown}{\underset{N\diagdown_O\diagup}{}}\!\!\!-CH_2-NH-\overset{\displaystyle C_6H_5}{\underset{}{C}}\!\!=\!NOH}
$$

13−15

35 *Grundmann, C.:* Fortschr. Chem. Forsch. **7**, 99 (1966).

Interesting are also the adducts *48* from nitrile oxides and *aziridines*, because they represent an elegant route, through an open-chain intermediate, to the little known ring system of 1,2,4-oxadiazine (*49*) (Eq. (25)).[18, 31]

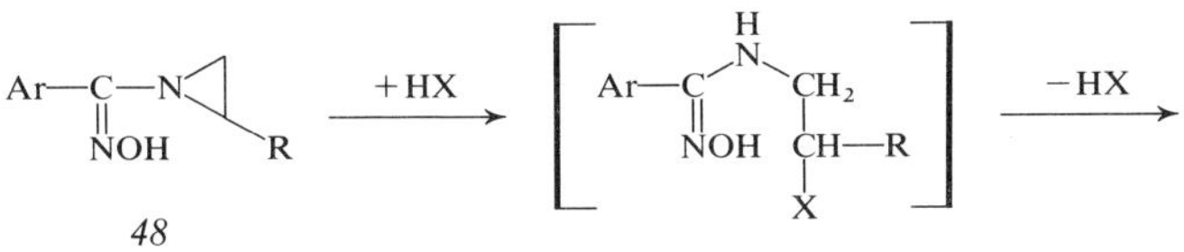

$$\text{(25)}$$

Cyanamide and phenyl-cyanamide could in principle react either with the amino group or with the C≡N triple bond (the C=N double bond of the tautomeric carbodiimide structure seems less plausible); in fact both compounds have been reported to add aromatic nitrile oxides through an initial nucleophilic attack of the amino group, followed by cyclization to 1,2,4-oxadiazolin-5-one imines (Eq. (26)).[17] N-cyano-

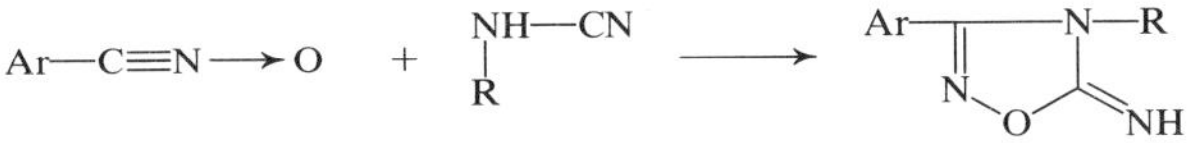

$$\text{(26)}$$

guanidine (dicyandiamide) on the contrary, owing to the high electron density on the nitrile nitrogen, adds to nitrile oxides with its C≡N triple bond.

Tertiary amines add to nitrile oxides to furnish usually very labile compounds, which are presumably inner salts of trisubstituted amidoximes (*50*); these can be stabilized in some cases by addition of hydrogen

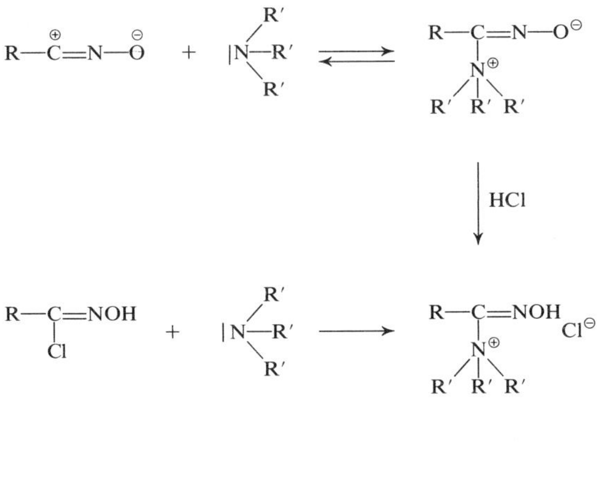

$$\text{(27)}$$

chloride as hydrochlorides of quaternary amidinium oximes (*51*).[5, 36] The same type of compounds have been occasionally obtained by the addition of pyridine to benzhydroximic acid chloride[37] or dichloroglyoxime.[38] The data presently available are insufficient to allow conclusions to be drawn regarding the structural requirements of either the nitrile oxide or the tertiary amine in the formation of stable adducts of type *50*.

The fact that many nitrile oxides have been prepared successfully, either in substance or *in situ* by the reaction of hydroximic acid chlorides and triethylamine suggests that in most cases the equilibrium between *50* and its components lies far on the left.

An interesting variation of this reaction is the addition of an indolizine, such as *51 a*, to an α-ketonitrile oxide (Eq. (27a)) to give the oxime *51 b* as main product, which is a useful intermediate in the synthesis of cycl[3,3,2]azine derivatives.[39]

$$\text{CH}_3\text{COC}{\equiv}\text{N}{\longrightarrow}\text{O}$$

(27 a)

As a powerful nucleophile, *phenylhydrazine* reacts readily with nitrile oxides, but the course of the reaction is complicated by the concomitant redox side reaction leading to the destruction of the hydrazine and conversion of the nitrile oxide to the nitrile (which may then further react with phenylhydrazine). Furthermore, the expected primary products of the addition reaction, the hydrazide-oximes (R—C(=NOH)—NH—NH—C$_6$H$_5$), (which have been synthesized by other routes) are particularly unstable compounds.[2] Only at $-40°$ to $-20°$, oxalonitrile-bis-N-oxide and phenylhydrazine yield the expected adduct, oxalyl-bis-phenylhydrazide-oxime (*52*), together with the product of its spontaneous dehydration, 3,4-di-phenylhydrazino-1,2,5-oxadiazole (*53*). Even under these conditions, appreciable amounts of oxalyl-bis-phenyl-hydrazide-imide (*54*), resulting from the known reaction of cyanogen

36 *Grundmann, C., Dean, J. M.:* Unpublished results.
37 *Wieland, H., Höchtlen, A.:* Liebigs Ann. Chem. **505**, 237 (1933).
38 *Quilico, A., Gaudiano, G., Ricca, A.:* Gazz. Chim. Ital. **87**, 638 (1957).
39 *Leaver, D.:* Personal communication.

with phenylhydrazine,[40] are formed. At 0° *54* is the only product which can be isolated (Eq. (28)).[5]

$$O \leftarrow N{\equiv}C{-}C{\equiv}N \rightarrow O \xrightarrow{\ C_6H_5NHNH_2\ } C_6H_5NHNH{-}\underset{\underset{HON}{\|}}{C}{-}\underset{\underset{NOH}{\|}}{C}{-}NHNHC_6H_5$$

$$\big\downarrow C_6H_5NHNH_2 \qquad\qquad\qquad\qquad\qquad\qquad\qquad\qquad\qquad\qquad\ \textit{52}$$

$$\big\downarrow -H_2O$$

$$N{\equiv}C{-}C{\equiv}N \qquad\qquad\qquad\qquad\qquad\qquad\qquad\qquad\qquad\qquad (28)$$

$$\big\downarrow C_6H_5NHNH_2 \qquad\qquad C_6H_5NHNH{-}\overset{\ }{\underset{N\diagdown_O\diagup N}{\text{(ring)}}}{-}NHNHC_6H_5$$

$$\qquad\qquad\qquad\qquad\qquad\qquad\qquad\qquad\qquad \textit{53}$$

$$C_6H_5{-}NHNH{-}\underset{\underset{HN}{\|}}{C}{-}\underset{\underset{NH}{\|}}{C}{-}NHNHC_6H_5$$

$$\textit{54}$$

Benzonitrile oxide yields a small amount of benzohydroximyl-phenyl-diimide (phenylazo-benzaldoxime, *55*) after reaction with phenylhydrazine and subsequent oxidation with ferric chloride. This result is undoubtedly due to the initial formation of a hydradine oxime intermediate (Eq. (29)).[14]

$$\begin{array}{c} C_6H_5{-}C{\equiv}N \rightarrow O \\ + \\ C_6H_5NHNH_2 \end{array} \longrightarrow \left[C_6H_5{-}\underset{\underset{NOH}{\|}}{C}{-}NHNHC_6H_5 \right] \xrightarrow{-H_2} C_6H_5{-}\underset{\underset{NOH}{\|}}{C}{-}N{=}N{-}C_6H_5$$

$$\qquad\qquad\qquad\qquad\qquad\qquad\qquad\qquad\qquad\qquad\qquad\qquad\qquad \textit{55}$$

$$\qquad\qquad\qquad\qquad\qquad\qquad\qquad\qquad\qquad\qquad\qquad\qquad\qquad (29)$$

---

40 *Rinman, E. L.:* Ber. **30**, 1193 (1897).

H. Addition of Azide, Cyanide and Thiocyanate Ions

Free *hydrazoic acid* does not seem to react with nitrile oxides, azide ion adds mesitonitrile oxide to give the rather unstable mesitohydroximic acid azide (*56*), which above room temperature splits off nitrogen and hyponitrous acid with quantitative yields of mesitonitrile (Eq. (30)).[1]

$$Mes-C\equiv N\rightarrow O \ + \ N_3^\ominus \longrightarrow$$

$$\longrightarrow \underset{\underset{NO^\ominus}{\|}}{Mes-C-N_3} \longrightarrow \underset{\underset{NOH}{\|}}{Mes-C-N_3} \longrightarrow \begin{array}{c} MesCN \\ + \ N_2 \ + \ (NOH)_2 \end{array} \tag{30}$$

$$56$$

$$Mes = 2,4,6\text{-trimethylphenyl}$$

This decomposition pattern is identical to that of benzhydroximic acid azide (*57*),[2] for which the isomeric cyclic structure *58* had also been proposed, but afterwards unequivocally refuted.[3]

$$\underset{\underset{NOH}{\|}}{C_6H_5-C-N_3} \qquad\qquad \underset{57}{} \qquad\qquad 58$$

While free *hydrogen cyanide* does not seem to react with nitrile oxides,[4] cyanide ion adds quickly mesitonitrile oxide[1] and oxalonitrile-bis-N-oxide[4] to form the corresponding cyanooximate ion, from which the free α-oximino-nitriles can be liberated (Eq. (31)):

$$R-C\equiv N\rightarrow O \ + \ CN^\ominus \longrightarrow \underset{\underset{N-O^\ominus}{\|}}{R-C-CN} \xrightarrow{H^+} \underset{\underset{NOH}{\|}}{R-C-CN} \tag{31}$$

Some earlier contradictory statements on the preparation of bis-oximino-succinonitrile (dicyano-glyoxime, *60*) from the reaction of oxalo-bis-hydroximic acid chloride (dichloro-glyoxime, *59*) with potassium cyanide can be explained by the different reactivity of the nitrile oxide toward cyanide ion and the free acid. If two moles of KCN are added to one mole of the dichloro-glyoxime (*59*), the cyanide acts only as a base to liberate the nitrile oxide, which cannot react with free hydrogen cyanide (Equation 32).[5] Addition of dichloroglyoxime, however, to a solution of two moles of KCN gives—as expected—an approximate

1 *Grundmann, C., Frommeld, H. D.:* J. Org. Chem. **31**, 157 (1966).

2 *Wieland, H.:* Ber. **42**, 4199 (1909).

3 *Eloy, F.:* J. Org. Chem. **26**, 952 (1961).

4 *Grundmann, C., Mini, V., Dean, J. M., Frommeld, H. D.:* Liebigs Ann. Chem. **687**, 191 (1965).

5 *Ungnade, H. E., Fritz, G., Kissinger, L. W.:* Tetrahedron **19**, Suppl. I, 235 (1963).

yield of 50% of dinitrile *60*, since, while half of the cyanide is used to liberate the nitrile oxide, the second half can add to the same nitrile oxide to yield after acidification the dicyanoglyoxime (*60*).[6] The use of

$$HON{=}C\text{---}C{=}NOH \;(\overset{Cl}{|}\;\overset{Cl}{|})\; + \; 2\,KCN \longrightarrow O{\leftarrow}N{\equiv}C{-}C{\equiv}N{\rightarrow}O \; + \; 2\,KCl \; + \; 2\,HCN$$

59

$$+\,2\,CN^-$$

$$HON{=}C\text{---}C{=}NOH\;(\overset{NC}{|}\;\overset{CN}{|}) \;\overset{H^+}{\longleftarrow}\; {}^{\ominus}O{-}N{=}C\text{---}C{=}N{-}O^{\ominus}\;(\overset{NC}{|}\;\overset{CN}{|})$$

60

$$(32)$$

four moles of KCN allowed to reach a yield of over 90% of *60*.[4] Other examples of formation of cyano-glyoximes from the corresponding chloro-glyoximes are known.[7] An excess of potassium cyanide was also needed to prepare several other α-isonitrosoacetonitriles, and the formation of the nitrile oxide as logical intermediate was expressly proposed, although no direct evidence was found.[8]

Some evidence has been reported that benzonitrile oxide reacts with *potassium thiocyanate* in dilute aqueous solution.[9] The most probable adduct is not the open-chain thiocyanate oxime (*61*), but the cyclic isomer, 5-phenyl-1,3,4-thiaoxazolin-2-one imine (*62a*). This latter compound indeed is the only product of the reaction between benzhydroximic acid chloride and potassium thiocyanate,[10] which too possibly proceeds via the nitrile oxide. The structure of *62a* has been assured by its easy thermal decomposition to phenyl isothiocyanate and compound *66a*. The formation of these products can be rationalized as follows (Eq. (33)).

$$(33)$$

61   *62*   *64*   *65*   *63*   *66*

a: R = phenyl,   b: R = mesityl,   c: R = 2,3,5,6-tetramethylphenyl

---

6 *Longo, G.:* Gazz. Chim. Ital. **61**, 575 (1931).

7 *Carbone, G.:* Gazz. Chim. Ital. **68**, 428 (1932).

8 *Poziomek, E.J., Melvin, A.R.:* J. Org. Chem. **26**, 3769 (1961).

9 *Armand, J.:* Bull. Soc. Chim. France 882 (1966).

10 *Musante, C.:* Gazz. Chim. Ital. **68**, 331 (1938).

Ring cleavage leads to the nitrene *63a*, which is stabilized by rearrangement to the isothiocyanate, and to cyanic acid, which adds to a second molecule of thiaoxazoline imine to give *66a*. More stable nitrile oxides, *e.g.* mesitonitrile oxide, add thiocyanate ion too, but here neither the open-chain hydroximyl isothiocyanates (*61b, c*) not the thiaoxazoline derivatives (*62b, c*) could be isolated. An almost quantitative yield of the isothiocyanates (*65b, c*) was obtained in boiling methanol, whereas at room temperature the reaction was slower and led to the methyl thiocarbamates (*64b, c*).[1] Clearly, in alkaline medium the thiaoxazoline imines are unstable and decompose into cyanate ion (which could be detected in the reaction mixture) and the nitrene intermediate. Since the isothiocyanate does not react with methanol under prevailing conditions, it must be inferred that the addition of the solvent depends on a concerted mechanism during the decomposition or the rearrangement of an intermediate like *62* or *63*. At 25° the lifetime of this intermediate is presumably long enough to allow this reaction to proceed to a major extend, while at 64° the decomposition is too fast, so that the isothiocyanate *65* is the main product. The proposed mechanism is consistent with the behaviour of the cyclic adducts from nitrile oxides and thiocarbonyl compounds (see Chapter V, D).

I. Addition of Metallo-Organic Compounds

Both aliphatic and aromatic Grignard reagents add to benzonitrile oxide to form complexes *67* from which the oximes are readily liberated with acids (Eq. (34)).[1]

$$C_6H_5-\overset{\displaystyle C}{\underset{N_{\searrow O}}{|||}} + RMgX \longrightarrow \left[C_6H_5-\overset{R}{\underset{NOMgX}{\overset{|}{C}}} \right] \xrightarrow[H_2O]{H^+} C_6H_5-\overset{R}{\underset{NOH}{\overset{|}{C}}} \quad (34)$$

$$\qquad\qquad\qquad\qquad\qquad 67 \qquad\qquad\qquad\qquad 68$$

Similar addition reaction are known starting from triphenylaceto-nitrile oxide[2] and oximinophenylacetonitrile oxide.[3] As by-products the corresponding ketones and benzonitrile have often been isolated. Most probably, the well-known reductive properties of the Grignard reagent provoke de-oxygenation of the nitrile oxide and the so formed nitrile can further react with a second molecule of alkylmagnesium halide.

Use of acetylenic Grignard reagents led to isoxazole derivatives, most probably through the unstable acetylenic oximes as intermediates.[4]

Direct reaction of an excess of Grignard reagent with the hydroximic acid chlorides is also well known and can in principle proceed through one of the following pathways: (a) nucleophilic substitution reaction of the Grignard reagent on the chlorine atom and subsequent formation of the ketoxime salt, (b) elimination—addition sequence involving an intermediate nitrile oxide, and (c) primary formation of the chloro-oxime salt, which can either eliminate to nitrile oxide or undergo a nucleophilic attack (Eq. 35). Although the by-products (ketones and nitriles) are

$$(35)$$

usually identical to those obtained in the reaction with nitrile oxides, no definite evidence is available to favour one or the other of the routes cited above.

1 *Wieland, H.:* Ber. **40**, 1667 (1907).
2 *Wieland, H., Rosenfeld, B.:* Liebigs Ann. Chem. **484**, 236 (1930).
3 *Ponzio, G.:* Gazz. Chim. Ital. **66**, 127 (1936).
4 *Palazzo, G.:* Gazz. Chim. Ital. **77**, 214 (1947).

Whatever the mechanism may be, the reaction has been largely applied to the synthesis of isoxazole derivatives[4-11] and has been a favorite one for building up di- and poly-isoxazole systems.[12-18]

The intermediate α,β-acetylenic ketoximes have also been isolated as a by-product[19] or as the only product of the reaction between hydroximic acid chlorides and acetylenic Grignard reagents.[20]

5 *Gaudiano, G., Ricca, A., Merlini, L.:* Gazz. Chim. Ital. **89**, 2466 (1959).
6 *Bravo, P., Gaudiano, G., Quilico, A., Ricca, A.:* Gazz. Chim. Ital. **91**, 47 (1961).
7 *Grünanger, P., Langella, M.R.:* Gazz. Chim. Ital. **91**, 1449 (1961).
8 *Bravo, P.:* Chim. Ind. (Milan) **45**, 1239 (1963).
9 *Langella, M.R., Vita Finzi, P.:* Chim. Ind. (Milan) **47**, 996 (1965).
10 *Feuer, H., Markovsky, S.:* J. Org. Chem. **29**, 935 (1964).
11 *Cabiddu, S., Solinas, V.:* Gazz. Chim. Ital. **99**, 1107 (1969).
12 *Quilico, A., Gaudiano, G., Ricca, A.:* Gazz. Chim. Ital. **87**, 638 (1957).
13 *Ricca, A.:* Gazz. Chim. Ital. **91**, 83 (1961).
14 *Gaudiano, G., Ricca, A.:* Gazz. Chim. Ital. **89**, 587 (1959).
15 *Ricca, A., Gaudiano, G.:* Rend. Accad. Naz. Lincei **26**, 240 (1959).
16 *Gaudiano, G., Quilico, A., Ricca, A.:* Tetrahedron **7**, 24 (1959).
17 *Gaudiano, G., Ricca, A., Quilico, A.:* Gazz. Chim. Ital. **90**, 1253 (1960).
18 *Ricca, A., Gaudiano, G.:* Rend. Accad. Naz. Lincei **28**, 211 (1960).
19 *Tronchet, J.M.J., Jotterand, A., Le-Long, N.:* Helv. Chim. Acta **52**, 2569 (1969).
20 *Hamlet, Z., Rampersad, M., Shearing, D.J.:* Tetrahedron Letters 2101 (1970).

K. Addition of Free Radicals

Recently, the reactivity of nitrile oxides toward nucleophilic radicals has been shown to follow a general pattern of 1,3-addition:

$$Ar-C\equiv N\rightarrow O \quad + \quad R\cdot \quad \longrightarrow \quad Ar-\underset{\underset{\underset{O\cdot}{|}}{\overset{\|}{N}}}{C}-R \quad \xrightarrow{H_2O} \quad Ar-\underset{NOH}{\overset{\|}{C}}-R$$

In the presence of the t-butylhydroperoxide-ferrous salt redox system, addition of aldehydes, dimethylformamide, methanol, dioxane or alkyl radicals has been carried out, with relatively fair yields in the case of p-chlorobenzonitrile oxide.[1]

---

1 *Caronna, T., Quilico, A., Minisci, F.:* Tetrahedron Letters 3633 (1970).

L. Other Addition Reactions

Alkyl chlorides, particularly prone to undergo nucleophilic attack, have occasionally been reported to add nitrile oxides. Treatment of silver fulminate with an excess of tritylchloride yielded *69* besides the stable triphenylacetonitrile oxide;[1] the most probable, although not the only explanation of it calls for a nucleophilic addition of the nitrile oxide itself to a second molecule of the chloride (Eq. (36)).

$$(C_6H_5)_3C—C\equiv N \longrightarrow O \ + \ (C_6H_5)_3\overset{\delta\oplus}{C}—\overset{\delta\ominus}{Cl} \longrightarrow (C_6H_5)_3—\underset{Cl}{C}=N—OC(C_6H_5)_3 \qquad (36)$$

$$69$$

Recently the reaction of benzonitrile oxide in moist ether with *benzylchloride* has been reported to give a 33% yield of benzyl benzhydroxamate, clearly produced by hydrolysis of the intermediate hydroximic acid chloride *70* (Eq. (37)).[2]

$$C_6H_5—C\equiv N \longrightarrow O$$
$$+$$
$$C_6H_5CH_2Cl$$
$$\longrightarrow \left[C_6H_5—\underset{Cl}{C}=NOCH_2C_6H_5 \right] \longrightarrow$$
$$70$$

$$\xrightarrow{H_2O} C_6H_5CONHOCH_2C_6H_5 \qquad (37)$$

The different sterical environment of the two chlorine atoms could explain the different stability of the two chlorides *69* and *70*, however, a more extensive investigation would be needed, in order to establish limits and mechanism of this interesting addition reaction.

The capacity of arylacetylenes to yield, along with the cycloaddition products, *viz.* the 3,5-disubstituted isoxazoles, the acetylenic oximes,[3] has already been mentioned previously, and reference is made to Chapter V, C.

Benzanilidoxime has been reported to react with benzonitrile oxide in ether to yield the O-benzoyl derivative;[4] the reaction can be visualized as a 1,3-addition of the acidic oxime to the nitrile oxide with subsequent

---

1 *Wieland, H., Rosenfeld, B.:* Liebigs Ann. Chem. **484**, 236 (1930).

2 *Awad, W.I., Sobhy, M.:* Can. J. Chem. **47**, 1473 (1969).

3 *Morrocchi, S., Ricca, A., Zanarotti, A., Bianchi, G., Gandolfi, R., Grünanger, P.:* Tetrahedron Letters 3329 (1969).

4 *Boyer, J.H., Frints, P.J.A.:* J. Org. Chem. **33**, 4554 (1968).

hydrolysis of the oximino group (Eq. (38)):

$$C_6H_5-\underset{\underset{\displaystyle NOH}{\|}}{C}-NHC_6H_5 \quad + \quad C_6H_5-C\equiv N\longrightarrow O \quad\longrightarrow$$

$$(38)$$

$$\longrightarrow \left[\; C_6H_5-\underset{\underset{\displaystyle HON}{\|}}{\overset{\|}{C}}-NHC_6H_5 \atop N-O-\underset{\|}{C}-C_6H_5 \;\right] \longrightarrow \quad C_6H_5-\underset{\underset{\displaystyle NOCOC_6H_5}{|}}{\overset{\|}{C}}-NHC_6H_5$$

Low yields of benzonitrolic acid have been obtained from the reaction of benzonitrile oxide with nitrite ion.[5]

---

5 *Speroni, G., Bartoli, M.:* Sopra gli ossidi di nitrili, VIII, Stab. Tip. Marzocco, Firenze, 1952.

VII. Miscellaneous Reactions

In this section a number of reactions of nitrile oxides are described, which can neither be classified as 1,3-additions leading to open-chain compounds nor as true 1,3-dipolar cycloadditions, although in some cases the reaction sequence may start with one of the above alternatives.

Carbethoxynitrene and aromatic nitrile oxides, both generated *in situ* by the action of triethylamine on an equimolecular mixture of N-4-nitrobenzenesulfonyl-oxyurethane and aroylhydroximic acid chloride, gave an extremely low yield of yellow compounds, for which structures *1* were assured. The proposed mechanism of formation is shown (Eq. (1)).[1]

$$Ar = C_6H_5, \text{ p-ClC}_6H_4, \text{ p-BrC}_6H_4 \tag{1}$$

The main products are, however, *diphenylfuroxan* in the first case and 3,5-diaryl-1,2,4-oxadiazoles in the other two. The principal pattern of the reaction is therefore dimerization or de-oxygenation of the nitrile oxide. The nitrile thus formed reacts easily with a second molecule of nitrile oxide to give the oxadiazole (see Chapter V, F).

Nitrosobenzene reacts already at $-20°$ with benzonitrile oxide. The reaction proceeds initially by a nucleophilic attack on the nitrile oxide; this stage is followed by a cyclization of the insoluble intermediate *2* through abstraction of a proton from the o-position of the nitrosobenzene to give 1-hydroxy-2-phenylbenzimidazole-3-oxide (*3*, R=H).[2] The iso-

1 *Rajagopalan, P., Talaty, C. N.:* Tetrahedron Letters 4877 (1966).

2 *Minisci, F., Galli, R., Quilico, A.:* Tetrahedron Letters 785 (1963).

lation of the intermediate is here undoubtely due to the stabilization performed by the prevalent contribution of the mesomeric form *2b*. p-Nitroso-N-dimethylaniline gives rise to the benzimidazole derivative *3* directly, without permitting the isolation of any intermediate of type *2*. With p-nitrosophenol on the other hand, the intermediate *2* (R=OH) stabilizes to the quinone imide derivative *4* (Eq. (2)).

$$C_6H_5-C\equiv N \longrightarrow O$$
$$+ \qquad \longrightarrow$$
$$R-C_6H_5-N=O$$

2a $\qquad\longleftrightarrow\qquad$ *2b*

$$\tag{2}$$

3 $\qquad\qquad$ R = H, N(CH$_3$)$_2$, OH $\qquad\qquad$ *4*

While diazoacetate and diazoacetophenone are reported to be inert toward nitrile oxides, *diazomethane* in excess reacts with benzonitrile oxide to afford with moderate yields a yellow product, identified through NMR spectroscopy, chemical degradation and comparison with a known sample, as 1-nitroso-3-phenyl-2-pyrazoline (6).[3] The proposed mechanism presumes an initial nucleophilic attack of diazomethane on the positively charged carbon of the nitrile oxide, followed by a nucleophilic displacement of nitrogen by a second molecule, rearrangement of the oxygen atom and ring closure (Eq. (3)), p. 180).

The reaction has been extended to other aromatic nitrile oxides, and the formation of very low yields of *3-phenyl-2-isoxazoline* (7) was explained, as a result of an intramolecular displacement of nitrogen from *5*. Interestingly enough, the same product was obtained by pyrolysis of the nitrosopyrazoline 6.[4]

3 *Lo Vecchio, G., Crisafulli, M., Aversa, M.C.:* Tetrahedron Letters 1909 (1966).
4 *Nagarajan, K., Rajagopalan, P.:* Tetrahedron Letters 5525 (1966).

$$C_6H_5-C\equiv N \rightarrow O$$
$$+$$
$$\overset{\ominus}{CH_2}-\overset{\oplus}{N}\equiv N$$

$$\left[C_6H_5-\underset{\underset{O^{\ominus}}{\diagdown N}}{\overset{\|}{C}}-CH_2-\overset{\oplus}{N}\equiv N \right] \xrightarrow[-N_2]{CH_2N_2} C_6H_5-\underset{\underset{O^{\ominus}}{\diagdown N}}{\overset{\|}{C}}-CH_2-\underset{\underset{\oplus}{N}\equiv N}{CH_2} \longleftrightarrow C_6H_5-\underset{\diagdown N}{\overset{\|}{C}}-CH_2-\underset{N}{CH_2}$$

$$\mathbf{5}$$

(3)

$$\left[C_6H_5-\overset{\oplus}{\underset{\ominus N}{C}}-\overset{CH_2}{\underset{CH_2}{\diagup}} \right] \longrightarrow C_6H_5-\underset{\underset{NO}{N}}{\diagup} \xrightarrow{\Delta} C_6H_5-\underset{\underset{O}{N}}{\diagup}$$

$$\mathbf{6} \qquad\qquad\qquad \mathbf{7}$$

A similar reaction has been observed between *dimethyloxosulfonium methylide* (*8*) and benzonitrile oxide, where the main final products were the *syn*-phenylvinylketoxime (*10*) and 3-phenyl-2-isoxazoline (*7*). The ylide *8* carries out two consecutive transfers of methylene to the nitrile oxide: the initial nucleophilic attack to the carbon atom is followed by a nucleophilic displacement of DMSO by a second molecule of ylide. The zwitterionic intermediate *9* can either give *10* through a β-elimination or yield *7* through an intramolecular displacement (Eq. (4)). It is probable that the choice between the two pathways depends on the configuration

$$C_6H_5C\equiv N \rightarrow O$$
$$+$$
$$\overset{\ominus}{CH_2}-\overset{\oplus}{\underset{\underset{O}{\|}}{S}}(CH_3)_2$$
$$\mathbf{8}$$

$$\longrightarrow \left[C_6H_5-\underset{\underset{O^{\ominus}}{\diagdown N}}{\overset{\|}{C}}-CH_2-\overset{\oplus}{S}(CH_3)_2 \right]$$

$$\Big\downarrow +8$$

(4)

$$C_6H_5-\underset{\underset{O}{N}}{\diagup} \xleftarrow{-DMSO} \left[C_6H_5-\underset{\underset{O^{\ominus}}{\diagdown N}}{\overset{\|}{C}}-\overset{\overset{H}{|}}{CH}-CH_2-\overset{\oplus}{\underset{\underset{O}{}}{S}}(CH_3)_2 \right] \xrightarrow{-DMSO} C_6H_5-\underset{\underset{HON}{}}{\overset{\|}{C}}-CH=CH_2$$

$$\mathbf{7} \qquad\qquad\qquad \mathbf{9} \qquad\qquad\qquad \mathbf{10}$$

$$C_6H_5-\underset{\underset{O}{N}}{\diagup}-\underset{\underset{NOH}{\|}}{\overset{\|}{C}}-C_6H_5 \qquad\qquad C_6H_5-\underset{\underset{NOH}{\|}}{\overset{\|}{N}-\underset{\underset{NOH}{}}{C}-C_6H_5}{\overset{\|}{C}}-CH=CH_2$$

$$\mathbf{11} \qquad\qquad\qquad\qquad\qquad \mathbf{12}$$

of the oxime ion in *9*, the cyclization process being favoured in the case of the *anti*-phenyl isomer.[5,6]

Several minor products have been isolated from the reaction mixture: the isoxazoline oxime *11* results from the cycloaddition of *10* to a second

$$
\begin{aligned}
&C_6H_5\!-\!C\!\equiv\!N\!\rightarrow\!O \\
&\qquad + \\
&\overset{\ominus}{C}H_2\!-\!\overset{\oplus}{A}s(C_6H_5)_3 \\
&\qquad 13
\end{aligned}
\longrightarrow
\left[
C_6H_5\!-\!\underset{\underset{O^{\ominus}}{\parallel}}{\overset{}{C}}\!=\!N,\; CH_2\!-\!\overset{\oplus}{A}s(C_6H_5)_3
\quad \text{or} \quad
\underset{N\diagdown O}{C_6H_5}\!\diagup As(C_6H_5)_3
\right]
$$

$-\,H^+$

$$
\left[
C_6H_5\!-\!\underset{\underset{O^{\ominus}}{\parallel}}{C}\!=\!N,\;\overset{\ominus}{C}H\!-\!\overset{\oplus}{A}s(C_6H_5)_3
\right]
\;\xleftarrow{\;+\,C_6H_5CNO\;}\;
\left[
C_6H_5\!-\!\underset{\underset{O^{\ominus}}{N}}{C}\!-\!CH\!-\!\underset{\overset{\oplus}{A}s(C_6H_5)_3}{C}\!-\!C_6H_5
\right]
$$

$+\,13$ $\qquad$ $-\,As(C_6H_5)_3$, [H]

$$
\left[
C_6H_5\!-\!C\!-\!CH\!-\!C\!-\!C_6H_5 \;\; (N,\,CH_2,\,O^{\ominus},\,\overset{\oplus}{A}s(C_6H_5)_3)
\right]
\qquad
\left[
C_6H_5\!-\!\underset{\underset{O^{\ominus}}{N}}{C}\!-\!CH_2\!-\!\underset{\underset{O^{\ominus}}{N}}{C}\!-\!C_6H_5
\right]
\qquad (5)
$$

$+\,H^+$ $\;\big|\;$ $-\,As(C_6H_5)_3$

$$
\begin{array}{c}
\overset{\displaystyle HON}{\underset{\displaystyle N\diagdown O}{C_6H_5\!-\!\!-\!\overset{\parallel}{C}\!-\!C_6H_5}} \\
14 \\
+ \\
\overset{\displaystyle NOH}{\underset{\displaystyle N\diagdown O}{C_6H_5\!-\!\!-\!\overset{\parallel}{C}\!-\!C_6H_5}} \\
15
\end{array}
\qquad
\underset{\underset{16}{}}{\underset{N\diagdown O}{C_6H_5\!-\!\!-\!C_6H_5}}
\qquad
\underset{17}{C_6H_5\!-\!\underset{NOH}{C}\!-\!CH_2\!-\!\underset{NOH}{C}\!-\!C_6H_5}
$$

5 *Umani-Ronchi, A., Bravo, P., Gaudiano, G.*: Tetrahedron Letters 3477 (1966).
6 *Bravo, P., Gaudiano, G., Umani-Ronchi, A.*: Gazz. Chim. Ital. **97**, 1664 (1967).

molecule of nitrile oxide, whereas the formation of *12* is more difficult to explain. The other by-products arose from hydrolytic or reductive side processes to be ascribed to the action of the strong basic medium on the nitrile oxide (see Chapter IV, B and VI, C).[7]

The reaction of *dimethylsulfonium methylide* toward benzonitrile oxide is very similar, compounds *7* and *10* are obtained as the major products.[8] The stable dimethylsulfonium phenacylide yields analogously the 4,5-dibenzoyl-2-isoxazoline.[9]

The reactivity of the *sulfur ylides* is thus completely different from that of the *phosphor ylides*, which enter a true cycloaddition with nitrile oxides (see Chapter V, G).

The behaviour of the *arsen-ylides* seems to be still different. Triphenyl-arsonium methylide (*13*), prepared in DMSO, reacts with benzonitrile oxide to yield a mixture of the *syn-* (*14*) and *anti-* (*15*) isomers of 3-phenyl-4-benzoximino-2-isoxazoline, along with 3,5-diphenyl-isoxazole (*16*) and a small amount of dibenzoylmethane dioxime (*17*). A possible mechanism is depicted in Eq. (5);[10] it should be noted that here too a reductive process must be involved, as already observed with sulfur ylides, to account for the formation of *16* and *17*. The influence of the solvent is here very strong; when the ylide was generated in ether from the arsonium salt and butyllithium, only *16* and *17* were obtained.[10]

$$C_6H_5C\equiv N \rightarrow O$$

$$+$$

$$\overset{\ominus}{C}H_2-SO-CH_3$$

$$\longrightarrow \quad \left[C_6H_5-\underset{\underset{O^\ominus}{\overset{\|}{N}}}{C}-CH_2SOCH_3 \right] \quad \xrightarrow{+H^+} \quad C_6H_5-\underset{\overset{\|}{NOH}}{C}-CH_2SOCH_3$$

18 *19*

$$\Big\downarrow -H^+$$

$$\left[C_6H_5-\underset{\underset{O^\ominus}{\overset{\|}{N}}}{C}-\underset{\overset{|}{SOCH_3}}{CH}-\underset{\underset{O^\ominus}{\overset{\|}{N}}}{C}-C_6H_5 \right] \quad \xleftarrow{+C_6H_5CNO} \quad \left[C_6H_5-\underset{\underset{O^\ominus}{\overset{\|}{N}}}{C}-\overset{\ominus}{C}H-SOCH_3 \right] \qquad (6)$$

$$\Big\downarrow$$

$$\underset{*20*}{C_6H_5-\underset{N \diagdown O}{\big|\big|}-\underset{C_6H_5}{\overset{SOCH_3}{\big|}}}$$

---

7 *Gaudiano, G., Ponti, P. P., Umani-Ronchi, A.:* Gazz. Chim. Ital. **98**, 48 (1968).

8 *Bravo, P., Gaudiano, G., Umani-Ronchi, A.:* Chim. Ind. (Milan) **49**, 286 (1967).

9 *Hayashi, Y., Watanabe, T., Oda, R.:* Tetrahedron Letters 605 (1970).

10 *Gaudiano, G., Ticozzi, C., Umani-Ronchi, A., Selva, A.:* Chim. Ind. (Milan) **50**, 1343 (1968).

Finally, the reaction of the conjugate base of *DMSO*, which is known to share ylide character, with benzonitrile oxide yielded, as expected, the oxime of methylphenacylsulfoxide (*19*), along with minor amounts of 3,5-diphenyl-4-methylsulfinyl-isoxazole (*20*), which might arise by further transformation of intermediate *18* according to Eq. (6).[6, 8]

The CNO group, like the CN group, is particularly unreactive towards *chlorine*, *bromine* and *iodine*, with the only exception being fulminic acid whose reaction with the halogens is treated on p. 39 ff. The behaviour of nitrile oxides towards *fluorine* has not been studied so far.

VIII. Physiological Properties and Applications of Nitrile Oxides

The parent compound, fulminic acid, is known to be comparable in toxicity to hydrocyanic acid,[1] but no data seem to be reported on the physiological effects of the lower aliphatic nitrile oxides. Their instability makes it unlikely that they would ever reach the site of biological action before being completely dimerized to furoxans.

Some furoxans, like the dichloro-, dibromo- and dicyano-furoxans,[2,3] are powerful *lachrymators*; 3,4-dicarbethoxyfuroxan is a vicious *skin irritant* and allergenic.[4] Inhalation of minute quantities of the vapors of the relatively stable oxalonitrile-bis-N-oxide causes long-lasting and painful irritations of the upper respiratory tracts.[5,6]

Stable aromatic nitrile oxides, like compounds No. 23, 25, 28–30, 46–48, 51 of Table III (p. 16), have shown *cytostatic activity* in cell culture tests (human epidermoid carcinoma of the naso-pharynx), but failed to perform significantly *in vivo*.[7]

The stable nitrile oxides No. 23, 25, 46 and 48 have a low toxicity and a moderate but interesting *anthelmintic activity*.[8]

Some halogen-substituted stable aromatic nitrile oxides exhibit *herbicidal*[9] *and pesticidal*[10] *activities*, although 2,6-dichloro-benzonitrile oxide is less active than the corresponding nitrile.

The unstable aromatic nitrile oxides are not especially dangerous, but their most frequent sources, *i.e.* hydroximic acid chlorides, have strong

---

1 *Schischkoff, L.:* Ann. d. Chem., Suppl. **1**, 109 (1861).

2 *Birckenbach, L., Sennewald, K.:* Liebigs Ann. Chem. **489**, 7 (1931).

3 *Wieland, H.:* Liebigs Ann. Chem. **424**, 107 (1921).

4 *Wieland, H., Semper, L., Gmelin, E.:* Liebigs Ann. Chem. **367**, 52 (1909).

5 *Grundmann, C.:* Angew. Chem. **75**, 450 (1963).

6 *Grundmann, C., Mini, V., Dean, J.M., Frommeld, H.D.:* Liebigs Ann. Chem. **687**, 191 (1965).

7 *Grundmann, C.:* Unpublished results (in collaboration with the National Cancer Institute, U.S. Public Health Service, Bethesda, Maryland); — see *Grundmann, C., Richter, R.:* J. Org. Chem. **32**, 2308 (1967).

8 *Hess, H.J.E., Mac Farland, J W.* (to Chas. Pfizer & Co., Inc.): U.S. Pat. 3,258,397 (1966); — Chem. Abstr. **65**, 8831d (1966).

9 *Koopman, H., Daams, J.:* Weed Res. **5**, 319 (1965); — Chem. Abstr. **64**, 10339a (1966).

10 *Hackmann, J.T., Harthoorn, P.A., Kidd, J.* (to Shell Research Ltd.): Brit. Pat. 949,372 (1964); — Chem. Abstr. **60**, 13199a (1964).

vescicant and lachrymatory properties and sometimes cause allergic symptons. Owing to its high volatility, benzhydroximic acid chloride is particularly dangerous; for its high persistance and solubility in water, pyruvohydroxamic acid chloride has even been evaluated as a chemical warfare agent.[11]

The labile addition products of tertiary amines and nitrile oxides (see Section VI, G) are powerful catalysts for the *deactivation of toxic fluoro-phosphorus compounds* (nerve gases), like isopropyl methylphosphono-fluoridate (GB, Sarin). Especially active, even at physiological pH, are the pyridinium salts obtained from cyanogen-bis-N-oxide or benzonitrile oxide.[12]

Some 1,3-dipolar cycloaddition reactions, involving nitrile oxides, have led to compounds of practical interest. Several steroidal isoxazolines belonging to the pregnane and androstane series have been prepared from aceto- or benzo-nitrile oxide and have shown *anabolic activity*.[13-15] Iso-xazolines derived from 5-nitro-2-furonitrile oxide are claimed to have antimicrobial activity.[16]

The reaction between hydroximic acid chlorides and sodium salts of activated carbonyl compounds (*e.g.* β-diketones, β-ketoesters and β-keto-nitriles), discovered by *Quilico* and *Fusco*,[17-19] has been shown to pass through the nitrile oxides[20] and is one of the most widely used methods of synthesis of isoxazole derivatives.[21-37] Several nitrofurylisoxazoles

11 *Milone, M.:* Ann. Chim. (Rome) **29**, 360 (1939).

12 *Grundmann, C., Dean, J. M.:* Unpublished results.

13 *Fritsch, W., Seidl, G.* (to Farbwerke Hoechst A.G.): Germ. Pat. 1,210,821 (1966); — Chem. Abstr. **64**, 17682f (1966).

14 *Fritsch, W., Stache, U.* (to Farbwerke Hoechst A.G.): Germ. Pat. 1,214,224 (1966); — Chem. Abstr. **65**, 12266 (1966).

15 *Fritsch, W., Stache, U.* (to Farbwerke Hoechst A.G.): Germ. Pat. 1,215,146 (1966); — Chem. Abstr. **65**, 15460e (1966).

16 *Hoyle, W., Howarth, A.G.* (to Geigy A.G.): S. African Pat. 5,484 (1969); — Chem. Abstr. **71**, 101846 (1969).

17 *Quilico, A., Fusco, R.:* Rend. Ist. Lombardo Sci. Lettere **69**, 1 (1936).

18 *Fusco, R.:* Rend. Ist. Lombardo Sci. Lettere **70**, 225 (1937).

19 *Quilico, A., Fusco, R.:* Gazz. Chim. Ital. **67**, 589 (1937).

20 *Quilico, A., Speroni, G.:* Gazz. Chim. Ital. **76**, 148 (1946).

21 *Speroni, G., Bartoli, M.:* Sopra gli ossidi di nitrili, IX, Stab. Tip. Marzocco, Firenze, 1952.

22 *Speroni, G., Bartoli, M.:* Sopra gli ossidi di nitrili, X, Stab. Tip. Marzocco, Firenze, 1952.

23 *Speroni, G., Scarpati, R.:* Nuovi metilbenzonitrilossidi. Centro Univ. Edit., Napoli, 1959.

24 *Klötzer, W.:* Monatsh. Chem. **95**, 265 (1964).

25 *Ajello, T., Sprio, V., Fabra, J.:* Ric. Sci. **34**, 575 (1964).

26 *Renzi, G., Dal Piaz, V.:* Gazz. Chim. Ital. **95**, 1478 (1965).

27 *Dal Piaz, V., Renzi, G.:* Chim. Ind. (Milan) **48**, 759 (1966).

28 *Caramella, P.:* Ric. Sci. **36**, 986 (1966).

thus prepared show antibacterial activity,[38-41] although none of them has so far received practical application.

Dichloxacillin (*1*, Ar = 2,6-dichlorophenyl), a broad spectrum *antibiotic*, especially valuable for its acid- and penicillinase-stability,[42-43] has been prepared starting from 2,6-dichlorobenzhydroximic acid chloride and sodium acetoacetate through the route depicted in Eq. (1).[44]

$$Ar-\underset{\underset{NOH}{\|}}{C}-Cl \;+\; Na^+(CH_3COCHCOOEt)^- \longrightarrow \left[\begin{array}{c} Ar-C\equiv N \longrightarrow O \\ + \\ CH_3COCH_2COOEt \end{array}\right] \tag{1}$$

Several other semisynthetic penicillin analogs, *e.g.* oxacillin (*1*, Ar = phenyl) and chloxacillin (Ar = 2-chlorophenyl), have been obtained in a similar manner.[45-48] The presence of a methyl group in 5-position of the isoxazole ring is essential for the biological activity.[40,47]

29 *Khisamutdinov, G.K., Pechenkin, A.G., Aitova, E.F.:* U.R.S.S. Pat. 181,659 (1966); — Chem. Abstr. **65**, 18592d (1966); — Zh. Obshchei Khim. **36**, 1856 (1966).

30 *Sprio, V., Ajello, E., Mazza, A.:* Ann. Chim. (Rome) **57**, 936 (1967).

31 *Rajagopalan, P., Talaty, C.N.:* Tetrahedron **23**, 3541 (1967).

32 *Renzi, G., Dal Piaz, V., Musante, C.:* Gazz. Chim. Ital. **98**, 656 (1968).

33 *Dal Piaz, V., Renzi, G.:* Gazz. Chim. Ital. **98**, 667 (1968).

34 *Cheney, L.C., Crast, B.L.* (to Bristol-Myers Co.): Fr. Pat. 1,535,810 (1968); — Chem. Abstr. **71**, 101850 (1969).

35 *Grundmann, C., Datta, S.K.:* J. Org. Chem. **34**, 2016 (1969).

36 *Tsujikawa, T., Aki, O., Tsujima, S.* (to Takeda Chem. Ind.): Japan Pat. 29460 (1969); — Chem. Abstr. **72**, 43646 (1970).

37 *Renzi, G., Dal Piaz, V., Pinzauti, S.:* Gazz. Chim. Ital. **99**, 753 (1969).

38 *Matsumoto, J., Minami, S.:* Chem. Pharm. Bull. (Tokyo) **15**, 1806 (1967).

39 *Giannella, M., Gualtieri, F., Pigini, M.:* Farmaco **22**, 333 (1967).

40 *Micetich, R.G.:* J. Med. Pharm. Chem. **12**, 611 (1969).

41 Dainippon Pharm. Co.: Brit. Pat. 1,162,257 (1969); — Chem. Abstr. **72**, 12709 (1970).

42 *Neumann, P., Kempf, B.:* Arzneimittel-Forsch. **15**, 139 (1965).

43 *Mössner, G., Maurer, H., Meisel, C.:* Arzneimittel-Forsch. **15**, 344 (1965).

44 Bristol-Myers Co.: Neth. Appl. Pat. 6,603,496 (1966); — Chem. Abstr. **66**, 37917 (1967).

45 *Doyle, F.P., Hanson, J.C., Long, A.A.W., Nayler, J.H.C., Stove, E.R.:* J. Chem. Soc. 5838 (1963).

46 *Doyle, F.P., Hanson, J.C., Long, A.A.W., Nayler, J.C.H.:* J. Chem. Soc. 5845 (1963).

47 *Hanson, J.C., Long, A.A.W., Nayler, J.H.C., Stove, E.R.:* J. Chem. Soc. 5976 (1965).

48 *Crast, L.B.* (to Bristol-Myers Co.): Brit. Pat. 1,160,565 (1969); — Chem. Abstr. **72**, 31785 (1970).

Aliphatic nitrile oxides, prepared *in situ* from nitroalkanes by the method patented by *Hoshino* and *Mukaiyama*,[49] have been used to synthesize 3,5-substituted 1,2,4-oxadiazoles useful as nematocides, insectides, fungicides and defoliants.[50]

Difunctional nitrile oxides, especially oxalo- and terephthalo-bis nitrile oxides have been proposed as building blocks for chain polymers by reaction with difunctional partners, such as olefins, acetylenes, nitriles or amines. A copolymer of tetramethyl-terephthalo-bis nitrile oxide and p-diethynylbenzene was stable up to 400°; oxalo-bis-nitrile oxide could be used to cross-link (vulcanize) a *polybutadiene rubber*; polymers from oxalo-bis-nitrile oxide and diamines have been recommended as chelating agents for transition metals; but the practical value of all the above suggestions is unknown.[6, 12, 51–54]

49 *Hoshino, T., Mukaiyama, M.:* Japan Pat. 9855 (1959); — Chem. Abstr. **54**, 7738h (1960).

50 *Eloy, F.* (to Union Carbide Corp.): U.S. Pat. 3,264,318 (1966); — Chem. Abstr. **65**, 15391f (1966).

51 *Iwakura, Y., Akiyama, M., Nagabuko, K.:* Bull. Chem. Soc. Japan **37**, 767 (1964).

52 *Overberger, C.G., Fujimoto, S.:* Polymer Letter **3**, 735 (1965); — J. Polymer Sci. (C) **16**, 4161 (1968).

53 *Overberger, C.G.:* Private communication.

54 *Eloy, F.:* Private communication.

IX. Tabular Survey

General Remarks. Tables XXVIII through XXXII include available data on *cycloaddition reactions of nitrile oxides to dipolarophiles* not covered already within the text of the corresponding Sections of Chapter V.

Within each table, the nitrile oxides are listed in the order aliphatic, cycloaliphatic, aromatic, heterocyclic; within each group, entries appear in the order of increasing number of carbon atoms and of increasing complexity of substitution, except for bifunctional terms, which are always located at the end of their group.

For the methods of preparation of the nitrile oxides the following abbreviations are used:
A – generated prior to reaction, but not necessarily isolated;
B – generated *in situ* from hydroximic acid halide and base;
C – generated *in situ* from nitroalkane;
D – by thermal reaction from hydroximic acid halide;
E – by thermal reaction from nitrolic acid;
F – fulminic acid prepared *in situ* from its sodium salt;
G – cycloaddition catalysed by Lewis acid;
H – miscellaneous methods.

Dipolarophiles are arranged following *Chemical Abstracts* indexing.

In the product column, the following abbreviations are used:
B = Bis-adduct;
Is = Isoxazole derivative (through elimination);
M = Mono-adduct;
Ox = 1,2,4-Oxadiazole derivative (through elimination);
P = Polymer;
T = Tris-adduct.

If two or more isomers are obtained, the number of isomers is bracketed.

The literature has been searched through 1969. Some later references are also included. Reference numbering refers to the appropriate chapter and section, to which the table relates. Additional references are placed at the end of each table.

The following Table XXVIII tabulates the available data on cyclo-additions of nitrile oxides to CC double bond derivatives (see Chapter V, B). The table is subdivided in the following sections:

XXVIII a – Cycloadditions of nitrile oxides to monosubstituted C=C compounds.

XXVIII b – Cycloadditions of nitrile oxides to 1,1-disubstituted C=C compounds.

XXVIII c – Cycloadditions of nitrile oxides to 1,2-disubstituted C=C compounds.

XXVIII d – Cycloadditions of nitrile oxides to poly-substituted C=C compounds.

XXVIII e – Cycloadditions of nitrile oxides to compounds containing more than one CC double bond.

The reaction of nitrile oxides with β-dicarbonyl derivatives (most probably in their enolic form)[108, 188, 189] is not covered by the table.

Note: References up to No. 144 are to be found on pages 96 to 111, while No. 145 to 200 are at page 212 to 214.

Table XXVIIIa. Cycloaddition of nitrile oxides to olefinic compounds

$$R-C{\equiv}N{\rightarrow}O \ + \ CH_2{=}CH-R' \longrightarrow \text{(I)} \quad \text{and/or} \quad \text{(II)}$$

Nitrile oxide		Dipolarophile R'	Product	Ref.
R	method			
H	F	$OCOCH_3$	Is	80
H	B	$COOCH_3$	I	34
H	B, F	C_6H_5	I	34, 2
CH_3	B	H	I	15
CH_3	C	CN	I	21
CH_3	B	$COCH_3$	I	86
CH_3	C	$OCOCH_3$	I	21, 77
CH_3	C	$COOCH_3$	I+II	2, 49, 66
CH_3	C	OC_2H_5	I	77
CH_3	B	$P(O)(OCH_3)_2$	I	182
CH_3	C	CH_2OCOCH_3	I	21
CH_3	C	$OC_4H_9\text{-}n$	I	21
CH_3	B	$P(O)[N(CH_3)_2]_2$	I	182
CH_3	C	$OCH_2CH_2N(CH_3)_2$	I	77
CH_3	A	$CH_2Si(CH_3)_3$	I	153
CH_3	A	$Si(OCOCH_3)_2CH_3$	I	153
CH_3	C	$p\text{-}ClC_6H_4$	I	33
CH_3	C, D	C_6H_5	I	21, 24

Table XXVIIIa. Cycloaddition of nitrile oxides to olefinic compounds

| Nitrile oxide | | Dipolarophile | Prod- | Ref. |
R	method	R'	uct	
CH_3	A	$Si(C_2H_5)_3$	I	153
CH_3	B	$p\text{-}NO_2C_6H_4CO$	I	88
CH_3	B	$p\text{-}CH_3C_6H_4CO$	I	88
CH_3	C	$n\text{-}C_8H_{17}$	I	2
CH_3	C	$\beta\text{-}C_{10}H_7$	I	33
CN	B	CN	I	25
CN	B	$COOCH_3$	I + II	66
CN	B	C_6H_5	I	25
CH=NOH	H	CH_2Br	I	140
CH=NOH	H	CH_2Cl	I	140
CH=NOH	H	CH_2I	I	140
CH=NOH	H	CH_2OH	I	140
CH=NOH	H	$COOCH_3$	I	55
CCl_3	D	C_6H_5	I	24
C_2H_5	C	CN	I	21
C_2H_5	C	$OCOCH_3$	I	21
C_2H_5	C	$COOCH_3$	I	2, 49
C_2H_5	C	OC_2H_5	I	77
C_2H_5	C	$OC_4H_9\text{-}n$	I	21
C_2H_5	C	$OCH_2CH_2N(CH_3)_2$	I	77
C_2H_5	C	C_6H_5	I	21
CH_3CO	D	C_6H_5	I	24
$COOCH_3$	B	$OCOCH_3$	Is	191
$n\text{-}C_3H_7$	C	$COOCH_3$	I	2, 49
$n\text{-}C_3H_7$	C	OC_2H_5	I	77
$i\text{-}C_3H_7$	H	$OCOCH_3$	I	74, 76
$COOC_2H_5$	B	H	I	16, 17
$COOC_2H_5$	B	Br	Is	191
$COOC_2H_5$	B	CH_2Cl	I	16, 17
$COOC_2H_5$	B	CH_3	I	16, 17
$COOC_2H_5$	B	$OCOCH_3$	I	191
$COOC_2H_5$	B	OC_2H_5	I	77
$COOC_2H_5$	B, D	C_6H_5	I	16, 22, 24
$(CH_3)_3C$	H	$OCOCH_3$	I	74, 76
$(CH_3)_3C$	B	$COOCH_3$	I	66
$CH_2CH_2COOCH_3$	C	OC_2H_5	I	77
C_9H_{17}	H	$COOCH_3$	I	197
$CH_3COOC_9H_{16}$	H	$COOCH_3$	I	197
$C\equiv N \rightarrow O$	E	H	B	181
$C\equiv N \rightarrow O$	A, D	C_6H_5	B(2)	41, 24
C_6H_5	A, B	Cl	Is	3, 55
C_6H_5	A, B	H	I	8, 55
C_6H_5	A	$SiCl_3$	I	153
C_6H_5	A	CN	I	3, 73
C_6H_5	A	CHO	I	83
C_6H_5	A	COOH	I	116

Table XXVIIIa (continued)

| Nitrile oxide | | Dipolarophile | Prod- | Ref. |
R	method	R′	uct	
C_6H_5	A	CH_2Br	I	3, 73
C_6H_5	A	CH_2Cl	I	3, 73
C_6H_5	A	CH_2I	I	3, 73
C_6H_5	A, B	CH_3	I	8, 55
C_6H_5	A	$SiCl_2CH_3$	I	153
C_6H_5	A	CH_2OH	I	3, 73
C_6H_5	A	SO_3CH_3	I or II	172
C_6H_5	A	CH_2CN	I	38
C_6H_5	A, B	$COCH_3$	I	84, 87
C_6H_5	A, B, C, H	$OCOCH_3$	I	3, 73, 74, 76, 77, 191
C_6H_5	A	$COOCH_3$	I	3, 109
	B		I + II	66
C_6H_5	A	C_2H_5	I	8
C_6H_5	A, C	OC_2H_5	I	77
C_6H_5	B	$P(O)(OCH_3)_2$	I	182
C_6H_5	A	COC_2H_5	I	13
C_6H_5	B	$COOC_2H_5$	II	18
C_6H_5	A	$n\text{-}C_3H_7$	I	8
C_6H_5	A	$i\text{-}C_3H_7$	I	8
C_6H_5	A	$\alpha\text{-}C_4H_3O$	I	35, 146
C_6H_5	A	$\alpha\text{-}C_4H_3S$	I	35
C_6H_5	A	5-methyl-4-isoxazolyl	I	38
C_6H_5	A	$(CH_2)_3CN$	I	156
C_6H_5	A	$n\text{-}C_4H_9$	I	2
C_6H_5	A	$CH_2Si(CH_3)_3$	I	153
C_6H_5	B	$P(O)[N(CH_3)_2]_2$	I	182
C_6H_5	B	$COC_4H_3S\text{-}\alpha$	I	87
C_6H_5	A	$\alpha\text{-}C_5H_4N$	I	172
C_6H_5	A	$\gamma\text{-}C_5H_4N$	I	172
C_6H_5	A	$CH(OC_2H_5)_2$	I	83
C_6H_5	A	$Si(CH_3)(C_2H_5)(OC_2H_5)$	I	153
C_6H_5	A	$p\text{-}BrC_6H_4$	I	30
C_6H_5	A	$o\text{-}ClC_6H_4$	I	30
C_6H_5	A	$p\text{-}ClC_6H_4$	I	148
C_6H_5	A	$o\text{-}NO_2C_6H_4$	I	31
C_6H_5	A	$m\text{-}NO_2C_6H_4$	I	31
C_6H_5	A	$p\text{-}NO_2C_6H_4$	I	31
C_6H_5	A, B, D	C_6H_5	I	3, 18, 20, 23, 64
C_6H_5	A	$o\text{-}HOC_6H_4$	I	32
C_6H_5	A	$p\text{-}HOC_6H_4$	I	30
C_6H_5	A	$Si(C_2H_5)_3$	I	153
C_6H_5	A	$Si(OC_2H_5)_3$	I	153
C_6H_5	B	$COC_6H_4Br\text{-}p$	I	87
C_6H_5	B	$COC_6H_4NO_2\text{-}p$	I	87
C_6H_5	B	COC_6H_5	I	87

Table XXVIIIa. Cycloaddition of nitrile oxides to olefinic compounds

Nitrile oxide R	method	Dipolarophile R′	Product	Ref.
C_6H_5	A	$C(=NOH)C_6H_5$	I + II	157, 158, 166
C_6H_5	A	$CH_2C_6H_5$	I	30
C_6H_5	A, B	$p\text{-}CH_3C_6H_4$	I	30, 64
C_6H_5	A	$p\text{-}CH_3OC_6H_4$	I	30
C_6H_5	B	$COC_6H_4CH_3\text{-}p$	I	87
C_6H_5	A	safrol	I	145
C_6H_5	A	eugenol	I	145
C_6H_5	A	$n\text{-}C_8H_{17}$	I	2
C_6H_5	A	$B(OC_4H_9\text{-}n)_2$	I	144
C_6H_5	A	3-phenyl-4-isoxazolyl	I	39
C_6H_5	B	indole-3-carbonyl	I	85
C_6H_5	B	$COC_6H_4OCH_3\text{-}p$	I	86
C_6H_5	A	$2,4,6\text{-}(CH_3)_3C_6H_2$	I	171
C_6H_5	A	$3,4\text{-}(CH_3O)_2C_6H_3CH_2$	I	145
C_6H_5	A	$n\text{-}C_9H_{19}$	I	8
C_6H_5	A	$\alpha\text{-}C_{10}H_7$	I	2
C_6H_5	A	$\beta\text{-}C_{10}H_7$	I	2
C_6H_5	A	3-phenyl-5-methyl-4-isoxazolyl	I	38
C_6H_5	A	apiol	I	145
C_6H_5	A	$H_2C\!-\!C(COOC_2H_5)_2$ $\mid$ $NHCOCH_3$	I	151
C_6H_5	B	$COC_{10}H_7\text{-}\beta$	I	87
C_6H_5	A	3-phenyl-5-dimethyl-amino-4-isoxazolyl	I	137
C_6H_5	A	1-carbazolyl	I	172
C_6H_5	B	$P(O)(C_6H_5)_2$	I	182
C_6H_5	B	$P(S)(C_6H_5)_2$	I	182
C_6H_5	B	3-phenyl-5-diethyl-amino-4-isoxazolyl	I	137
C_6H_5	A	$n\text{-}C_{20}H_{41}$	I	8
$o\text{-}ClC_6H_4$	B	$OCOCH_3$	I	191
$m\text{-}ClC_6H_4$	A	C_6H_5	I	29, 159
$m\text{-}ClC_6H_4$	A	3-phenyl-5-methyl-4-isoxazolyl	I	38
$p\text{-}ClC_6H_4$	A	C_6H_5	I	29, 159
$p\text{-}BrC_6H_4$	B	$COCH_3$	I	86
$p\text{-}BrC_6H_4$	A	$OCOCH_3$	I	79
$p\text{-}BrC_6H_4$	A	C_6H_5	I	44
$p\text{-}BrC_6H_4$	B	COC_6H_5	I	86
$o\text{-}NO_2C_6H_4$	A	1-methyl-5-tetrazolyl	I	40
$o\text{-}NO_2C_6H_4$	A	2-methyl-5-tetrazolyl	I	40
$m\text{-}NO_2C_6H_4$	A	H	I	13
$m\text{-}NO_2C_6H_4$	A	1-methyl-5-tetrazolyl	I	40
$m\text{-}NO_2C_6H_4$	A	2-methyl-5-tetrazolyl	I	40

Table XXVIIIa (continued)

Nitrile oxide R	method	Dipolarophile R'	Product	Ref.
m-$NO_2C_6H_4$	A	$OCOCH_3$	Is	[78]
p-$NO_2C_6H_4$	A, B	H	I	[13,14]
p-$NO_2C_6H_4$	A	1-methyl-5-tetrazolyl	I	[40]
p-$NO_2C_6H_4$	A	2-methyl-5-tetrazolyl	I	[40]
p-$NO_2C_6H_4$	B	$COCH_3$	I	[86]
p-$NO_2C_6H_4$	A	$OCOCH_3$	Is	[78]
p-$NO_2C_6H_4$	B	1-morpholinyl	I	[136]
2,6-$Cl_2C_6H_3$	A	$OCOCH_3$	I	[191]
2,4-$(NO_2)_2C_6H_3$	A	1-methyl-5-tetrazolyl	I	[40]
2,4-$(NO_2)_2C_6H_3$	A	2-methyl-5-tetrazolyl	I	[40]
C_6Cl_5	A	C_6H_5	I	[195]
C_6F_5	H	C_6H_5	I	[195]
p-$CH_3C_6H_4$	A	$OCOCH_3$	I	[79]
p-$CH_3C_6H_4$	A	C_6H_5	I	[29,159]
p-$CH_3OC_6H_4$	A	C_6H_5	I	[29,159]
C_6H_5CO	B	OC_2H_5	I	[191]
C_6H_5CO	D	C_6H_5	I	[24]
C_6H_5NHCO	H	$OCOCH_3$	I	[12]
C_6H_5NHCO	C	OC_2H_5	I	[12]
C_6H_5NHCO	H	$n-C_3H_7$	I	[12]
C_6H_5NHCO	C	$n-C_4H_9$	I	[12]
$C_6H_5CH=CH$	A	$CONH_2$	I	[27]
$C_6H_5CH=CH$	A	p-ClC_6H_4	I	[27]
$C_6H_5CH=CH$	A	C_6H_5	I	[27]
2,4,6-$(CH_3)_3C_6H_2$	A	H	I	[15]
2,4,6-$(CH_3)_3C_6H_2$	A	$COOCH_3$	I+II	[66]
2,4,6-$(CH_3)_3C_6H_2$	A	m-ClC_6H_4	I	[29,159]
2,4,6-$(CH_3)_3C_6H_2$	A	p-ClC_6H_4	I	[29,159]
2,4,6-$(CH_3)_3C_6H_2$	A	p-$NO_2C_6H_4$	I	[29,159]
2,4,6-$(CH_3)_3C_6H_2$	A	C_6H_5	I	[29,159]
2,4,6-$(CH_3)_3C_6H_2$	A	p-$CH_3C_6H_4$	I	[29,159]
2,4,6-$(CH_3)_3C_6H_2$	A	p-$CH_3OC_6H_4$	I	[29,159]
2,4,6-$(CH_3O)_3C_6H_2$	A	$COOCH_3$	I+II	[66]
16-nor-O-methylpodocarp-4-yl	A	$OCOCH_3$	I	[74,76]
m-ONC-C_6H_4	A, B	$CONH_2$	B	[162]
m-$ONCC_6H_4$	A, B	CH_2OH	B	[162]
m-$ONCC_6H_4$	A, B	$COOCH_3$	B(2)	[162]
p-$ONCC_6H_4$	A, B	CH_2OH	B	[152,162]
p-$ONCC_6H_4$	A, B	$OCOCH_3$	B	[162]
p-$ONCC_6H_4$	A, B	$COOCH_3$	B(2)	[42,162,174]
α-C_4H_3O	A	H	I	[11]
α-C_4H_3O	A	$CONH_2$	I	[175]
α-C_4H_3O	A	γ-C_5H_4N	I	[175]
α-C_4H_3O	A, B	C_6H_5	I	[175,26]
α-C_4H_3O	A	p-$CH_3C_6H_4$	I	[175]

Table XXVIII a. Cycloaddition of nitrile oxides to olefinic compounds

| Nitrile oxide | | Dipolarophile | Prod- | Ref. |
R	method	R′	uct	
5-NO$_2$-2-furyl	B	CH$_2$Cl	I	183
5-NO$_2$-2-furyl	B, D	CONH$_2$	I	26, 183
5-NO$_2$-2-furyl	B	CH$_2$CN	I	183
5-NO$_2$-2-furyl	B	COCH$_3$	I	28, 183
5-NO$_2$-2-furyl	B	OCOCH$_3$	I, Is	191, 183
5-NO$_2$-2-furyl	B	OC$_2$H$_5$	I	28, 183
5-NO$_2$-2-furyl	B	COOC$_2$H$_5$	I	183
5-NO$_2$-2-furyl	B	OCH$_2$CH(CH$_2$)O	I	183
5-NO$_2$-2-furyl	B	α-C$_5$H$_4$N	I	28, 183
5-NO$_2$-2-furyl	B	γ-C$_5$H$_4$N	I	28, 193
5-NO$_2$-2-furyl	B	N(CH$_2$)$_5$	Is	183
5-NO$_2$-2-furyl	B, D	C$_6$H$_5$	I	26, 28, 183
5-NO$_2$-2-furyl	B	2-CH$_3$-3-pyridyl	I	28
5-NO$_2$-2-furyl	B	2-CH$_3$-5-pyridyl	I	183
α-C$_4$H$_3$S	B	CONH$_2$	I	165
α-C$_4$H$_3$S	B	CH$_2$OH	I	165
α-C$_4$H$_3$S	B	COOCH$_3$	I	165
5-Cl-2-thienyl	B	COOCH$_3$	I	165
α-C$_5$H$_4$N	B	OCOCH$_3$	I	77
α-C$_5$H$_4$N	B	OC$_2$H$_5$	I	77
β-C$_5$H$_4$N	B	OCOCH$_3$	I	77
β-C$_5$H$_4$N	B	OC$_2$H$_5$	I	77
γ-C$_5$H$_4$N	B	OCOCH$_3$	I	77
γ-C$_5$H$_4$N	B	OC$_2$H$_5$	I	77
α-C$_4$H$_3$OCH=CH	A	H	I	9
α-C$_4$H$_3$OCH=CH	A	CONH$_2$	I	27
α-C$_4$H$_3$OCH=CH	A	C$_6$H$_5$	I	27
5-NO$_2$-2-C$_4$H$_2$O—CH=CH	A	CONH$_2$	I	27
5-NO$_2$-2-C$_4$H$_2$O—CH=CH	A	C$_6$H$_5$	I	27
α-C$_4$H$_3$SCH=CH	A	H	I	10

Table XXVIII b. Cycloaddition of nitrile oxides to olefinic compounds

$$R\text{—}C\equiv N\rightarrow O \;+\; CH_2\text{=}C\diagdown_{R''}^{R'} \longrightarrow \text{III}$$

Nitrile oxide		Dipolarophile		Product	Ref.
R	method	R′	R″		
$COOCH_3$	B	CH_3	$OCOCH_3$	Is	191
$(CH_3)_3C$	A	1-piperidinyl	1-piperidinyl	Is	128
C_6H_5	B	Cl	Cl	Is	193
C_6H_5	A	Br	CH_2Br	Is	67
C_6H_5	A	Cl	CH_3	Is	48
C_6H_5	A	Br	CH_3	Is	2
C_6H_5	A		$-OCOCH_2-$	B	70, 71
C_6H_5	A	CH_3	CH_3	III	8
C_6H_5	A		$-CH_2COOCO-$	III	121
C_6H_5	B	CH_3	$OCOCH_3$	Is	75, 191
C_6H_5	A, B	CH_3	$COOCH_3$	III	3, 109, 55
C_6H_5	B		$-CH_2CH_2CH_2CO-$	III	187
C_6H_5	A		$-CH_2CH_2CH_2CH_2-$	III	45, 50
C_6H_5	A	OC_2H_5	OC_2H_5	III	125
C_6H_5	A	SC_2H_5	SC_2H_5	Is	127
C_6H_5	B		3-methylenenorbornan-2-one	III	187
C_6H_5	A	$COOCH_3$	CH_2COOCH_3	III	121
C_6H_5	A		$-(CH_2)_5-$	III	45, 50
C_6H_5	A		$-CH(OH)(CH_2)_4-$	III	91
C_6H_5	A	Br	C_6H_5	Is	30
C_6H_5	B		$-(CH_2)_5CO-$	III	187
C_6H_5	A		$-(CH_2)_6-$	III	45, 50
C_6H_5	A		3-phenyl-4-methyleneisoxazolin-5-one	III	96, 98
C_6H_5	A	Cl	$p\text{-}CH_3C_6H_4$	Is	30
C_6H_5	A, B	CH_3	C_6H_5	III	30, 64
C_6H_5	B		$-(CH_2)_6CO-$	III	187
C_6H_5	B		2-methyleneindanone	III	187
C_6H_5	B		3-methylenopinone	III	187
C_6H_5	B		pinocarvone	III	187
C_6H_5	B		3-methylenesabinaketone	III	187
C_6H_5	B		β-pinene	III	45
C_6H_5	B		$-(CH_2)_7CO-$	III	187
C_6H_5	A	1-morpholinyl	1-morpholinyl	Is	128
C_6H_5	A		$-CH_2CON(C_6H_5)CO-$	III	161
C_6H_5	B	Cl	3-phenyl-5-isoxazolinyl	Is	178
C_6H_5	A	CH_3	3-phenyl-1,2,4-oxadiazol-5-yl	III	37
C_6H_5	B		2-methylenetetral-1-one	III	187
C_6H_5	B		3-methylenecamphor	III	187

Table XXVIII b. Cycloaddition of nitrile oxides to olefinic compounds

Nitrile oxide		Dipolarophile		Prod-uct	Ref.
R	method	R'	R''		
C_6H_5	B	$-(CH_2)_8CO-$		III	[187]
C_6H_5	A	CH_3	3-phenyl-5-isoxazolyl	III	[36]
C_6H_5	B	$-(CH_2)_{10}CO-$		III	[187]
C_6H_5	A	C_6H_5	p-BrC_6H_4	III	[30]
C_6H_5	A	C_6H_5	p-ClC_6H_4	III	[30]
C_6H_5	A, B	C_6H_5	C_6H_5	III	[30, 55, 192]
C_6H_5	A	C_6H_5	p-$CH_3C_6H_4$	III	[30]
C_6H_5	A	methyl gibberellate		III	[150]
C_6H_5	A	atractiligenin methylester		III	[149, 150]
C_6H_5	A	phyllocladene		III	[150]
o-ClC_6H_4	B	Cl	Cl	Is	[193]
o-ClC_6H_4	A	OC_2H_5	OC_2H_5	III	[126]
p-ClC_6H_4	A	SC_2H_5	SC_2H_5	Is	[127]
p-ClC_6H_4	A	i-C_3H_7	1-morpholinyl	III	[132]
p-ClC_6H_4	A	1-morpholinyl	1-morpholinyl	Is	[128]
p-ClC_6H_4	A	i-C_4H_9	1-morpholinyl	III	[132]
p-ClC_6H_4	A	C_6H_5	1-morpholinyl	III	[129]
p-ClC_6H_4	A	1-piperidinyl	1-piperidinyl	Is	[128]
p-ClC_6H_4	A	4-methylpipera-zinyl	4-methylpipera-zinyl	Is	[128]
p-ClC_6H_4	A	C_6H_5	$N(CH_3)C_6H_5$	Is	[129]
p-BrC_6H_4	A	Br	CH_3	Is	[48]
p-BrC_6H_4	A	CH_3	$OCOCH_3$	Is	[88]
p-BrC_6H_4	A	CH_3	$COOCH_3$	III	[88]
p-BrC_6H_4	A	OC_2H_5	OC_2H_5	III	[126]
p-BrC_6H_4	A	$-(CH_2)_5-$		III	[50]
p-BrC_6H_4	B	CH_3	COC_6H_5	III	[88]
m-$NO_2C_6H_4$	A	$-OCOCH_2-$		B	[70]
m-$NO_2C_6H_4$	A	$-(CH_2)_5-$		III	[50]
m-$NO_2C_6H_4$	A	1-morpholinyl	1-morpholinyl	Is	[128]
p-$NO_2C_6H_4$	B	$-OCOCH_2-$		B	[70, 71]
p-$NO_2C_6H_4$	A	$-(CH_2)_5-$		III	[50]
p-$NO_2C_6H_4$	B	i-C_3H_7	1-morpholinyl	III	[132]
p-$NO_2C_6H_4$	B	C_6H_5	$N(CH_3)C_6H_5$	Is	[129]
2,6-$Cl_2C_6H_3$	A	Cl	Cl	Is	[193, 194]
2,6-$Cl_2C_6H_3$	A	Cl	CN	Is	[194]
2,6-$Cl_2C_6H_3$	A	Cl	CH_2Cl	Is	[194]
2,6-$Cl_2C_6H_3$	A	Br	CH_3	Is	[194]
p-$CH_3C_6H_4$	A	SC_2H_5	SC_2H_5	Is	[127]
$C_6H_5CH=CH$	A	C_6H_5	1-morpholinyl	III	[27]
2,4,6-$(CH_3)_3C_6H_2$	A	$-OCOCH_2-$		B	[70]
2,4,6-$(CH_3)_3C_6H_2$	A	CH_3	CH_3	III	[89]
m-$ONCC_6H_4$	A, B	CH_3	$CONH_2$	B	[162]
p-$ONCC_6H_4$	A, B	CH_3	$COOCH_3$	B(2)	[152, 162]
5-NO_2-2-furyl	D	$-OCOCH_2-$		Is	[69]

Table XXVIII b (continued)

Nitrile oxide		Dipolarophile		Prod-uct	Ref.
R	method	R′	R″		
$5\text{-}NO_2\text{-}2\text{-furyl}$	B	$CONH_2$	CH_3	III	[183]
$5\text{-}NO_2\text{-}2\text{-furyl}$	B	C_2H_5	1-piperidinyl	III	[28,183]
$5\text{-}NO_2\text{-}2\text{-furyl}$	B	$i\text{-}C_4H_9$	1-pyrrolidinyl	III	[183]
$5\text{-}NO_2\text{-}2\text{-furyl}$	B	$\beta\text{-}C_5H_4N$	1-morpholinyl	III	[183]
$5\text{-}NO_2\text{-}2\text{-furyl}$	D	C_6H_5	1-morpholinyl	III	[164]
$5\text{-}NO_2\text{-}2\text{-furyl}$	B	$\alpha\text{-}C_5H_4N$	1-piperidinyl	III	[28]
$5\text{-}NO_2\text{-}2\text{-furyl}$	B	$\beta\text{-}C_5H_4N$	1-piperidinyl	III	[28,183]
$5\text{-}NO_2\text{-}2\text{-furyl}$	B	$\gamma\text{-}C_5H_4N$	1-piperidinyl	III	[28,183]
$5\text{-}NO_2\text{-}2\text{-furyl}$	B	C_6H_5	1-piperidinyl	III	[28,183]
$\alpha\text{-}C_4H_3S$	B	CH_3	$CONH_2$	III	[165]
$\alpha\text{-}C_4H_3S$	B	CH_3	$COOCH_3$	III	[165]
5-Cl-2-thienyl	B	CH_3	$COOCH_3$	III	[165]
$5\text{-}NO_2\text{-}2\text{-}C_4H_2O\text{—}CH{=}CH$	A	C_6H_5	1-morpholinyl	III	[27]

Table XXVIII c. Cycloadditions of nitrile oxides to olefinic compounds

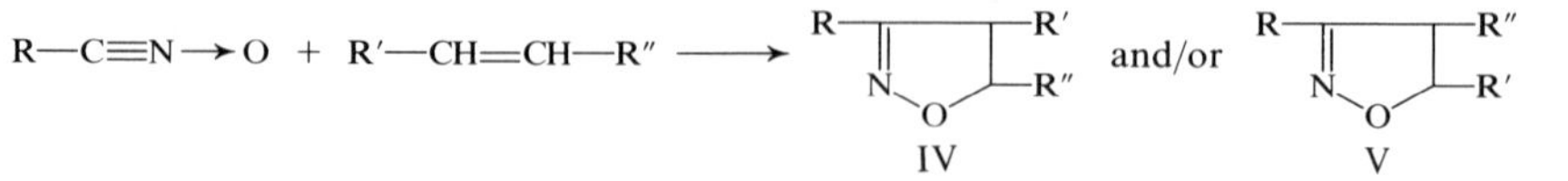

$$R-C\equiv N\rightarrow O \;+\; R'-CH=CH-R'' \longrightarrow \underset{IV}{\text{(isoxazoline IV)}} \;\text{and/or}\; \underset{V}{\text{(isoxazoline V)}}$$

Nitrile oxide		Dipolarophile		Product	Ref.
R	method	R'	R''		
H	B	CH_3	$COOCH_3$	IV + V	34, 66
H	B	$-(CH_2)_3-$		IV	34
H	B	$COOCH_3$ (trans)	$COOCH_3$	IV	55
H	B	norbornene		IV	34
H	B	$COOCH_3$	C_6H_5	IV + V	34, 66
CH_3	C	$-CH_2CH_2O-$		IV	77
CH_3	C	$-(CH_2)_3-$		IV	198
CH_3	C	$-CH_2CH_2CH_2O-$		IV	77
CH_3	C	$COOCH_3$	CH_3	IV + V	66
CH_3	B	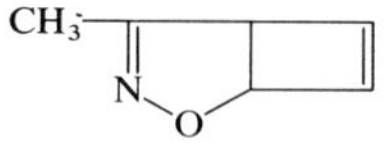 (isoxazoline bicyclic structure)		IV + V (3)	48
CH_3	C	$-(CH_2)_4-$		IV	198
CH_3	C	$-(CH_2)_5-$		IV	198
CH_3	C	$COOC_2H_5$ (cis)	$COOC_2H_5$	IV	21
CH_3	B, C	indene		IV + V	53
CH_3	B, C	C_6H_5	$COCH_3$	IV + V	89
CH_3	C	$COOCH_3$	C_6H_5	IV + V	21, 66
CH_3	C	acenaphthylene		IV	49
CH_3	C	C_6H_5	C_6H_5	IV	192
CH_3	B	17-acetoxyandro-1-sten-3-one		IV or V	155
CN	B	$COOCH_3$	CH_3	IV + V	66
CN	B	$COOCH_3$	C_6H_5	IV + V	66
$CH=NOH$	H	$COOCH_3$	CH_3	IV + V	34, 66
$CH=NOH$	H	norbornene		IV	55
$CH=NOH$	H	$COOCH_3$	C_6H_5	IV + V	34
$CONH_2$	B	$-CH_2CH_2O-$		IV	81
$CONH_2$	B	$-CH_2OCH_2-$		IV	81
$CONH_2$	B	$-CH_2CH_2CH_2O-$		IV	81
$CONH_2$	B	$-(CH_2)_3N(COCH_3)-$		IV	81
C_2H_5	C	$-CH_2CH_2O-$		IV	77
C_2H_5	C	$-CH_2OCH_2-$		IV	77
C_2H_5	C	$-(CH_2)_3O-$		IV	77
C_2H_5	C	$COOC_2H_5$	C_6H_5	IV	21
C_2H_5	C	acenaphthylene		IV	49
$n\text{-}C_3H_7$	C	$-CH_2CH_2O-$		IV	77
$n\text{-}C_3H_7$	C	acenaphthylene		IV	49
$COOC_2H_5$	B	$-CH_2CH_2O-$		IV	77
$COOC_2H_5$	B	C_6H_5	$NHCOOCH_3$	IV	22

Table XXVIII c (continued)

Nitrile oxide		Dipolarophile		Product	Ref.
R	method	R'	R''		
$(CH_3)_3C$	B, C	$COOCH_3$	CH_3	IV + V	55, 66
$(CH_3)_3C$	B	$-(CH_2)_4-$		IV	46
$(CH_3)_3C$	B	$COOCH_3$	C_6H_5	IV + V	66
$CH_2CH_2COOCH_3$	C	$-CH_2CH_2O-$		IV	77
$CH_2CH_2COOCH_3$	C	$-CH_2OCH_2-$		IV	77
$n-C_6H_{13}$	C	acenaphthylene		IV	49
$n-C_7H_{15}$	C	acenaphthylene		IV	49
$C\equiv N\rightarrow O$	A	$-COOCO-$		B(2)	24
$C\equiv N\rightarrow O$	A	C_6H_5	C_6H_5	B(2)	24
$CH_2CH_2C\equiv N\rightarrow O$	C	acenaphthylene		B	49
C_6H_5	B	$-OCOO-$		IV	82
C_6H_5	A, D	$-COOCO-$		IV	3, 23, 109
C_6H_5	B	$-CH(Cl)CHCl-$		IV(2)	48
C_6H_5	B	$-COOCH_2-$		IV	200
C_6H_5	A	COOH (cis)	COOH	B	117
C_6H_5	A	Cl	$COCH_3$	Is	141
C_6H_5	A, B	$-CH_2CH_2O-$		IV	77, 81
C_6H_5	A, B	$-CH_2OCH_2-$		IV	77, 81
C_6H_5	B	$-CH_2SO_2CH_2-$		IV	48
C_6H_5	A	CH_3 (trans)	CH_3	IV	8
C_6H_5	A	$-COCH_2CO-$		IV	45
C_6H_5	A	$-CHBrCH_2CHBr-$		IV	45
C_6H_5	A	$-COCH_2CH_2-$		IV	45, 47, 48
C_6H_5	A	Cl	COC_2H_5	Is	141
C_6H_5	A	$-CH_2CH_2CH_2NH-$		IV	186
C_6H_5	A	$-CH_2CH_2CH_2-$		IV	45, 47
C_6H_5	A, B	$-(CH_2)_3O-$		IV	77, 81
C_6H_5	B	$COOCH_3$	CH_3	IV + V	66
C_6H_5	A	$-CO(CH_2)_3-$		IV or V	45, 47
C_6H_5	A	$COOCH_3$ (cis)	$COOCH_3$	IV	3, 109, 122
C_6H_5	A	$COOCH_3$ (trans)	$COOCH_3$	IV	3, 109
C_6H_5	A	Cl	COC_3H_7-n	Is	141
C_6H_5	B	$COOCH_3$	$N(CH_3)_2$	Is	55
C_6H_5	B	$-(CH_2)_4-$		IV	46
C_6H_5	B	norbornene		IV	18
C_6H_5	B	$-C(OCH_2)_2CH_2CH_2-$		IV	48
C_6H_5	A	Cl	COC_4H_9-i	Is	141
C_6H_5	B	$-(CH_2)_3N(COCH_3)-$		IV	81
C_6H_5	A	NO_2	$o-NO_2C_6H_4$	IV	139
C_6H_5	A	NO_2	$m-NO_2C_6H_4$	IV	139
C_6H_5	A	NO_2	$p-NO_2C_6H_4$	IV	139
C_6H_5	A	1,4-endoxotetrahydrophthalic anhydride		IV	45, 47
C_6H_5	A	Br (trans)	C_6H_5	IV	3, 20
	B			Is(2)	55

Table XXVIII c. Cycloadditions of nitrile oxides to olefinic compounds

Nitrile oxide		Dipolarophile		Product	Ref.
R	method	R'	R''		
C_6H_5	A	NO_2	C_6H_5	IV	30
	B			Is(2)	55
C_6H_5	A	$COCH_3$	α-C_4H_3S	IV	35
C_6H_5	A	$COCH_3$	α-C_4H_3O	IV	35, 90
C_6H_5	A	$COOCH_3$	α-C_4H_3S	IV	35
C_6H_5	A	$COOCH_3$	α-C_4H_3O	IV	148
C_6H_5	A	\-$CH(COOCH_3)CH(COOCH_3)$\-		IV	65
C_6H_5	A	Cl	COC_5H_{11}-n	Is	141
C_6H_5	B	CH_3	1-piperidinyl	Is	201
C_6H_5	B	C_2H_5	1-pyrrolidinyl	IV	131
C_6H_5	A	Cl	COC_6H_4Br-p	IV, Is	143
C_6H_5	A	Cl	$COC_6H_4NO_2$-p	Is	143
C_6H_5	A	Cl	COC_6H_4Cl-p	IV, Is	143
C_6H_5	A	1-bromoindene		IV	53
C_6H_5	A	Cl	COC_6H_5	IV	143
	A			Is	143
C_6H_5	A	COOH	p-ClC_6H_4	IV	116
C_6H_5	A	COOH	o-$NO_2C_6H_4$	IV	118
C_6H_5	A	COOH	p-$NO_2C_6H_4$	IV	118
C_6H_5	A, B, D	indene		IV	3, 20, 52, 64, 190
C_6H_5	A	$CONH_2$	p-$NO_2C_6H_4$	IV	118
C_6H_5	A	COOH (trans)	C_6H_5	IV	54
C_6H_5	A	COOH	o-HOC_6H_4	IV	32
C_6H_5	A	1,4-endomethylenetetra-hydrophthalic anhydride		IV	121
C_6H_5	A	CH_3	C_6H_5	IV	192
C_6H_5	A	$COOC_2H_5$	α-C_4H_3S	IV	35
C_6H_5	A	$COOC_2H_5$	α-C_4H_3O	IV	35
C_6H_5	A	Cl	COC_6H_{13}-n	Is	141
C_6H_5	A	\-CON(p-ClC_6H_4)CO\-		IV	161
C_6H_5	A, B	1,4-naphthoquinone		Is	100, 101
	G			IV	101
C_6H_5	A	\-$CON(C_6H_5)CO$\-		IV	161
C_6H_5	A	Cl	$COC_6H_4OCH_3$-p	Is	143
C_6H_5	A	$COOCH_3$	p-ClC_6H_4	IV	148
C_6H_5	A	$COCH_3$	p-$NO_2C_6H_4$	IV + V	90
C_6H_5	A	$COOCH_3$	o-$NO_2C_6H_4$	IV	118
C_6H_5	A	$COOCH_3$	p-$NO_2C_6H_4$	IV	118
C_6H_5	B	3,4-dihydronaphthalene		IV + V	48
C_6H_5	A, B	$COCH_3$	C_6H_5	IV + V	3, 20, 89
C_6H_5	A	$COOCH_3$ (cis)	C_6H_5	IV + V	54, 119
C_6H_5	A	$COOCH_3$ (trans)	C_6H_5	IV	120
	B			IV + V	66
	D			IV	23
C_6H_5	B	C_6H_5	$OCOCH_3$	Is	201

Table XXVIIIc (continued)

Nitrile oxide		Dipolarophile		Product	Ref.
R	method	R'	R''		
C_6H_5	A	$-CH_2PO(OC_6H_5)CH_2-$		IV	[160]
C_6H_5	B	p-BrC$_6$H$_4$— (fused isoxazoline–cyclopentene)		IV + V(3)	[48]
C_6H_5	B	COC_4H_3S-α	α-C_4H_3O	IV + V	[89]
C_6H_5	B	COC_4H_3O-α	α-C_4H_3S	IV + V	[89]
C_6H_5	B	C_6H_5— (fused isoxazoline–cyclopentene)		IV + V(2)	[48]
C_6H_5	A	$-CON(p-CH_3C_6H_4)CO-$		IV	[161]
C_6H_5	A	$COOC_2H_5$	o-$NO_2C_6H_4$	IV	[118]
C_6H_5	A	$COOC_2H_5$	p-$NO_2C_6H_4$	IV	[118]
C_6H_5	B	COC_2H_5	C_6H_5	IV + V	[89]
C_6H_5	A	$COOC_2H_5$ (cis)	C_6H_5	IV + V	[54, 119]
C_6H_5	A	$COOC_2H_5$ (trans)	C_6H_5	IV	[120]
C_6H_5	B	C_6H_5	$CH(CH_3)_2$	IV + V	[55]
C_6H_5	A, D	acenaphthylene		IV	[45, 47, 23, 190]
C_6H_5	A	COOH	3-phenyl-5-isoxazolyl	IV	[147]
C_6H_5	A	COOH	3-phenyl-2-isoxazolin-5-yl	IV	[147]
C_6H_5	B	COC_4H_3S-α	C_6H_5	IV + V	[89]
C_6H_5	B	COC_6H_5	C_4H_3S-α	IV + V	[89]
C_6H_5	B	$COOC_2H_5$	3-indolyl	IV	[85]
C_6H_5	A	isoapiol		IV or V	[145]
C_6H_5	B	C_6H_5	1-pyrrolidinyl	IV	[55]
C_6H_5	A	$SO_2C_4H_9$-n (cis)	C_6H_5	IV	[123]
C_6H_5	A	$COOCH_3$	3-phenyl-5-isoxazolyl	IV	[147]
C_6H_5	A	$COC(CH_3)_3$	C_6H_5	IV	[90]
C_6H_5	A, B	C_6H_5 (cis)	C_6H_5	IV	[54, 55]
C_6H_5	A, B	C_6H_5 (trans)	C_6H_5	IV	[3, 20, 55]
C_6H_5	A	$SO_2C_6H_5$ (trans)	C_6H_5	IV	[123]
C_6H_5	A	C_6H_5— (fused isoxazoline–norbornene)		IV + V(4)	[62, 170]
C_6H_5	A	$COOC_2H_5$	3-phenyl-2-isoxazolin-5-yl	IV	[147]
C_6H_5	A	COC_6H_5	p-$NO_2C_6H_4$	IV + V	[90]
C_6H_5	B	C_6H_5	COC_6H_5	IV + V	[89]
C_6H_5	B	lumisantonin		IV	[168]
C_6H_5	B	COC_6H_5 (trans)	COC_6H_5	IV	[86]

Table XXVIII c. Cycloadditions of nitrile oxides to olefinic compounds

Nitrile oxide		Dipolarophile		Product	Ref.
R	method	R′	R″		
C_6H_5	B	(spiro structure: $N{=}\text{C—}C_6H_5$, naphthalenone)		IV	173
C_6H_5	B	(spiro structure: naphthalenone, $O{-}N{=}\text{C—}C_6H_5$)		B	173
C_6H_5	A	$COOCH_2C_6H_5$ (cis)	$COOCH_2C_6H_5$	IV	121
C_6H_5	A	$COOCH_2C_6H_5$ (trans)	$COOCH_2C_6H_5$	IV	121
C_6H_5	B	17-acetoxyandro-1-sten-3-one		IV or V	155
$o\text{-}ClC_6H_4$	A	—COOCO—		IV	117
$o\text{-}ClC_6H_4$	A	—CON(p-ClC_6H_4)CO—		IV	161
$o\text{-}ClC_6H_4$	A	—CON(C_6H_5)CO—		IV	161
$o\text{-}ClC_6H_4$	A	—CON(p-CH_3C_6H_4)CO—		IV	161
$m\text{-}ClC_6H_4$	A	Cl	$COC_3H_7\text{-}n$	IV	142
$p\text{-}ClC_6H_4$	A	D	C_6H_5	IV	29, 159
$p\text{-}BrC_6H_4$	A	—COOCO—		IV	117
$p\text{-}BrC_6H_4$	B	—CH_2CH_2—		IV + V(3)	48
$p\text{-}BrC_6H_4$	A	$COOCH_3$ (cis)	$COOCH_3$	IV	117
$p\text{-}BrC_6H_4$	A	$COOCH_3$ (trans)	$COOCH_3$	IV	117
$p\text{-}BrC_6H_4$	A	Cl	$COC_3H_7\text{-}n$	Is	141
$p\text{-}BrC_6H_4$	A	—CH(COOCH_3)CH(COOCH_3)—		IV	65
$p\text{-}BrC_6H_4$	B	$COOC_2H_5$ (trans)	$COOC_2H_5$	IV	86
$p\text{-}BrC_6H_4$	B	$COCH_3$	C_6H_5	IV + V	89
$p\text{-}BrC_6H_4$	B	(bicyclic isoxazoline: C_6H_5, N—O)		IV	48
$m\text{-}NO_2C_6H_4$	A	—COOCO—		IV	117
$m\text{-}NO_2C_6H_4$	A	COOH (cis)	COOH	IV	117
$m\text{-}NO_2C_6H_4$	A	$COOCH_3$ (cis)	$COOCH_3$	IV	117
$m\text{-}NO_2C_6H_4$	A	$COOCH_3$ (trans)	$COOCH_3$	IV	117
$m\text{-}NO_2C_6H_4$	D	CN	1-morpholinyl	Is	164
$m\text{-}NO_2C_6H_4$	A	—CH(COOCH_3)CH(COOCH_3)—		IV	65
$m\text{-}NO_2C_6H_4$	B	—(CH_2)_6—		IV	48
$m\text{-}NO_2C_6H_4$	A	Cl	$COC_6H_4Cl\text{-}o$	Is	143
$m\text{-}NO_2C_6H_4$	D	indene		IV	190
$m\text{-}NO_2C_6H_4$	A	$COOC_2H_5$ (trans)	C_6H_5	IV	126
$m\text{-}NO_2C_6H_4$	D	acenaphthylene		IV	190
$p\text{-}NO_2C_6H_4$	B	—OCOO—		IV	82
$p\text{-}NO_2C_6H_4$	A	—COOCO—		IV	117
$p\text{-}NO_2C_6H_4$	A	COOH (cis)	COOH	IV	117
$p\text{-}NO_2C_6H_4$	A	$COOCH_3$ (cis)	$COOCH_3$	IV	117

Table XXVIIIc (continued)

Nitrile oxide		Dipolarophile		Product	Ref.
R	method	R′	R″		
p-NO$_2$C$_6$H$_4$	A	COOCH$_3$ (trans)	COOCH$_3$	IV	[117]
p-NO$_2$C$_6$H$_4$	D	CN	1-morpholinyl	Is	[164]
p-NO$_2$C$_6$H$_4$	D	indene		IV	[190]
p-NO$_2$C$_6$H$_4$	B	COCH$_3$	C$_6$H$_5$	IV + V	[89]
p-NO$_2$C$_6$H$_4$	A	—PO(OC$_6$H$_5$)CH$_2$CH$_2$—		IV	[100]
p-NO$_2$C$_6$H$_4$	B	n-C$_5$H$_{11}$	1-morpholinyl	Is	[129]
p-NO$_2$C$_6$H$_5$	D	acenaphthylene		IV	[190]
2,6-Cl$_2$C$_6$H$_3$	A	Cl (trans)	Cl	IV	[194]
2,6-Cl$_2$C$_6$H$_3$	A	Cl (cis)	Cl	IV	[194]
2,6-Cl$_2$C$_6$H$_3$	A	Br	CH$_3$	Is	[194]
2,6-Cl$_2$C$_6$H$_3$	A	CN	OC$_2$H$_5$	Is	[194]
p-CH$_3$C$_6$H$_4$	B	COCH$_3$	C$_6$H$_5$	IV + V	[89]
p-CH$_3$OC$_6$H$_4$	B	COCH$_3$	C$_6$H$_5$	IV + V	[89]
C$_6$H$_5$CO	B	—CH$_2$CH$_2$O—		IV	[77]
C$_6$H$_5$CO	B	—CH$_2$OCH$_2$—		IV	[77]
C$_6$H$_5$CO	B	COCH$_3$	C$_6$H$_5$	IV + V	[24]
2-Cl-4-CH$_3$OC$_6$H$_3$	A	—CON(C$_6$H$_5$)CO—		IV	[161]
2-Cl-4-CH$_3$OC$_6$H$_3$	A	—CON(p-CH$_3$C$_6$H$_4$)CO—		IV	[161]
C$_6$H$_5$NHCO	H	—CH$_2$CH$_2$O—		IV	[12]
C$_6$H$_5$CH=CH	A	CONH$_2$	C$_6$H$_5$	IV	[27]
2,4,6-(CH$_3$)$_3$C$_6$H$_2$	A	—COOCH$_2$—		IV	[200]
2,4,6-(CH$_3$)$_3$C$_6$H$_2$	A	COOCH$_3$	CH$_3$	IV + V	[66]
2,4,6-(CH$_3$)$_3$C$_6$H$_2$	A	D	C$_6$H$_5$	IV	[29, 159]
2,4,6-(CH$_3$)$_3$C$_6$H$_2$	A	C$_6$H$_5$	COCH$_3$	IV + V	[89]
2,4,6-(CH$_3$)$_3$C$_6$H$_2$	A	COOCH$_3$ (trans)	C$_6$H$_5$	IV + V	[66]
2,4,6-(CH$_3$)$_3$C$_6$H$_2$	B	COC$_4$H$_3$O-α	α-C$_4$H$_3$O	IV + V	[89]
2,4,6-(CH$_3$)$_3$C$_6$H$_2$	B	COC$_4$H$_3$S-α	C$_6$H$_5$	IV + V	[89]
2,4,6-(CH$_3$)$_3$C$_6$H$_2$	B	COC$_4$H$_3$O-α	C$_6$H$_5$	IV + V	[89]
2,4,6-(CH$_3$)$_3$C$_6$H$_2$	B	COC$_6$H$_5$	C$_4$H$_3$O-α	IV + V	[89]
2,4,6-(CH$_3$)$_3$C$_6$H$_2$	A	COC$_6$H$_5$	C$_6$H$_5$	IV	[89]
m-ONCC$_6$H$_4$	A, B	—CON(C$_6$H$_5$)CO—		B	[162]
p-ONCC$_6$H$_4$	A, B	—CON(C$_6$H$_5$)CO—		B	[162]
p-ONCC$_6$H$_4$	B	N(C$_2$H$_5$)$_2$	N(C$_2$H$_5$)$_2$	B	[167]
α-C$_4$H$_3$O	A	—CH$_2$SO$_2$CH$_2$—		IV	[175]
α-C$_4$H$_3$O	D	C$_6$H$_5$ (trans)	C$_6$H$_5$	IV	[26]
5-NO$_2$-2-furyl	B	—(CH$_2$)$_3$O—		IV	[28]
5-NO$_2$-2-furyl	B	—CH$_2$OCH$_2$CH$_2$—		IV	[183]
5-NO$_2$-2-furyl	D	—(CH$_2$)$_4$—		IV	[26]
5-NO$_2$-2-furyl	B	C$_2$H$_5$	N(C$_2$H$_5$)$_2$	IV	[28, 183]
5-NO$_2$-2-furyl	D	CN	1-pyrrolidinyl	Is	[164]
5-NO$_2$-2-furyl	D	CN	1-morpholinyl	Is	[164]
5-NO$_2$-2-furyl	D	CN	1-piperidinyl	Is	[164]
5-NO$_2$-2-furyl	B	CH$_3$	1-piperidinyl	IV	[28]
5-NO$_2$-2-furyl	B, D	—(CH$_2$)$_6$—		IV	[26]
5-NO$_2$-2-furyl	B	COOC$_2$H$_5$ (cis)	COOC$_2$H$_5$	IV	[183]

Table XXVIII c. Cycloadditions of nitrile oxides to olefinic compounds

Nitrile oxide		Dipolarophile		Product	Ref.
R	method	R'	R''		
$5\text{-}NO_2\text{-}2\text{-furyl}$	D	indene		IV	[190]
$5\text{-}NO_2\text{-}2\text{-furyl}$	D	1,4-naphthoquinone		Is	[51]
$5\text{-}NO_2\text{-}2\text{-furyl}$	B	$-CON(C_6H_5)CO-$		IV	[28, 183]
$5\text{-}NO_2\text{-}2\text{-furyl}$	B	$COOC_2H_5$	C_6H_5	IV	[183]
$5\text{-}NO_2\text{-}2\text{-furyl}$	D	acenaphthylene		IV	[190]
$5\text{-}NO_2\text{-}2\text{-furyl}$	B	C_6H_5	1-piperidinyl	IV	[28]
$5\text{-}NO_2\text{-}2\text{-furyl}$	D	C_6H_5 (trans)	C_6H_5	IV	[26]
$\alpha\text{-}C_4H_3S$	B	$-CON(C_6H_5)CO-$		IV	[165]
$\alpha\text{-}C_5H_4N$	B	$-CH_2CH_2O-$		IV	[77]
$\alpha\text{-}C_5H_4N$	B	$-CH_2OCH_2-$		IV	[77]
$\beta\text{-}C_5H_4N$	B	$-CH_2CH_2O-$		IV	[77]
$\beta\text{-}C_5H_4N$	B	$-CH_2OCH_2-$		IV	[77]
$\gamma\text{-}C_5H_4N$	B	$-CH_2CH_2O-$		IV	[77]
$\gamma\text{-}C_5H_4N$	B	$-CH_2OCH_2-$		IV	[77]

Table XXVIII d. Cycloadditions of nitrile oxides to olefinic compounds

$$R{-}C{\equiv}N{\rightarrow}O \quad + \quad R^1R^2C{=}CR^3R^4 \quad \longrightarrow$$

Nitrile oxide		Dipolarophile				Product	Ref.
R	method	R^1	R^2	R^3	R^4		
CH_3	C	H	$COOCH_3$	CH_3	CH_3	VI	55
CH_3	C	H	$COOC_2H_5$	H	1-pyrrolidinyl	Is	134
CH_3	C	H	$COOC_2H_5$	CH_3	1-pyrrolidinyl	Is	134
CH_3	B		2-pyrrolidinylindene			VI	53
CH_3	B		3-pyrrolidinylindene			VI	53
CH_3	C		4-benzal-3-methyl-1-phenylpyrazolin-5-one			VI	99a
CH_3	C	H	$COOC_2H_5$	C_6H_5	1-pyrrolidinyl	Is	134
CH_3	C		1,3-diphenyl-4-benzalpyrazolin-5-one			VI (3)	99a
CH_3	B		2-benzalandrostan-17-ol-3-one			VI	93, 95
C_2H_5	C	H	$COOC_2H_5$	CH_3	1-pyrrolidinyl	Is	134
CH_3CO	B		4-benzal-3-methyl-1-phenylpyrazolin-5-one			VI (2)	99a
CH_3CO	B		1,3-diphenyl-4-benzalpyrazolin-5-one			VI (2)	99a
$n\text{-}C_3H_7$	C	H	$COOC_2H_5$	CH_3	1-pyrrolidinyl	Is	134
$COOC_2H_5$	B	H	$-(CH_2)_4-$		1-morpholinyl	Is	130
$COOC_2H_5$	B		4-benzal-3-methyl-1-phenylpyrazolin-5-one			VI(2)	99a
$COOC_2H_5$	B		1,3-diphenyl-4-benzalpyrazolin-5-one			VI	99a
$(CH_3)_3C$	B	H	$COOCH_3$	CH_3	CH_3	VI	66
$(CH_3)_3C$	A	H	C_6H_5	OCH_3	OCH_3	VI	125
$(CH_3)_3C$	A	C_6H_5	C_6H_5	$=O$		VI	58, 124
$CH_3C(OCH_2)_2(CH_2)_3$	B	H	$COOC_2H_5$	CH_3	1-pyrrolidinyl	Is	134
C_6H_5	A	H	$-COOCO-$		CH_3	VI	3, 109
C_6H_5	A	H	$-COOCH_2-$		CH_3	VI	200
C_6H_5	B	H	$-CH_2CH_2O-$		CH_3	VI	81

Table XXVIIId. Cycloadditions of nitrile oxides to olefinic compounds

Nitrile oxide		Dipolarophile				Product	Ref.
R	method	R^1	R^2	R^3	R^4		
C_6H_5	A	H	$-(CH_2)_3-$		CH_3	VI	45, 50
C_6H_5	B	H	$COCH_3$	CH_3	CH_3	VI	89
C_6H_5	C	H	$COOCH_3$	CH_3	CH_3	VI	66
C_6H_5	A	H	$-(CH_2)_3-$		$COCH_3$	VI	102
C_6H_5	A	H	$COOCH_3$ (cis)	$COOCH_3$	CH_3	VI	3, 121
C_6H_5	A	H	$COOCH_3$ (trans)	$COOCH_3$	CH_3	VI	3, 121
C_6H_5	C	H	$COOCH_3$	CH_3	$N(CH_3)_2$	Is	184
C_6H_5	A	H	CH_3	$-CO(CH_2)_4-$		VI	91
C_6H_5	B	CH_3	CH_3	1-morpholinyl	H	Is	55
C_6H_5	B, C	H	$COOC_2H_5$	CH_3	$N(CH_3)_2$	Is	129, 184
C_6H_5	A	H	$COOCH_3$	$COOCH_3$	CH_2COOCH_3	VI	121
C_6H_5	A	H	C_2H_5	$-CO(CH_2)_4-$		VI	91
C_6H_5	A	H	$COOC_2H_5$ (cis)	$COOC_2H_5$	CH_3	VI	121
C_6H_5	A	H	$COOC_2H_5$ (trans)	$COOC_2H_5$	CH_3	VI	121
C_6H_5	A, B	H	$-(CH_2)_3-$		1-pyrrolidinyl	VI	133, 131
C_6H_5	B	H	$-(CH_2)_3-$		1-morpholinyl	VI	130
C_6H_5	A, B	H	$-COOCH_2-$		C_6H_5	VI	200
C_6H_5	B	H	COC_6H_5	NH_2	CH_3	Is	138
C_6H_5	A	H	C_6H_5	OCH_3	OCH_3	VI	125
C_6H_5	A	H	n-C_3H_7	$-CO(CH_2)_4-$		VI	91
C_6H_5	A, B	H	$-(CH_2)_4-$		1-pyrrolidinyl	VI	130, 131
C_6H_5	A	H	$-(CH_2)_4-$		1-morpholinyl	VI	135
	B					Is	131
C_6H_5	B	H	$-(CH_2)_2N(CH_3)CH_2-$		1-pyrrolidinyl	VI	131
C_6H_5	A	H	$-CON(C_6H_5)CO-$		CH_3	VI	161
C_6H_5	A		3-methyl-4-benzalisoxazolin-5-one			VI	97
C_6H_5	A	H	α-C_4H_3O	$-CO(CH_2)_4-$		VI	91

C_6H_5	A	H	$-(CH_2)_3-$	C_6H_5	VI	45, 47
C_6H_5	A	H	$-PO(C_6H_5)CH_2CH_2-$	CH_3	VI	160
C_6H_5	A	H	$-PO(OC_6H_5)CH_2CH_2-$	CH_3	VI	160
C_6H_5	B	H	$-(CH_2)_4-$	1-piperidinyl	VI	131
C_6H_5	B	H	$-(CH_2)_5-$	1-pyrrolidinyl	VI	131
C_6H_5	A, B	H	$-CH_2CH(CH_3)(CH_2)_2$	1-pyrrolidinyl	VI	135, 131
C_6H_5	A	H	C_6H_5 $-CO(CH_2)_3-$		VI	91
C_6H_5	A	H	C_6H_5 $-CO(CH_2)_4-$		VI	91
C_6H_5	A		1-benzoylcyclopentene		VI(2)	102
C_6H_5	A		2-pyrrolidinylindene		VI	53
C_6H_5	A		3-pyrrolidinylindene		VI	53
C_6H_5	A	H	$-(CH_2)_3-$	3-phenyl-5-isoxazolyl	VI	36
C_6H_5	A	H	C_6H_5 $-CO(CH_2)_5-$		VI	91
C_6H_5	A	H	$p\text{-}CH_3OC_6H_4$ $-CO(CH_2)_4-$		VI	92
C_6H_5	A		2-piperidinylindene		VI	53
C_6H_5	A		1-pyrrolidinyl-3,4-dihydronaphthalene		VI	135
C_6H_5	A		1-morpholinyl-3,4-dihydronaphthalene		VI	135
C_6H_5	A		3-phenyl-4-benzalisoxazolin-5-one		VI	96, 97
C_6H_5	A		4-benzal-3-methyl-1-phenylpyrazolin-5-one		VI(2)	99
C_6H_5	A		2-benzal-1-tetralone		VI	94
C_6H_5	B	H	C_6H_5 C_6H_5	1-pyrrolidinyl	VI	131
C_6H_5	B		1,3-diphenyl-4-benzalpyrazolin-5-one		VI(3)	99a
C_6H_5	B		2-benzal-androstan-17β-ol-3-one		VI	93, 95
$o\text{-}ClC_6H_4$	A	H	$-CON(C_6H_5)CO-$	CH_3	VI	161
$p\text{-}ClC_6H_4$	B	H	$-(CH_2)_3-$	1-morpholinyl	VI	130
$p\text{-}ClC_6H_4$	B	H	$-(CH_2)_4-$	1-piperidinyl	VI	130
$p\text{-}ClC_6H_4$	B	H	$-(CH_2)_3-$	1-CH_3-piperazinyl	VI	130
$p\text{-}NO_2C_6H_4$	B	H	$-(CH_2)_3-$	1-morpholinyl	VI	130
$p\text{-}NO_2C_6H_4$	B	H	CH_3 C_2H_5	1-morpholinyl	VI	129
$p\text{-}NO_2C_6H_4$	B	H	$-(CH_2)_4-$	1-morpholinyl	VI	130
$p\text{-}NO_2C_6H_4$	B	H	$-(CH_2)_5-$	1-morpholinyl	VI	130
$p\text{-}NO_2C_6H_4$	B		2-morpholinylindene		VI	130

Table XXVIII d. Cycloadditions of nitrile oxides to olefinic compounds

Nitrile oxide		Dipolarophile				Product	Ref.
R	method	R^1	R^2	R^3	R^4		
p-NO$_2$C$_6$H$_4$	B	H	COCH$_3$	C$_6$H$_5$	1-morpholinyl	Is	129
p-NO$_2$C$_6$H$_4$	B	H	COC$_6$H$_5$	CH$_3$	1-morpholinyl	Is	129
2,6-Cl$_2$C$_6$H$_3$	C	H	COOCH$_3$	CH$_3$	N(CH$_3$)$_2$	Is	184
2,6-Cl$_2$C$_6$H$_3$	C	H	COOC$_2$H$_5$	CH$_3$	N(CH$_3$)$_2$	Is	184
2,6-Cl$_2$C$_6$H$_3$	C	H	COOC$_2$H$_5$	CH$_2$COOC$_2$H$_5$	N(CH$_3$)$_2$	Is	185
2,4,6-(CH$_3$)$_3$C$_6$H$_2$	A	H	—COOCH$_2$—		CH$_3$	VI	200
2,4,6-(CH$_3$)$_3$C$_6$H$_2$	A	H	COCH$_3$	CH$_3$	CH$_3$	VI	89
2,4,6-(CH$_3$)$_3$C$_6$H$_2$	A	H	COOCH$_3$	CH$_3$	CH$_3$	VI	66
2,4,6-(CH$_3$)$_3$C$_6$H$_2$	A	CH$_3$	CH$_3$	CH$_3$	CH$_3$	VI	19
2,4,6-(CH$_3$)$_3$C$_6$H$_2$	A	H	—COOCH$_2$—		C$_6$H$_5$	VI	200
5-NO$_2$-2-furyl	B	H	CH$_3$	H	C$_2$H$_5$	VI	183
5-NO$_2$-2-furyl	B	H	CN	OCH$_3$	N(C$_2$H$_5$)$_2$	Is (2)	196
5-NO$_2$-2-furyl	B	H	CH$_3$	H	N(CH$_2$)$_5$	VI	183
5-NO$_2$-2-furyl	B	H	—(CH$_2$)$_3$—		N(CH$_2$)$_4$	VI	28, 183
5-NO$_2$-2-furyl	B	H	C$_2$H$_5$	H	N(CH$_2$)$_5$	Is	183
5-NO$_2$-2-furyl	B	CH$_3$	CH$_3$	H	N(CH$_2$)$_5$	VI	183
5-NO$_2$-2-furyl	B	H	CH$_3$	C$_2$H$_5$	1-morpholinyl	VI	183
5-NO$_2$-2-furyl	D		α-pinene			VI	51
5-NO$_2$-2-furyl	B	H	—(CH$_2$)$_4$—		N(CH$_2$)$_4$	VI	183
5-NO$_2$-2-furyl	B	H	—(CH$_2$)$_4$—		1-morpholinyl	VI	183
5-NO$_2$-2-furyl	B	H	—(CH$_2$)$_4$—		N(C$_2$H$_5$)$_2$	VI	183
5-NO$_2$-2-furyl	B	H	CN	1-morpholinyl	1-morpholinyl	Is	196
5-NO$_2$-2-furyl	B	H	—(CH$_2$)$_4$—		N(CH$_2$)$_5$	VI	28, 183
5-NO$_2$-2-furyl	B	H	CN	N(C$_2$H$_5$)$_2$	N(C$_2$H$_5$)$_2$	Is	196
5-NO$_2$-2-furyl	B	H	C$_6$H$_5$	H	N(CH$_2$)$_5$	VI	183
5-NO$_2$-2-furyl	B	H	CH$_2$C$_6$H$_5$	H	1-morpholinyl	VI	183

Table XXVIII e. Cycloadditions of nitrile oxides to dienes and polyenes

Nitrile oxide		Dipolarophile	Product	Ref.
R	method			
H	B	norbornadiene	M	34
H	B	1,3,5,7-cyclooctatetraene	M	34
CH_3	B	$CH_2=C=CH_2$	B	68
CH_3	C	1,3,5,7-cyclooctatetraene	M	48
CH_3	B	pregna-4,16-dien-3,11,20-trione	M	93, 106
CH_3	B	pregna-4,16-dien-21-ol-3,11,20-trione	M	93, 106
CH_3	B	pregna-4,16-dien-3,20-dione	M	93, 106
CH_3	B	pregna-5,16-dien-3β-ol-20-one	M	93, 106
CH_3	B, C	3β-acetoxypregna-5,16-dien-20-one	M	93, 104, 199
CH_3	B	3β,17-diacetoxyandrosta-5,16-diene	M	104
CH_3	B	3β-acetoxy-6-methylpregna-5,16-dien-20-one	M	93, 106
CN	B	$CH_2=CHCH=CH_2$	M	25
CH=NOH	H	$(CH_2=CHCH_2)_2$	B	140
$(CH_2)_3NO_2$	C	$(CH_2=CHCH_2)_2Si(CH_3)_2$	M	154
$(CH_2)_3NO_2$	C	$[CH_2=CHSi(CH_3)_2]_2O$	M	154
$COOC_2H_5$	B	$CH_2=CHCH=CH_2$	M	57
$COOC_2H_5$	B	$CH_3CH=CHCH=CH_2$	M	57
$COOC_2H_5$	B	$CH_2=C(CH_3)CH=CH_2$	M	57
$COOC_2H_5$	B	3-methoxyestrone-17-enolacetate	M	107
$COOC_2H_5$	B	pregna-4,16-dien-3,20-dione	M	103, 105
$COOC_2H_5$	B	6-methylpregna-4,16-dien-3,20-dione	M	103
$COOC_2H_5$	B	6-methylpregna-5,16-dien-3β-ol-20-one	M	103
$COOC_2H_5$	B	3β-acetoxypregna-5,16-dien-20-one	M	103, 104, 105
$COOC_2H_5$	B	3β,17-diacetoxyandrosta-5,16-diene	M	104, 107
$(CH_3)_3C$	A	$CH_2=CHCH=CH_2$	M/B	58
$C\equiv N\rightarrow O$	A	$CH_2=CHCH=CH_2$	B	41
$C\equiv N\rightarrow O$	A	cyclopentadiene	B	41
$C\equiv N\rightarrow O$	A	1,4-cyclohexadiene	B	41
$C\equiv N\rightarrow O$	B	$[CH_2=CHSi(CH_3)_2]_2O$	B	154
C_6H_5	A	$CH_2=C=CH_2$	B	67
C_6H_5	B	$ClCH=C(Cl)CH=CH_2$	M	177
C_6H_5	B	$CH_2=C(Cl)C(Cl)=CH_2$	Is(M/B)	177
C_6H_5	B	furan	B	72
C_6H_5	B	$CH_2=C(Cl)CH=CH_2$	M/B	178
C_6H_5	A	$CH_3CH=C=CH_2$	B(2)	67
C_6H_5	A	$CH_2=CHCH=CH_2$	M/B	56
C_6H_5	A	cyclopentadiene	M/B	45, 47, 56
C_6H_5	B	$CH_3CH=CHCH=CH_2$	M	178

Table XXVIII e. Cycloadditions of nitrile oxides to dienes and polyenes

Nitrile oxide		Dipolarophile	Product	Ref.
R	method			
C_6H_5	A	$CH_2{=}C(CH_3)CH{=}CH_2$	M(2)/B	59
C_6H_5	A	tetrabromo-o-benzoquinone	B	169
C_6H_5	A	o-benzoquinone	T	173
C_6H_5	A, B, D	p-benzoquinone	Is(B)	100, 101, 51
	G		B	101
C_6H_5	A	$CH_2{=}C(CH_3)C(CH_3){=}CH_2$	B	56
C_6H_5	A	$(CH_2{=}CHCH_2)_2$	M/B	56
C_6H_5	A	$(CH_2{=}CHCH_2)_2O$	B	3, 73
C_6H_5	A	$(CH_2{=}CHCH_2)_2S$	B	3, 73
C_6H_5	A	norbornadiene	M(2) B(4)	62, 170, 176
C_6H_5	B	$CH_2{=}CHCH{=}CHN(CH_3)_2$	M/B(2)	137
C_6H_5	A, B	1,3,5,7-cyclooctatetraene	M	65, 66
C_6H_5	A	dimethylfulvene	M/B	56
C_6H_5	B	$CH_2{=}CHCH{=}CHN(CH_2)_4$	B(2)	137
C_6H_5	B	$CH_2{=}CHCH{=}CHN(C_2H_4)_2O$	B(2)	137
C_6H_5	A, B	$CH_2{=}CHCH{=}CHN(C_2H_5)_2$	M/B(2)	137
C_6H_5	A	$[CH_2{=}CHSi(CH_3)_2]_2O$	B	153
C_6H_5	A	$C_6H_5CH{=}C{=}CH_2$	B(2)	67
C_6H_5	A	methylethylfulvene	B	56
C_6H_5	B	$CH_2{=}CHCH{=}CHN(CH_2)_5$	B(2)	137
C_6H_5	A	$o\text{-}NO_2C_6H_4CH{=}CHCH{=}CH_2$	M	60
C_6H_5	A	$m\text{-}NO_2C_6H_4CH{=}CHCH{=}CH_2$	M	60
C_6H_5	A	$p\text{-}NO_2C_6H_4CH{=}CHCH{=}CH_2$	M	60
C_6H_5	A	$C_6H_5CH{=}CHCH{=}CH_2$	M	56
C_6H_5	A, B	dicyclopentadiene	M	3, 56, 64
C_6H_5	A	$H_5C_2OOCCH{=}CHCH{=}CHCOOC_2H_5$	B	121
C_6H_5	B	2,8-dimethylenecyclooctanone	B	187
C_6H_5	B	myrcene	B(4)	61
C_6H_5	A	$o\text{-}NO_2C_6H_4CH{=}CHCH{=}CHCOOH$	M	60
C_6H_5	A	$p\text{-}NO_2C_6H_4CH{=}CHCH{=}CHCOOH$	M	60
C_6H_5	A	$C_6H_5CH{=}CHCH{=}CHCOOH$	M	60
C_6H_5	B	2,9-dimethylenecyclononanone	B	187
C_6H_5	A	$CH_2{=}CH(CH_2)_7CH{=}CH_2$	B	8
C_6H_5	A	aldrine	M	62
C_6H_5	A	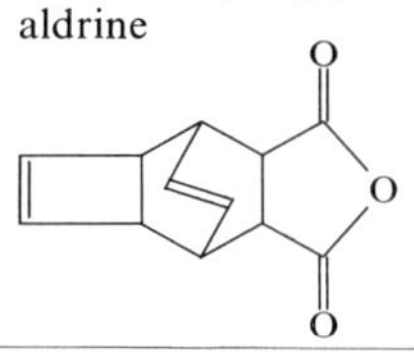	M	65

Table XXVIIIe (continued)

Nitrile oxide		Dipolarophile	Product	Ref.
R	method			
C_6H_5	A	$[CH_2\!=\!CHCH_2CH_2\cdot(CN)CH]_2Ni$	B	156
C_6H_5	B	2,10-dimethylenecyclodecanone	B	187
C_6H_5	B	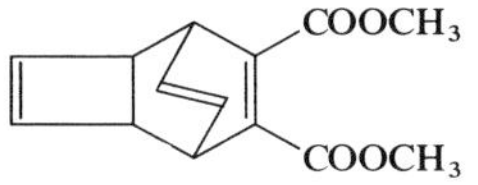 (bicyclic diester) $COOCH_3$ / $COOCH_3$	M(2)	48
C_6H_5	A	$CH_2\!=\!C(C_6H_5)C(C_6H_5)\!=\!CH_2$	B	56
C_6H_5	A	diphenylfulvene	M/B	56
C_6H_5	B	androsta-5,16-dien-3β-ol	M	104
C_6H_5	B	4,16-pregnadien-3,20-dione	M	93, 106
C_6H_5	B	5,16-pregnadien-3β-ol-20-one	M	93, 104, 106
C_6H_5	B	4,16-pregnadien-3,11,20-trione	M	93, 106
C_6H_5	B	4,16-pregnadien-21-ol-3,11,20-trione	M	93, 106
C_6H_5	B	3β-acetoxy-17-cyanoandrosta-5,16-diene	M	104
C_6H_5	B	3β-formoxy-17-acetoxyandrosta-5,16-diene	M	104
C_6H_5	B	3β-hydroxy-6-methyl-5,16-pregnadien-20-one	M	93
C_6H_5	B	3β-acetoxy-5,16-pregnadien-20-one	M	93, 104
C_6H_5	B	3β,17-diacetoxyandrosta-5,16-diene	M	104
C_6H_5	B	3β-acetoxy-17-acetamido androsta-5,16-diene	M	104
C_6H_5	B	3β-acetoxy-6-methyl-5,16-pregnadien-20-one	M	93, 106
o-ClC_6H_4	A	tetrabromo-o-benzoquinone	B	169
p-BrC_6H_4	A, B	1,3,5,7-cyclooctatetraene	M	65
p-BrC_6H_4	B	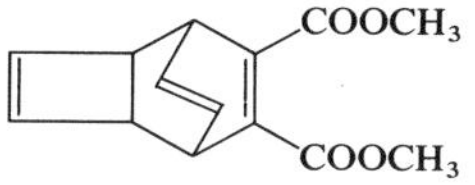 (bicyclic diester) $COOCH_3$ / $COOCH_3$	M(2)	48
m-$NO_2C_6H_4$	B	$ClCH\!=\!C(Cl)CH\!=\!CH_2$	M	177
m-$NO_2C_6H_4$	B	$CH_2\!=\!C(Cl)C(Cl)\!=\!CH_2$	Is(M/B)	177
m-$NO_2C_6H_4$	B	furan	M	72
m-$NO_2C_6H_4$	A	1,3,5,7-cyclooctatetraene	M	65
p-$NO_2C_6H_4$	A	$CH_2\!=\!C\!=\!CH_2$	B	71
p-$NO_2C_6H_4$	A	furan	B	72
p-$NO_2C_6H_4$	D	1,5-cyclooctadiene	M	51
3-Cl-4,5-$CH_2O_2C_6H_2$	A	tetrabromo-o-benzoquinone	B	169
2,4,6-$(CH_3)_3C_6H_2$	A	furan	B	179
m-$ONCC_6H_4$	A, B	$[CH_2\!=\!C(CH_3)COOCH_2]_2$	P	162

Table XXVIII e. Cycloadditions of nitrile oxides to dienes and polyenes

Nitrile oxide		Dipolarophile	Product	Ref.
R	method			
m-ONCC$_6$H$_4$	A, B	(CH$_2$=CHCH$_2$OCONHCH$_2$CH$_2$)$_2$	P	[163]
m-ONCC$_6$H$_4$	A, B	m-phenylenedimaleinimide	P	[162]
m-ONCC$_6$H$_4$	A, B	p-phenylenedimaleinimide	P	[162]
m-ONCC$_6$H$_4$	A, B	(CH$_2$=CHCH$_2$OCONHC$_6$H$_4$)$_2$CH$_2$	P	[163]
p-ONCC$_6$H$_4$	A	CH$_2$=C=CH$_2$	P	[42, 174]
p-ONCC$_6$H$_4$	A	cyclopentadiene	B(2)+P	[42, 174]
p-ONCC$_6$H$_4$	A	p-benzoquinone	P	[42, 174]
p-ONCC$_6$H$_4$	A, B	[CH$_2$=C(CH$_3$)COOCH$_2$]$_2$	P	[162]
p-ONCC$_6$H$_4$	A, B	(CH$_2$=CHCH$_2$OCONHCH$_2$)$_2$	P	[163]
p-ONCC$_6$H$_4$	A, B	m-phenylenedimaleinimide	P	[162]
p-ONCC$_6$H$_4$	A, B	p-phenylenedimaleinimide	P	[162]
p-ONCC$_6$H$_4$	A, B	(CH$_2$=CHCH$_2$OCONHC$_6$H$_4$)$_2$CH$_2$	P	[163]
5-NO$_2$-2-furyl	D	p-benzoquinone	Is(M)	[51]
5-NO$_2$-2-furyl	D	1,5-cyclooctadiene	M	[51]
5-NO$_2$-2-furyl	A, D	dicyclopentadiene	M	[26]
5-NO$_2$-2-furyl	D	1,5,9-cyclododecatriene	M	[51]

145 *Stagno d'Alcontres, G.:* Gazz. Chim. Ital. **82**, 823 (1952).

146 *Bianchi, G., Cogoli, A., Gandolfi, R.:* Gazz. Chim. Ital. **98**, 74 (1968).

147 *Stagno d'Alcontres, G., De Giacomo, G.:* Atti Soc. Peloritana **5**, 169 (1958/59).

148 *Grünanger, P., Vita Finzi, P.:* Gazz. Chim. Ital. **89**, 1771 (1959).

149 *Ajello, T., Piozzi, F., Quilico, A., Sprio, V.:* Rend. Accad. Naz. Lincei **28**, 545 (1960).

150 *Ajello, T., Piozzi, F., Quilico, A., Sprio, V.:* Gazz. Chim. Ital. **93**, 867 (1963).

151 *Lo Vecchio, G., Polimeni Conti, M., Cum, G.:* Biochim. Appl. **10**, 192 (1963).

152 *Iwakura, Y., Akiyama, M., Nagakubo, K.:* Bull. Chem. Soc. Japan **37**, 767 (1964).

153 *Fritsch, C., Lefort, M.* (to Rhone-Poulenc S.A.): Fr. Pat. 1,371,325 (1964); — Chem. Abstr. **62**, 1689b (1965).

154 Rhone-Poulenc S.A.: Fr. Pat. Addn. 84,686 (1965); — Chem. Abstr. **63**, 4332d (1965).

155 *Fritsch, W., Stache, U.* (to Farbwerke Hoechst A.G.): Ger. Pat. 1,215,146 (1966); — Chem. Abstr. **65**, 15460e (1966).

156 *Dubini, M., Montino, F.:* J. Organomet. Chem. **6**, 188 (1966).

157 *Umani-Ronchi, A., Bravo, P., Gaudiano, G.:* Tetrahedron Letters 3477 (1966).

158 *Bravo, P., Gaudiano, G., Umani-Ronchi, A.:* Gazz. Chim. Ital. **97**, 1664 (1967).

159 *Battaglia, A., Dondoni, A.:* Ric. Sci. **38**, 201 (1968).

160 *Arbuzov, B.A., Vizel, A.O., Rakov, A.P., Samitov, Y.Y.:* Doklady Akad. Nauk SSSR **172**, 1075 (1967).

161 *Awad, W.I., Omran, S., Sobhy, M.:* J. Chem. U.A.R. **10**, 1 (1967).

162 *Iwakura, Y., Shiraishi, S., Akiyama, M., Yuyama, M.:* Bull. Chem. Soc. Japan **41**, 1648 (1968).

163 *Iwakura, Y., Shiraishi, S., Asakawa, Z., Akiyama, M., Tanaka, N.:* Bull. Chem. Soc. Japan **41**, 1654 (1968).

164 *Sasaki, T., Yoshioka, T.:* Bull. Chem. Soc. Japan **41**, 2212 (1968).

165 *Iwakura, Y., Uno, K., Shiraishi, S., Hongu, T.:* Bull. Chem. Soc. Japan **41**, 2954 (1968).

166 *Gaudiano, G., Ticozzi, C., Umani-Ronchi, A., Selva, A.:* Chim. Ind. (Milan) **50**, 1343 (1968).

167 *Halleux, A., Viehe, H.G.:* J. Chem. Soc. (C) 1726 (1968).

168 *Sasaki, T., Eguchi, S.:* J. Org. Chem. **33**, 4389 (1968).

169 *Awad, W.I., Sobhy, M.:* Can. J. Chem. **47**, 1473 (1969).

170 *Lazar, R., Cocu, F.G., Barbulescu, N.:* Riv. Chim. (Bucaresti) **20**, 3 (1969).

171 *Morrocchi, S., Ricca, A., Zanarotti, A., Bianchi, G., Gandolfi, R., Grünanger, P.:* Tetrahedron Letters 3329 (1969).

172 *Grünanger, P.:* Unpublished results.

173 *Morrocchi, S., Ricca, A., Selva, A., Zanarotti, A.:* Gazz. Chim. Ital. **99**, 565 (1969).

174 *Overberger, C.G., Fujimoto, S.:* J. Polymer Sci. (C) 4161 (1968).

175 *Sasaki, T., Yoshioka, T.:* Nippon Kagaku Zasshi **88**, 1122 (1967); — Chem. Abstr. **69**, 43703 (1968).

176 *Cocu, F.G., Lazar, R., Barbulescu, N.:* Rev. Chim. (Bucaresti) **19**, 625 (1968).

177 *Chistokletov, V.N., Vabayan, V.O.:* Zh. Organ. Khim. **2**, 201 (1966).

178 *Chistokletov, V.N., Troshchenko, A.T.T.:* Izv. Sibirsk. Otd. Akad. Nauk SSSR, Ser. Khim. Nauk 147 (1963).

179 *Corsico Coda, A., Grünanger, P.:* Unpublished results.

180 *Ricca, A.:* Personal communication.

181 *Batigne, D., Boichard, J., Gay, M., Janin, R.:* Fr. Pat. 1,523,291 (1968); — Chem. Abstr. **72**, 43650 (1970).

182 *Kolokot'tseva, I.G., Chistokletov, V.N., Ionin, B.I., Petrov, A.A.:* Zh. Obshchei Khim. **38**, 1248 (1968).

183 Dainippon Pharm. Co.: Brit. Pat. 1,162,257 (1969); — Chem. Abstr. **72**, 12709 (1970).

184 *Nojima, S., Tsujikawa, T., Aki, O.:* Japan Pat. (1969) 18,226; — Chem. Abstr. **71**, 101843 (1969).

185 *Aki, O., Tsujikawa, T.* (to Takeda Chem. Ind.): Japan Pat. (1969) 29,654; — Chem. Abstr. **72**, 43645 (1970).

186 *Kano, H., Adachi, I., Kido, R., Hirose, K.* (to Shionogi Co.): Japan Pat. (1969) 29,656; — Chem. Abstr. **72**, 43643 (1970).

187 *Müller, G., Frischleder, H., Mühlstädt, M.:* J. Prakt. Chem. **311**, 118 (1969).

188 *Aki, O., Tsujikawa, T.* (to Takeda Chem. Ind.): Japan Pat. (1969) 29,655; — Chem. Abstr. **72**, 43649 (1970).

189 *Plemmons, D.C.* (to American Home Products Corp.): U.S. Pat. 3,466,296 (1969); — Chem. Abstr. **72**, 12706 (1970).

190 *Sasaki, T., Yoshioka, T., Susuki, Y.:* Yuki Gosei Kagaku Kyokai Shi **27**, 998 (1969); — Chem. Abstr. **72**, 43542 (1970).

191 *Micetich, R.G.:* Can. J. Chem. **48**, 467 (1970).

192 *Kotera, K., Takano, Y., Matsuura, A., Kitahonoki, K.:* Tetrahedron **26**, 539 (1970).

193 *Micetich, R.G.:* Can. J. Chem. **48**, 1371 (1970).

194 *Micetich, R.G.:* Can. J. Chem. in press.

195 *Wakefield, B.J., Wright, D.J.:* J. Chem. Soc. (C) 1165 (1970).

196 *Sasaki, T., Kojima, A.:* J. Chem. Soc. (C) 476 (1970).

197 *Just, G., Dahl, K.:* Can. J. Chem. **48**, 966 (1970).

198 *Park, K.P., Shine, C., Clapp, L.B.:* J. Org. Chem. **35**, 2065 (1970).

199 *Ius, A.* and coworkers, private communication.

200 *Metelli, R., Bettinetti, G.F.:* Synthesis 365 (1970).

201 *Adachi, I., Kano, H.:* Chem. Pharm. Bull. **17**, 2201 (1969).

Available data on the cycloaddition reaction of nitrile oxides with CC triple bond derivatives (see Section V C) are collected in table XXIX, which is subdivided into the following sections:

XXIX a – Cycloadditions of nitrile oxides to C≡C compounds.

XXIX b – Cycloadditions of nitrile oxides to enynes.

XXIX c – Cycloadditions of nitriles oxides to diynes.

The direct reaction of hydroximic acid chlorides with sodium acetylides [15, 18, 34, 35, 94, 95] or with acetylenic Grignard reagents (see Section VI, I) to yield isoxazoles is not covered by the table.

Note: References up to No. 77 are to be found on pages 112 to 119, while No. 78 to 107 are at page 224.

Table XXIX a. Cycloadditions of nitrile oxides to acetylenes

$$R-C\equiv N\rightarrow O \ + \ R'-C\equiv C-R'' \ \longrightarrow \ \text{I} \quad \text{and/or} \quad \text{II}$$

Nitrile oxide		Dipolarophile		Product	Ref.
R	method	R'	R''		
H	B, F	H	H	I	9, 8
H	F	H	CH_2OH	I	12, 88
H	B	H	$COOCH_3$	I + II	14, 36
H	F	H	$CH(OH)CH_3$	I + II	42
H	F	H	$C(CH_3)_2OH$	I	12
H	B	$COOCH_3$	$COOCH_3$	I	9
H	B, F	H	C_6H_5	I	8, 14
H	B	C_6H_5	$COOC_2H_5$	I + II	9
CH_3	A	H	CHO	I	79
CH_3	B	H	COOH	I	29
CH_3	A	H	CH_2Cl	I	43
CH_3	B	H	CH_2OH	I	29
CH_3	C	H	$p\text{-}ClC_6H_4$	I	22
CH_3	B, C, D	H	C_6H_5	I	29, 22, 31
CH_3	A	$Si(CH_3)_3$	$Si(CH_3)_3$	I	63
CH_3	C	H	$p\text{-}CH_3C_6H_4$	I	22
CH_3	C	H	$p\text{-}CH_3OC_6H_4$	I	22
CH_3	C	H	$n\text{-}C_8H_{17}$	I	22
CH_3	C	H	$\alpha\text{-}C_{10}H_7$	I	22
CH_3	C	H	$\beta\text{-}C_{10}H_7$	I	22
CH_3	B, C		3-methoxy-17α-ethynylestra-1,3,5(10)-trien-17β-ol	I	100, 102
CH_3	C		3β-acetoxy-17α-ethynylandro-5-sten-17β-ol	I	102
CN	B	H	$COOCH_3$	I + II	36
CN	B	$COOCH_3$	$COOCH_3$	I	26
CN	B	H	C_6H_5	I	26
CH=NOH	H	H	H	I	9, 13
CH=NOH	H	H	CH_2OH	I	12
CH=NOH	H	H	$C(CH_3)_2OH$	I	12
CH=NOH	H	H	C_6H_5	I	13
CCl_3	D	H	C_6H_5	I	31
C_2H_5	B	H	COOH	I	29
C_2H_5	B	H	CH_2OH	I	29
C_2H_5	C	H	$p\text{-}ClC_6H_4$	I	22
C_2H_5	C	H	C_6H_5	I	22
C_2H_5	C	H	$p\text{-}CH_3C_6H_4$	I	22
C_2H_5	C	H	$p\text{-}CH_3OC_6H_4$	I	22
CH_3CO	E	H	H	I	15, 78
CH_3CO	E	$COOCH_3$	$COOCH_3$	I	61
CH_3CO	D, E	H	C_6H_5	I	31, 18
CH_3CO	E	H	$n\text{-}C_6H_{13}$	I	18
$n\text{-}C_3H_7$	B	H	COOH	I	29

Table XXIX a. Cycloadditions of nitrile oxides to acetylenes

Nitrile oxide		Dipolarophile		Prod-	Ref.
R	method	R′	R″	uct	
n-C$_3$H$_7$	B	H	CH$_2$OH	I	29
n-C$_3$H$_7$	B	H	C$_6$H$_5$	I	29
i-C$_3$H$_7$	B	H	COOH	I	29
i-C$_3$H$_7$	B	H	CH$_2$OH	I	29
i-C$_3$H$_7$	B	H	C$_6$H$_5$	I	29
C(CH$_3$)$_2$OH	H	H	H	I	10, 11
C(CH$_3$)$_2$OH	H	H	C$_6$H$_5$	I	8
COOC$_2$H$_5$	E	H	H	I	15
COOC$_2$H$_5$	D, E	H	C$_6$H$_5$	I	31, 15
COOC$_2$H$_5$	B	COC$_6$H$_5$	CH$_3$	I	58, 59
COOC$_2$H$_5$	B	17α-ethynylestra-1,3,5(10)-trien-3,17β-diol		I	102
COOC$_2$H$_5$	B	17α-ethynylestra-4-en-3β,17β-diol		I	102
COOC$_2$H$_5$	B	3-methoxy-17α-ethynylestra-1,3,5(10)-trien-17β-ol		I	100
n-C$_4$H$_9$	B	H	COOH	I	29
n-C$_4$H$_9$	B	H	CH$_2$OH	I	29
sec-C$_4$H$_9$	B	H	H	I	16
(CH$_3$)$_3$C	B	H	COOCH$_3$	I + II	36
CH$_2$CH$_2$COOCH$_3$	C	H	CH$_2$CH$_2$COCH$_3$	I	48 a
CH$_2$CH$_2$COOCH$_3$	C	H	CH$_2$CH$_2$COOCH$_3$	I	48 a
CH$_2$CH$_2$COOCH$_3$	C	H	CH$_2$CH(CH$_3$)COCH$_3$	I	48 a
(CH$_3$)$_2$C(OH)C(=NOH)	H	H	H	I	11
(CH$_3$)$_2$C(OH)C(=NOH)	H	H	C$_6$H$_5$	I	8
n-C$_5$H$_{11}$	B	H	COOH	I	29
n-C$_5$H$_{11}$	B	H	CH$_2$OH	I	29
(CH$_3$)$_2$C(CH$_2$COCH$_3$)	C	H	CH$_2$C(CN)(CH$_3$)$_2$	I	48 b
(CH$_3$)$_2$ · C[CH(COOCH$_3$)$_2$]	C	H	CH$_2$C(COCH$_3$)(CH$_3$)$_2$	I	48 b
[(CH$_3$)$_3$C]$_2$CH	A	H	C$_6$H$_5$	I	92
C≡N→O	E	H	H	B	82
C≡N→O	A, D	H	C$_6$H$_5$	B	31, 64, 65
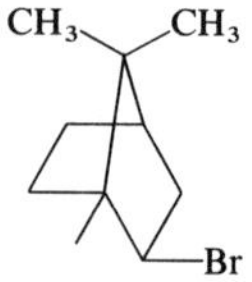 *(bicyclic structure, CH$_3$, CH$_3$, —Br)*	A	COOCH$_3$	COOCH$_3$	I	108
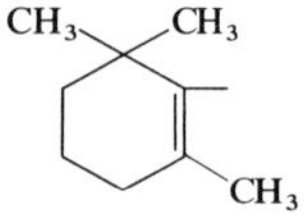 *(cyclohexene structure, CH$_3$, CH$_3$, CH$_3$)*	A	H	C$_6$H$_5$	I	92
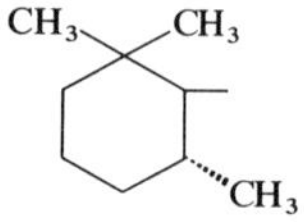 *(cyclohexane structure, CH$_3$, CH$_3$, CH$_3$)*	A	H	C$_6$H$_5$	I	92

Table XXIX a (continued)

Nitrile oxide	method	R′	R″	Product	Ref.
C_6H_5	A, B	H	H	I	8, 9
C_6H_5	A	H	CHO	I	79
C_6H_5	A	H	COOH	I	17
C_6H_5	A, H	H	CH_2Br	I	44, 85, 91
C_6H_5	A	H	CH_2Cl	I	43
C_6H_5	A, B	H	CH_2OH	I	37, 41
C_6H_5	A	H	OCH_3	I	50
C_6H_5	A	H	CH_2NH_2	I	46
C_6H_5	A	COOH	COOH	I	17
C_6H_5	B, D	H	$COOCH_3$	I + II	14, 36, 30
C_6H_5	A	COOH	CH_3	I	17
C_6H_5	B	H	C_2H_5	I	27
C_6H_5	B	H	CH_2CH_2OH	I	41, 87
C_6H_5	B	H	$CH(OH)CH_3$	I	86, 87
C_6H_5	A	H	OC_2H_5	I	49
C_6H_5	A	CN	CN	I	57
C_6H_5	B	$COOCH_3$	CH_3	I + II	36
C_6H_5	A	H	$CH_2NHCOCH_3$	I	46
C_6H_5	A, B	H	$C(CH_3)_2OH$	I	39, 41
C_6H_5	A, B	H	$CH_2N(CH_3)_2$	I	45
C_6H_5	A	H	5-Br-2-furyl	I	90
C_6H_5	A	benzyne		I	68
C_6H_5	B	$COOCH_3$	$COOCH_3$	I	60
C_6H_5	A, B	H	$n\text{-}C_4H_9$	I	84, 9
C_6H_5	A, B	H	$CH_2CH_2N(CH_3)_2$	I	45, 87
C_6H_5	A, B	H	$C(CH_2)_4(OH)$	I	39, 41
C_6H_5	A, B	H	$CH_2N(CH_2)_4$	I	45
C_6H_5	A	H	$CH(OC_2H_5)_2$	I	79
C_6H_5	A, B	H	$CH_2N(C_2H_5)_2$	I	45, 87
C_6H_5	A	H	$m\text{-}BrC_6H_4$	I	107
C_6H_5	A	H	$p\text{-}ClC_6H_4$	I	107
C_6H_5	A	H	$o\text{-}NO_2C_6H_4$	I	20
C_6H_5	A	H	$m\text{-}NO_2C_6H_4$	I	20, 107
C_6H_5	A	H	$p\text{-}NO_2C_6H_4$	I	20, 107
C_6H_5	A, B, D, E	H	C_6H_5	I	17, 9, 81 30, 18, 107
C_6H_5	A, B	H	$C(CH_2)_5(OH)$	I	39, 41
C_6H_5	B	H	$CH_2C(CH_2)_4(OH)$	I	41
C_6H_5	A, B	H	$CH_2N(CH_2)_5$	I	45, 87
C_6H_5	B	H	$CH_2CH_2N(CH_2)_4$	I	87
C_6H_5	B	H	$CH_2CH_2N(CH_2CH_2)_2O$	I	87
C_6H_5	B	H	$CH_2CH_2N(C_2H_5)_2$	I	87
C_6H_5	A	COOH	$o\text{-}NO_2C_6H_4$	I	20
C_6H_5	A	COOH	$m\text{-}NO_2C_6H_4$	I	20
C_6H_5	A	COOH	$p\text{-}NO_2C_6H_4$	I	20
C_6H_5	A	H	COC_6H_5	I	40

Table XXIX a. Cycloadditions of nitrile oxides to acetylenes

Nitrile oxide		Dipolarophile		Product	Ref.
R	method	R'	R''		
C_6H_5	A, E	COOH	C_6H_5	I	17, 18
C_6H_5	A	H	p-$CH_3C_6H_4$	I	107
C_6H_5	A	H	$CH_2C_6H_5$	I	44
C_6H_5	A	H	p-$CH_3OC_6H_4$	I	107
C_6H_5	A	H	$CH(OH)C_6H_5$	I	40
C_6H_5	B	H	$CH_2C(CH_2)_5(OH)$	I	41
C_6H_5	A, B	H	$CH_2N(CH_2)_6$	I	45
C_6H_5	B	H	$CH_2CH_2N(CH_2)_5$	I	87
C_6H_5	A, B	H	$CH_2CH_2CH_2N(CH_2)_4$	I	45
C_6H_5	A, B	H	$CH_2CH_2CH_2N(CH_2CH_2)_2O$	I	45
C_6H_5	A, B	H	$CH_2CH_2CH_2N(C_2H_5)_2$	I	45
C_6H_5	B	COC_6H_5	CH_3	I	59
C_6H_5	B	$COOCH_3$	C_6H_5	I+II	36
C_6H_5	B	H	$C(C_6H_5)(CH_3)OH$	I	41
C_6H_5	A, B	H	$CH_2CH_2CH_2N(CH_2)_5$	I	45, 87
C_6H_5	A	H	n-C_8H_{17}	I	84
C_6H_5	A	H	$B(OC_4H_9-n)_2$	I	54, 55
C_6H_5	A	H	3-phenyl-5-isoxazolyl	I	38
C_6H_5	A	H	(3-phenyl-1,2,4-oxadiazolin-5-on-4-yl)methyl	I	46
C_6H_5	A	H	3-phenyl-5-isoxazolinyl	I	72
C_6H_5	A	H	2,4,6-$(CH_3)_3C_6H_2$	I	2
C_6H_5	B	H	$CH_2C(C_6H_5)(CH_3)OH$	I	41
C_6H_5	A	H	3-phenyl-5-methyl-5-isoxazolinyl	I	75
C_6H_5	B	C_6H_5	$N(C_2H_5)_2$	I	53
C_6H_5	A	H	$CH_2C(COOC_2H_5)_2NHCOCH_3$	I	83
C_6H_5	B	17α-ethynylestra-1,3,5(10)-trien-3,17β-diol		I	102
C_6H_5	B	17α-ethynylestra-4-en-3β,17β-diol		I	102
C_6H_5	B	3-methoxy-17β-ethynylestra-1,3,5(10)-trien-17β-ol		I	100
C_6H_5	B	3β-acetoxy-17α-ethynylandro-5-sten-17β-ol		I	102
m-FC_6H_4	A	H	CH_2Br	I	85
p-FC_6H_4	A	H	CH_2Br	I	85
o-ClC_6H_4	A	H	OCH_3	I	50
o-ClC_6H_4	A	H	OC_2H_5	I	50
m-ClC_6H_4	A	H	CH_2OH	I	38
m-ClC_6H_4	A	H	OCH_3	I	50
m-ClC_6H_4	A	H	OC_2H_5	I	49
m-ClC_6H_4	A	H	C_6H_5	I	107
m-ClC_6H_4	A	H	3-m-chlorophenyl-5-isoxazolyl	I	38
m-ClC_6H_4	A	H	3-phenyl-5-isoxazolyl	I	38
p-ClC_6H_4	A	H	CH_2Br	I	85
p-ClC_6H_4	A	H	CH_2OH	I	85

Table XXIX a (continued)

| Nitrile oxide | | Dipolarophile | | Prod-uct | Ref. |
R	method	R'	R''		
p-ClC$_6$H$_4$	A	H	OCH$_3$	I	50
p-ClC$_6$H$_4$	A	H	OC$_2$H$_5$	I	50
p-ClC$_6$H$_4$	A	H	COOC$_2$H$_5$	I + II	107
p-ClC$_6$H$_4$	A	H	n-C$_4$H$_9$	I	107
p-ClC$_6$H$_4$	B	H	CH$_2$CH$_2$N(CH$_3$)$_2$	I	87
p-ClC$_6$H$_4$	B	H	CH$_2$N(CH$_2$CH$_2$)$_2$O	I	87
p-ClC$_6$H$_4$	B	H	CH$_2$N(C$_2$H$_5$)$_2$	I	87
p-ClC$_6$H$_4$	A	H	C$_6$H$_5$	I	19
p-ClC$_6$H$_4$	B	H	CH$_2$N(CH$_2$)$_5$	I	87
p-ClC$_6$H$_4$	B	H	CH$_2$CH$_2$N(C$_2$H$_4$)$_2$O	I	87
p-ClC$_6$H$_4$	B	H	CH$_2$CH$_2$N(CH$_2$)$_5$	I	87
p-ClC$_6$H$_4$	B	H	CH$_2$CH$_2$CH$_2$N(CH$_2$)$_5$	I	87
p-ClC$_6$H$_4$	A	H	2,4,6-(CH$_3$)$_3$C$_6$H$_2$	I	107
p-BrC$_6$H$_4$	A	H	CH$_2$Br	I	85
p-BrC$_6$H$_4$	A	H	OCH$_3$	I	50
p-BrC$_6$H$_4$	A	H	OC$_2$H$_5$	I	49
p-BrC$_6$H$_4$	A	H	B(OC$_4$H$_9$-n)$_2$	I	54, 55
p-BrC$_6$H$_4$	A, B	H	α-C$_{10}$H$_7$	I	2
o-NO$_2$C$_6$H$_4$	B	H	C$_6$H$_5$	I	25
m-NO$_2$C$_6$H$_4$	A	H	CN	I	56
m-NO$_2$C$_6$H$_4$	A	H	CH$_2$Br	I	85
m-NO$_2$C$_6$H$_4$	A	H	OC$_2$H$_5$	I	50
m-NO$_2$C$_6$H$_4$	B	benzyne		I	69
2,4-Cl$_2$C$_6$H$_3$	A	H	CH$_2$Br	I	85
3,4-Cl$_2$C$_6$H$_3$	A	H	CH$_2$Br	I	85
3,4-Cl$_2$C$_6$H$_3$	A	H	CH$_2$OH	I	85
3-NO$_2$-4-ClC$_6$H$_3$	A	H	CH$_2$Br	I	85
3,4-Cl$_2$-5-NO$_2$C$_6$H$_2$	A	H	CH$_2$Br	I	85
C$_6$Cl$_5$	A	H	C$_6$H$_5$	I	105
C$_6$F$_5$	H	H	C$_6$H$_5$	I	105
p-CH$_3$C$_6$H$_4$	A	H	C$_6$H$_5$	I	19, 107
p-CH$_3$C$_6$H$_4$	A	H	B(OC$_4$H$_9$-n)$_2$	I	54, 55
C$_6$H$_5$CO	D	H	C$_6$H$_5$	I	31
p-CH$_3$OC$_6$H$_4$	B	H	CH$_2$CH$_2$N(CH$_3$)$_2$	I	87
p-CH$_3$OC$_6$H$_4$	A, B	H	(CH$_2$)$_3$N(CH$_3$)$_2$	I	45
p-CH$_3$OC$_6$H$_4$	A	H	C$_6$H$_5$	I	107
p-CH$_3$OC$_6$H$_4$	B	H	CH$_2$CH$_2$N(C$_2$H$_4$)$_2$O	I	87
p-CH$_3$OC$_6$H$_4$	B	H	CH$_2$CH$_2$N(C$_2$H$_5$)$_2$	I	87
p-CH$_3$OC$_6$H$_4$	B	H	CH$_2$CH$_2$N(CH$_2$)$_5$	I	87
p-CH$_3$OC$_6$H$_4$	A, B	H	(CH$_2$)$_3$N(C$_2$H$_4$)$_2$O	I	45
p-CH$_3$OC$_6$H$_4$	A, B	H	(CH$_2$)$_3$N(C$_2$H$_5$)$_2$	I	45
p-CH$_3$OC$_6$H$_4$	A, B	H	CH$_2$CH$_2$CH$_2$N(CH$_2$)$_5$	I	45
2,4,6-(CH$_3$)$_3$C$_6$H$_2$	B	COOCH$_3$	H	I + II	36

Table XXIX a. Cycloadditions of nitrile oxides to acetylenes

Nitrile oxide		Dipolarophile		Product	Ref.
R	method	R'	R''		
2,4,6-$(CH_3)_3C_6H_2$	A	H	C_6H_5	I	21,23,80
2,4,6-$(CH_3O)_3C_6H_2$	A	H	C_6H_5	I	21,23
2,4,6-$(CH_3)_3$-3,5-Cl_2C_6	A	H	C_2H_5	I	27
2,4,6-$(CH_3)_2$-3,5-Cl_2C_6	A	H	p-BrC_6H_4	I	27
2,4,6-$(CH_3)_3$-3,5-Cl_2C_6	A	H	p-ClC_6H_4	I	27
2,4,6-$(CH_3)_3$-3,5-Cl_2C_6	A	H	p-$NO_2C_6H_4$	I	27
2,4,6-$(CH_3)_3$-3,5-Cl_2C_6	A	H	C_6H_5	I	27,80
2,4,6-$(CH_3)_3$-3,5-Cl_2C_6	A	H	p-$CH_3C_6H_4$	I	27
2,4,6-$(CH_3)_3$-3,5-Cl_2C_6	A	H	p-$CH_3OC_6H_4$	I	27
2,3,5,6-$(CH_3)_4C_6H$	A	H	C_6H_5	I	21,23,80
2,6-$(CH_3)_2$-4-$(CH_3)_2NC_6H_2$	A	H	C_6H_5	I	28
9-anthracenyl	A	H	C_6H_5	I	23
m-$ONCC_6H_4$	A, B	H	CH_2OH	B	89
m-$ONCC_6H_4$	A, B	H	$COOCH_3$	B	89
p-$ONCC_6H_4$	A, B	H	CH_2OH	B	89
p-$ONCC_6H_4$	A, B	H	$COOCH_3$	B	89
p-$ONCC_6H_4$	A, D	H	C_6H_5	B	66,93
p-$ONCC_6H_4$	A	H	p-$HC\equiv CC_6H_4$	P	93
p-$ONCC_6H_4$	A	NHC_4H_9-i	NHC_4H_9-i	B	52
p-$ONCC_6H_4$	A	C_6H_5	$N(C_2H_5)_2$	B	51,52
p-$ONCC_6H_4$	A	$N(C_4H_9$-i$)_2$	$N(C_4H_9$-i$)_2$	B	67
4-ONC-2,3,5,6-$(CH_3)_4C_6$	A	H	C_6H_5	B	21,23
2,4-$(CH_3)_2$-3-ONC-6-$[(CH_3)_2N]C_6H$	A	H	C_6H_5	B	28
2-furyl	A, B, D	H	CH_2Br	I	32,104
2-furyl	B	H	CH_2OH	I	104
2-furyl	B	H	CH_2NH_2	I	104
2-furyl	B, D	H	C_6H_5	I	32,104
5-NO_2-2-furyl	D	H	CH_2Br	I	32
5-NO_2-2-furyl	A, B, D	H	CH_2OH	I	32,104
5-NO_2-2-furyl	B	H	CH_2NH_2	I	104
5-NO_2-2-furyl	A, D	H	$COOCH_3$	I	32
5-NO_2-2-furyl	B	benzyne		I	69
5-NO_2-2-furyl	D	$COOCH_3$	$COOCH_3$	I	32
5-NO_2-2-furyl	B	CN	$N(CH_2CH_2)_2O$	I	106
5-NO_2-2-furyl	B	CN	$N(C_2H_5)_2$	I	106
5-NO_2-2-furyl	D	H	$CONHC_4H_9$-t	I	56
5-NO_2-2-furyl	A, B, D	H	C_6H_5	I	32,104
5-NO_2-2-furyl	D	H	$C(CH_2)_5(OH)$	I	32
5-NO_2-2-furyl	D	C_6H_5	C_6H_5	I	32
2-thienyl	B	H	$COOCH_3$	I	62
2-thienyl	B	$COOCH_3$	$COOCH_3$	I	62

Table XXIX a (continued)

Nitrile oxide		Dipolarophile		Product	Ref.
R	method	R'	R''		
2-thienyl	B	H	C_6H_5	I	62
5-Cl-2-thienyl	B	H	$COOCH_3$	I	62
5-Cl-2-thienyl	B	$COOCH_3$	$COOCH_3$	I	62
2-pyridyl	B	H	$CH_2CH_2N(CH_3)_2$	I	87
2-pyridyl	A, B	H	$(CH_2)_3N(CH_3)_2$	I	45
2-pyridyl	B	H	$CH_2CH_2N(C_2H_5)_2$	I	87
2-pyridyl	B	H	$CH_2CH_2N(CH_2)_5$	I	87
2-pyridyl	A, B	H	$(CH_2)_3N(C_2H_5)_2$	I	45
2-pyridyl	A, B	H	$CH_2CH_2CH_2N(CH_2)_5$	I	45
3-pyridyl	B	H	$CH_2CH_2N(C_2H_5)_2$	I	87
3-pyridyl	A, B	H	$(CH_2)_3N(C_2H_5)_2$	I	45
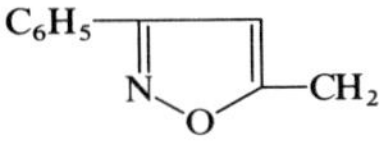 C_6H_5 isoxazolyl–CH_2	B	H	CH_3	I	24
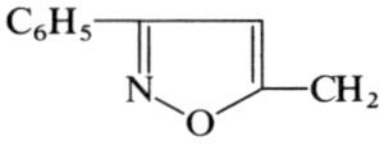 C_6H_5 isoxazolyl–CH_2	B	H	CH_2OH	I	24

Table XXIX b. Cycloadditions of nitrile oxides to enynes

Nitrile oxide		Dipolarophile	Reactive bond	Ref.
R	method			
H	F	$CH \equiv CCH = CHCH_2OH$	$C \equiv C$	88
CH_3	B	$CH_2 = CHC \equiv CH$	$C = C$	74
			$C = C$ both	
CH_3	B	$CH_2 = CHC \equiv CCH_3$	$C = C$	74
CH_3	B	$CH \equiv CCH = CHCH_3$	$C \equiv C$	74
CH_3	B	$CH_2 = C(CH_3)C \equiv CH$	$C \equiv C$	74
			$C = C$ both	
CH_3	B	$CH_2 = CHC \equiv CCOOCH_3$	$C = C$	74
CH_3	B	$CH_2 = CHC \equiv CC_2H_5$	$C = C$	74
CH_3	B	$CH_2 = CHC \equiv CCH(OH)CH_3$	$C = C$	74
CH_3	B	$CH_2 = CHC \equiv CCH_2CH = CH_2$	$C = C$	74
CH_3	B	$CH_2 = CHC \equiv CC_3H_7\text{-n}$	$C = C$	74
CH_3	B	$CH_2 = CHC \equiv CC_3H_7\text{-i}$	$C = C$	74
CH_3	B	1-ethynylcyclohexene	$C \equiv C$	74
C_6H_5	A	$CH_2 = CHC \equiv CH$	$C = C$	72, 73
			both	72
C_6H_5	A	$CH \equiv CCH = CHCHO$	$C \equiv C$	76
C_6H_5	A	$CH \equiv CCOCH = CH_2$	$C = C$	77
			$C \equiv C$ both	
C_6H_5	A	$CH_2 = CHC \equiv CCH_3$	$C = C$	73
C_6H_5	A	$CH \equiv CCH = CHCH_3$	$C \equiv C$	76
C_6H_5	A	$CH_2 = C(CH_3)C \equiv CH$	both	39
	B		$C = C$	75
			$C \equiv C$ both	
C_6H_5	A	$CH \equiv CCH = CHCH_2OH$	$C \equiv C$	72, 76
C_6H_5	A, B	$CH \equiv CCH = CHOCH_3$	$C \equiv C$	24
C_6H_5	A	$CH_2 = CHC \equiv CSCH_3$	$C = C$	96
C_6H_5	A	$CH_2 = CHC \equiv CSeCH_3$	$C = C$	96
C_6H_5	A	$CH_2 = CHC \equiv CTeCH_3$	$C = C$	96
C_6H_5	A	$CH_2 = CHC \equiv CCH = CH_2$	$C = C(B)$	98
C_6H_5	A	$CH_2 = CHCOC \equiv CCH_3$	$C = C$	77
C_6H_5	A	$CH \equiv CCOCH = CHCH_3$	$C \equiv C$	77
C_6H_5	B	$CH_2 = CHC \equiv CCOOCH_3$	$C = C$ both	97
C_6H_5	A	$CH \equiv CCH = CHCOOCH_3$	$C \equiv C$	72
C_6H_5	A	$CH_2 = CHC \equiv CC_2H_5$	$C = C$	73
C_6H_5	B	$CH_2 = CHCH(OH)C \equiv CCH_3$	$C = C$	77
C_6H_5	B	$CH \equiv CCH(OH)CH = CHCH_3$	$C \equiv C$	77
C_6H_5	A	$CH \equiv CCH = CHOC_2H_5$	$C \equiv C$	24
C_6H_5	A	$CH_2 = CHC \equiv CSC_2H_5$	$C = C$	96
C_6H_5	B	$CH_2 = CHC \equiv CCH_2CH = CH_2$	$C = C(B)$	97
			$C = C(M)$	

Table XXIX b (continued)

Nitrile oxide		Dipolarophile	Reactive bond	Ref.
R	method			
C_6H_5	A	1-ethynylcyclopentene	$C{\equiv}C$ both	[75]
C_6H_5	A	$CH_2{=}CHC{\equiv}CC_3H_7\text{-n}$	$C{=}C$	[97]
C_6H_5	A	$CH_2{=}CHC{\equiv}CC(CH_3)_2OH$	$C{=}C$	[97]
C_6H_5	B	$CH_2{=}CHC{\equiv}CSi(CH_3)_3$	$C{=}C$	[101]
C_6H_5	A	1-ethynylcyclohexene	$C{\equiv}C$	[39]
C_6H_5	A	$CH_2{=}CHC{\equiv}CC(CH_3)_3$	$C{=}C$	[97]
C_6H_5	B	$CH_2{=}C[C(CH_3)_3]C{\equiv}CH$	$C{=}C$ both	[75]
C_6H_5	B	$CH_2{=}C(CH_3)C{\equiv}CSi(CH_3)_3$	$C{=}C$	[101]
C_6H_5	A	$CH{\equiv}CCH{=}CHC_6H_5$	$C{\equiv}C$	[76]
C_6H_5	B	1-trimethylsilylethynylcyclopentene	$C{=}C$	[101]
C_6H_5	B	$CH_2{=}CHC{\equiv}CSi(C_6H_5)_3$	$C{=}C$	[101]
$m\text{-}NO_2C_6H_4$	A	$CH_2{=}CHCOC{\equiv}CH$	$C{=}C$ both	[77]
$m\text{-}NO_2C_6H_4$	A	$CH_2{=}CHCOC{\equiv}CCH_3$	$C{=}C$	[77]

Table XXIX c. Cycloaddition of nitrile oxides to diynes

Nitrile oxide		Dipolarophile	Product	Ref.
R	method			
H	F	$CH{\equiv}CC{\equiv}CH$	B	[99]
CH_3	B	$CH{\equiv}CC{\equiv}CCH_3$	M	[70]
CH_3	B	$CH{\equiv}CC{\equiv}CC_2H_5$	M	[70]
C_6H_5	A	$CH{\equiv}CC{\equiv}CH$	M, B	[38]
C_6H_5	B	$CH{\equiv}CCH_2C{\equiv}CH$	B	[24]
C_6H_5	B	$CH{\equiv}CC{\equiv}CCH_3$	M	[70]
C_6H_5	B	$CH{\equiv}CC{\equiv}CC_2H_5$	M	[70]
C_6H_5	A	$CH{\equiv}C(CH_2)_4C{\equiv}CH$	B	[38]
C_6H_5	A	$CH{\equiv}C(CH_2)_5C{\equiv}CH$	B	[38]
C_6H_5	B	$CH{\equiv}CC{\equiv}CC_6H_5$	M	[70]
C_6H_5	A	p-dihydroxy-p-diethynyldihydrobenzene	B	[39]
C_6H_5	A	$CH{\equiv}C(CH_2)_6C{\equiv}CH$	B	[38]
C_6H_5	A	1,4-dihydroxy-1,4-diethynyldihydronaphthalene	B	[39]
$m\text{-}ClC_6H_4$	A	$CH{\equiv}CC{\equiv}CH$	M/B	[38]
$p\text{-}BrC_6H_4$	A	$CH{\equiv}C(CH_2)_4C{\equiv}CH$	B	[38]
$m\text{-}NO_2C_6H_4$	B	$CH{\equiv}CC{\equiv}CCH_3$	M	[70]
$m\text{-}NO_2C_6H_4$	B	$CH{\equiv}CC{\equiv}CC_2H_5$	M	[70]
$m\text{-}NO_2C_6H_4$	B	$CH{\equiv}CC{\equiv}CC_6H_5$	M	[70]
$p\text{-}ON{\equiv}C{-}C_6H_4$	A	$CH{\equiv}CC_6H_4C{\equiv}CH\text{-p}$	P	[66, 93]

78 *Quilico, A., Freri, M.:* Gazz. Chim. Ital. **76**, 3 (1946).

79 *Stagno d'Alcontres, G., De Giacomo, G.:* Atti Soc. Peloritana **5**, 159 (1958/59).

80 *Speroni, G., Scarpati, R.:* Nuovi metilbenzonitrilossidi. Centro Univ. Edit., Napoli, 1959.

81 *Huisgen, R.:* Angew. Chem. **75**, 604 (1963); — Intern. Ed. Engl. **2**, 565 (1963).

82 *Cramer, R., McClellan, W.R.:* J. Org. Chem. **26**, 2976 (1961).

83 *LoVecchio, G., Polimeni Conti, M., Cum, G.:* Biochim. Appl. **10**, 192 (1963).

84 *Bianchi, G., Grünanger, P.:* Tetrahedron **21**, 817 (1965).

85 *Sen, H.G., Seth, D., Joshi, U.N., Rajagopalan, P.:* J. Med. Chem. **9**, 431 (1966).

86 *Kano, H., Adachi, I.* (to Shionogi Co.): Japan Pat. 9147 (1967); — Chem. Abstr. **68**, 78272 (1968).

87 *Kano, H., Adachi, I., Kido, R., Hirose, K.:* J. Med. Chem. **10**, 411 (1967).

88 *Adembri, G., Sarti-Fantoni, P., De Sio, F., Franchini, P.F.:* Tetrahedron **23**, 4697 (1967).

89 *Iwakura, Y., Shiraishi, S., Akiyama, M., Yuyama, M.:* Bull. Chem. Soc. Japan **41**, 1648 (1968).

90 *Bianchi, G., Cogoli, A., Gandolfi, R.:* Gazz. Chim. Ital. **98**, 74 (1968).

91 *Caramella, P., Grünanger, P.:* Unpublished results.

92 *Grundmann, C., Datta, S.K.:* J. Org. Chem. **34**, 2016 (1969).

93 *Overberger, C.G., Fujimoto, S.:* J. Polymer. Sci.(C) 4161 (1968).

94 *Quilico, A., Panizzi, L:* Gazz. Chim. Ital. **72**, 458 (1942).

95 *Sprio, V., Ajello, E.:* Ann. Chim. (Rome) **58**, 128 (1968).

96 *Radchenko, S.I., Chistokletov, V.N., Petrov, A.A.:* Zh. Obshchei Khim. **35**, 1735 (1965).

97 *Chistokletov, V.N., Troshchenko, A.T., Petrov, A.A.:* Zh. Obshchei Khim. **33**, 2555 (1963).

98 *Quilico, A., Grünanger, P., Mazzini, R.:* Gazz. Chim. Ital. **82**, 349 (1952).

99 *Grünanger, P., Fabbri, E.:* Rend. Accad. Naz. Lincei **26**, 235 (1959).

100 *Culbertson, P.T.* (to Parke, Davis and Co.): U.S. Pat. 3,402,172 (1968); — Chem. Abstr. **70**, 4378 (1969).

101 *Kolokol'tseva, I.G., Chistokletov, V.N., Stadnichuk, M.D., Petrov, A.A.:* Zh. Obshchei Khim. **38**, 1820 (1968).

102 *Laurent, H., Schulz, G.:* Chem. Ber. **102**, 3324 (1969).

103 *Micetich, R.G.:* Can. J. Chem. **48**, 467 (1970).

104 *Yoshioka, T., Sasaki, T.:* Japan Pat. 32,782 (1969); — Chem. Abstr. **72**, 79076 (1970).

105 *Wakefield, B.J., Wright, D.J.:* J. Chem. Soc. (C) 1165 (1970).

106 *Sasaki, T., Kojima, A.:* J. Chem. Soc. (C) 476 (1970).

107 *Battaglia, A., Dondoni, A., Taddei, F.:* J. Heterocycl. Chem. **7**, 721 (1970).

108 *Ranganathan, S., Singh, B.B., Panda, C.S.:* Tetrahedron Letters 1225 (1970).

In table XXX the available data on the cycloaddition reaction of nitrile oxides to C=X bonds are collected (see Section V, D). The table is subdivided into the following two sections:

XXXa – Cycloadditions of nitrile oxides to carbonyl compounds.

XXXb – Cycloadditions of nitrile oxides to thiocarbonyl compounds.

Note: References are to be found on pages 120 to 124.

Table XXX a. Cycloadditions of nitrile oxides to carbonyl compounds

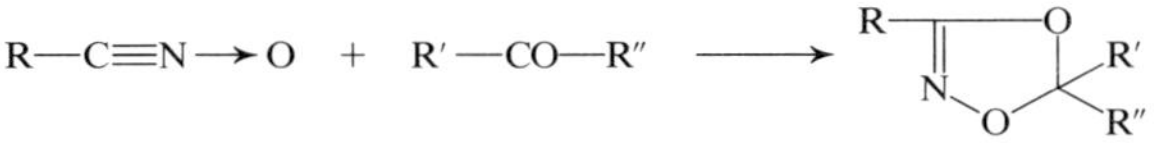

$$R-C\equiv N\rightarrow O \;+\; R'-CO-R'' \longrightarrow$$

Nitrile oxide		Dipolarophile		Ref.
R	method	R'	R''	
C_6H_5	A	H	CCl_3	1
C_6H_5	G	H	CH_3	5
C_6H_5	G	NH_2	CH_3	10
C_6H_5	B	CF_3	CF_2NO_2	3
C_6H_5	B	CF_3	CF_3	3
C_6H_5	G	CH_2Cl	CH_2Cl	5
C_6H_5	G	CH_3	CH_3	5
C_6H_5	A	CH_3	$COCH_3$	1
C_6H_5	B	H	$COOC_2H_5$	1
C_6H_5	G	H	$n\text{-}C_3H_7$	5
C_6H_5	G	CH_3	C_2H_5	5
C_6H_5	D	H	$5\text{-}NO_2\text{-}2\text{-furyl}$	6
C_6H_5	B	H	$\alpha\text{-}C_4H_3O$	1
C_6H_5	A	CH_3	$COOC_2H_5$	1
C_6H_5	A	tetrabromo-o-benzoquinone		13
C_6H_5	A	o-benzoquinone		11
C_6H_5	B, G	p-benzoquinone		15
C_6H_5	B	H	$\alpha\text{-}C_5H_4N$	1
C_6H_5	G	$-(CH_2)_5-$		5
C_6H_5	G	$N(C_2H_5)_2$	CH_3	10
C_6H_5	B	H	$o\text{-}ClC_6H_4$	1
C_6H_5	D	H	$p\text{-}NO_2C_6H_4$	6
C_6H_5	A, B	H	C_6H_5	1
C_6H_5	G	NH_2	C_6H_5	10
C_6H_5	B	$COOC_2H_5$	$COOC_2H_5$	1
C_6H_5	B	CF_3	C_6H_5	3
C_6H_5	G	CH_3	C_6H_5	5
C_6H_5	G	NHC_6H_5	CH_3	10
C_6H_5	B	o-naphthoquinone		11
C_6H_5	B, G	p-naphthoquinone		15
C_6H_5	A	CH_3	3-phenyl-1,2,4-oxadiazol-5-yl	2
C_6H_5	A, B	phenanthrenequinone		12, 11
C_6H_5	G	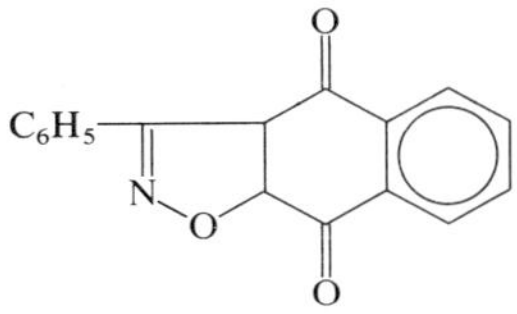		15
C_6H_5	B	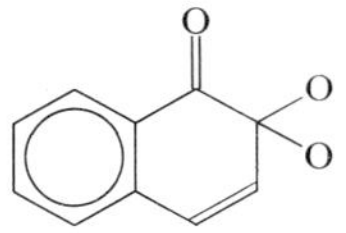		11

Table XXX a. Cycloadditions of nitrile oxides to carbonyl compounds

Nitrile oxide		Dipolarophile		Ref.
R	method	R′	R″	
C_6H_5	A		chrysenequinone	12
C_6H_5	B	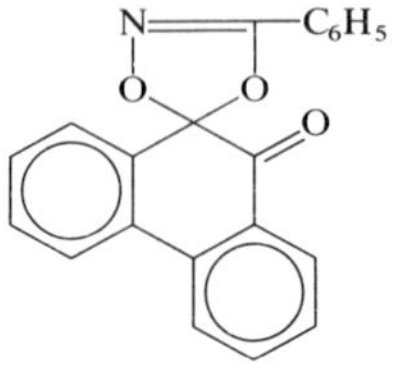		11
$o\text{-}ClC_6H_4$	A		tetrabromo-o-benzoquinone	13
$o\text{-}ClC_6H_4$	A		phenanthrenequinone	13
$o\text{-}ClC_6H_4$	A		chrysenequinone	13
$p\text{-}ClC_6H_4$	A	CH_3	$COCH_3$	1
$p\text{-}ClC_6H_4$	B	$COOC_2H_5$	$COOC_2H_5$	1
$m\text{-}NO_2C_6H_4$	D	H	$5\text{-}NO_2\text{-}2\text{-furyl}$	6
$p\text{-}NO_2C_6H_4$	B	$COOC_2H_5$	$COOC_2H_5$	1
$2\text{-}Cl\text{-}4\text{-}CH_3OC_6H_3$	A		phenanthrenequinone	13
$2\text{-}Cl\text{-}4\text{-}CH_3OC_6H_3$	A		chrysenequinone	13
$3\text{-}Cl\text{-}4,5\text{-}CH_2O_2C_6H_2$	A		tetrabromo-o-benzoquinone	13
$3\text{-}Cl\text{-}4,5\text{-}CH_2O_2C_6H_2$	A		phenanthrenequinone	13
$3\text{-}Cl\text{-}4,5\text{-}CH_2O_2C_6H_2$	A		chrysenequinone	13

Table XXX b. Cycloaddition of nitrile oxides to thiocarbonyl compounds

$$R-C\equiv N \rightarrow O \;+\; R'-CS-R'' \longrightarrow$$

Nitrile oxide		Dipolarophile		Ref.
R	method	R'	R''	
CH_3	B	α-$C_{10}H_7$	CH_3S	16
C_2H_5	B	α-$C_{10}H_7$	CH_3S	16
C_6H_5	B	$(CH_3OOC)(CN)C{=}C{=}S$		20
C_6H_5	B	Ar$=$p-ClC_6H_4		17
C_6H_5	B	Ar$=$$C_6H_5$		17
C_6H_5	B	Ar$=$p-$CH_3OC_6H_4$		17
C_6H_5	B	C_6H_5NH	C_6H_5NH	16
C_6H_5	B	C_6H_5	$HOOCCH_2S$	16
C_6H_5	B	$C_6H_5CH_2$	$HOOCCH_2S$	16
C_6H_5	B	α-$C_{10}H_7$	CH_3S	16
C_6H_5	B	OC_6H_5	OC_6H_5	16
C_6H_5	B	C_6H_5	C_6H_5	16
C_6H_5	B	SC_6H_5	SC_6H_5	16
p-ClC_6H_4	B	$(NC)_2C{=}C{=}S$		20
p-ClC_6H_4	B	$(CH_3OOC)(NC)C{=}C{=}S$		20
p-$NO_2C_6H_4$	B	C_6H_5	OC_2H_5	16
p-$NO_2C_6H_4$	B	p-$(CH_3)_2NC_6H_4$	NH_2	16
p-$NO_2C_6H_4$	B	α-$C_{10}H_7$	SCH_3	16
p-$NO_2C_6H_4$	B	p-$CH_3OC_6H_4$	p-$CH_3OC_6H_4$	16
2,4,6-$(CH_3)_3C_6H_2$	A	$=S$		19

Table XXXI. Cycloadditions of nitrile oxides to C=N compounds

$$R-C\equiv N\rightarrow O + {R' \atop R''}\!\!>\!\!C=N-R''' \longrightarrow \text{I} \quad \left(\xrightarrow{-R''R'''} \text{II}\right)$$

Note: References up to No. 21 are to be found on pages 125 to 129, while Nos. 22 – 24 are at page 230.

Nitrile oxide		Dipolarophile			Product	Ref.
R	method	R'	R''	R'''		
CH_3	C	$p\text{-}NO_2C_6H_4$	H	$t\text{-}C_4H_9$	I	24
CH_3	C	$p\text{-}ClC_6H_4$	H	C_6H_5	I	24
CH_3	C	C_6H_5	D	C_6H_5	I	24
CH_3	C	$m\text{-}NO_2C_6H_4$	H	C_6H_5	I	24
CH_3	C	C_6H_5	H	C_6H_5	I	5, 24
CH_3	C	$m\text{-}NO_2C_6H_4$	CH_3	C_6H_5	I	24
CH_3	C	C_6H_5	H	$p\text{-}CH_3OC_6H_4$	I	24
$C(=NOH)Cl$	D	C_6H_5	OC_2H_5	H	II	11
C_2H_5	C	C_6H_5	H	C_6H_5	I	5
$(CH_3)_3C$	A	$p\text{-}ClC_6H_4$	OCH_3	H	II	9
$(CH_3)_3C$	A	C_6H_5	H	CH_3	I	22
$COOC_2H_5$	B	C_6H_5	NH_2	H	II	13
C_6H_5	G	CH_3	H	OH	I	14
C_6H_5	B	CH_3	NH_2	H	II	12
C_6H_5	B	SCH_3	NH_2	H	II	12
C_6H_5	B	CF_3	CF_3	H	I	23
C_6H_5	G	CH_3	CH_3	OH	I	14
C_6H_5	A, B	CH_3	OC_2H_5	H	II	4, 12
C_6H_5	B	CH_3	CH_2COCH_3	H	M/B	4
C_6H_5	D	$5\text{-}NO_2\text{-}2\text{-furyl}$	OCH_3	H	II	11
C_6H_5	B	$\alpha\text{-}C_4H_3O$	H	CH_3	I	2
C_6H_5	A, B	CH_3	H	$n\text{-}C_4H_9$	I	3, 2
C_6H_5	G	C_6H_5	H	OH	I	14
C_6H_5	B	C_6H_5	NH_2	H	II	12, 13
C_6H_5	A, B	C_6H_5	H	CH_3	I	1, 4, 22
C_6H_5	G	C_6H_5	CH_3	OH	I	14
C_6H_5	A	C_6H_5	OCH_3	H	II	4
C_6H_5	B	CF_3	CF_3	C_6H_5	I	23
C_6H_5	A	C_6H_5	H	C_2H_5	I	4
C_6H_5	B	C_6H_5	H	CH_2CH_2OH	I	3
C_6H_5	A, D	C_6H_5	OC_2H_5	H	II	4, 11
C_6H_5	B	CF_3	CF_3	COC_6H_5	I	23
C_6H_5	A	$n\text{-}C_3H_7$	H	C_6H_{11}	I	3
C_6H_5	B	$\alpha\text{-}C_4H_3O$	H	C_6H_5	I	1, 22
C_6H_5	B	mesityl	H	CH_3	I	22
C_6H_5	A	$=N-C_6H_4Cl\text{-}p$		$p\text{-}ClC_6H_4$	I	1
C_6H_5	A	C_6H_5	H	$p\text{-}ClC_6H_4$	I	1, 22
C_6H_5	B	$p\text{-}ClC_6H_4$	H	C_6H_5	I	1, 22
C_6H_5	A	$=N-C_6H_5$		C_6H_5	I	1

Table XXXI (continued)

Nitrile oxide		Dipolarophile			Prod-uct	Ref.
R	method	R'	R''	R'''		
C_6H_5	A, B	C_6H_5	H	C_6H_5	I	3, 1, 10, 22
C_6H_5	A, B	C_6H_5	C_6H_5	H	I	1, 4
C_6H_5	B	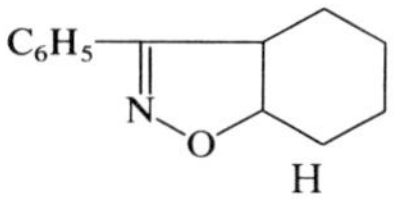 C_6H_5—(isoxazoline fused to cyclohexane ring)			I	21
C_6H_5	B	C_6H_5	H	$CH_2COC_4H_9$-n	B	4
C_6H_5	A	phenanthrenequinone-imine			I	7
C_6H_5	A	phenanthrenequinone-oxime			I	8
C_6H_5	B	p-$NO_2C_6H_4$	H	o-$CH_3OC_6H_4$	I	1
C_6H_5	B	p-$NO_2C_6H_4$	H	p-$CH_3OC_6H_4$	I	1, 22
C_6H_5	B	p-$CH_3OC_6H_4$	H	C_6H_5	I	1, 22
C_6H_5	A	=N—$C_6H_4CH_3$-p			I	1
C_6H_5	B	$CH=CHC_6H_5$	H	C_6H_5	I	6
C_6H_5	A	chrysenequinone-imine			I	7
o-ClC_6H_4	A	phenanthrenequinone-imine			I	8
o-ClC_6H_4	A	chrysenequinone-imine			I	8
p-ClC_6H_4	A	CH_3	OCH_3	H	II	9
p-ClC_6H_4	B	CH_3	H	n-C_4H_9	I	3
p-ClC_6H_4	A	$CH_2CH_2COOC_2H_5$	OCH_3	H	II	9
p-ClC_6H_4	A	C_6H_5	OCH_3	H	II	9
p-ClC_6H_4	A	$CH_2C_6H_4Cl$-p	OCH_3	H	II	9
p-ClC_6H_4	A	$CH_2C_6H_5$	OCH_3	H	II	9
p-ClC_6H_4	A	$CH_2CH_2C_6H_5$	OCH_3	H	II	9
p-ClC_6H_4	B	n-C_3H_7	H	C_6H_{11}	I	3
p-ClC_6H_4	B	C_6H_5	H	n-C_4H_9	I	3
p-ClC_6H_4	B	C_6H_5	H	C_6H_5	I	3
o-$NO_2C_6H_4$	A	C_6H_5	H	CH_3	I	22
m-$NO_2C_6H_4$	B	CH_3	NH_2	H	II	12
m-$NO_2C_6H_4$	B	SCH_3	NH_2	H	II	12
m-$NO_2C_6H_4$	B	CH_3	OC_2H_5	H	II	12
m-$NO_2C_6H_4$	B, D	C_6H_5	NH_2	H	II	12
m-$NO_2C_6H_4$	G	=N—C_6H_5		C_6H_5	B	17
m-$NO_2C_6H_4$	G	3-m-$NO_2C_6H_4$-4-C_6H_5-1,2,4-oxadiazol-on-5-phenylimine			I	17
p-$NO_2C_6H_4$	B	CH_3	NH_2	H	II	12
p-$NO_2C_6H_4$	B	SCH_3	NH_2	H	II	12
p-$NO_2C_6H_4$	B	CH_3	OC_2H_5	H	II	12
p-$NO_2C_6H_4$	B	CH_3	H	n-C_4H_9	I	3
p-$NO_2C_6H_4$	B, D	C_6H_5	NH_2	H	II	12
p-$NO_2C_6H_4$	B	CH_3	CH_3	n-C_4H_9	I	3
p-$NO_2C_6H_4$	B	n-C_3H_7	H	C_6H_{11}	I	3
p-$NO_2C_6H_4$	B	—$(CH_2)_5$—		n-C_4H_9	I	3
p-$NO_2C_6H_4$	B	C_6H_5	H	n-C_4H_9	I	3
p-$NO_2C_6H_4$	B	—$(CH_2)_5$—		C_6H_5	I	3
p-$NO_2C_6H_4$	B	p-$NO_2C_6H_4$	H	C_6H_5	I	3

Table XXXI. Cycloadditions of nitrile oxides to C=N compounds

Nitrile oxide		Dipolarophile			Prod-uct	Ref.
R	method	R'	R''	R'''		
p-$NO_2C_6H_4$	B	C_6H_5	H	C_6H_5	I	[3]
2,4-$Cl_2C_6H_3$	A	C_6H_5	H	CH_3	I	[22]
2,6-$Cl_2C_6H_3$	A	C_6H_5	H	CH_3	I	[22]
C_6H_5CO	D	C_6H_5	OC_2H_5	H	II	[11]
2-Cl-4-$CH_3OC_6H_3$	A	phenanthrenequinone-imine			I	[8]
2-Cl-4-$CH_3OC_6H_3$	A	chrysenequinone-imine			I	[8]
3-Cl-4,5-$CH_2O_2C_6H_2$	A	phenanthrenequinone-imine			I	[8]
3-Cl-4,5-$CH_2O_2C_6H_2$	A	chrysenequinone-imine			I	[8]
2,4,6-$(CH_3)_3C_6H_2$	A	1,5-diazabicyclo [4.3.0]non-5-ene			I	[19]
2,4,6-$(CH_3)_3C_6H_2$	A	1,5-diazabicyclo [5.4.0]undec-5-ene			I	[19]
2,4,6-$(CH_3)_3C_6H_2$	G	=N—C_6H_5		C_6H_5	B	[17]
2,4,6-$(CH_3)_3C_6H_2$	G	3-p-$NO_2C_6H_4$-4-C_6H_5-1,2,4-oxadiazol-on-5-phenylimine			I	[17]
α-C_4H_3O	B	C_6H_5	NH_2	H	II	[12]
5-NO_2-2-furyl	B	CH_3	NH_2	H	II	[12]
5-NO_2-2-furyl	B	NH_2	SCH_3	H	II	[12]
5-NO_2-2-furyl	B	CH_3	OC_2H_5	H	II	[12]
5-NO_2-2-furyl	D	5-NO_2-2-C_4H_2O	OCH_3	H	II	[11]
5-NO_2-2-furyl	B, D	C_6H_5	NH_2	H	II	[12]
5-NO_2-2-furyl	D	C_6H_5	OC_2H_5	H	II	[11]
5-NO_2-2-furyl	D	=N—C_6H_{11}		C_6H_{11}	*	[18]

* Rearranged product.

22 *De Sarlo, F., Fabbrini, L. Ruggini, U.:* Unpublished data.

23 *Simonyan, L.A., Zeifman, U.V., Gambaryan, N.P.:* Izv. Akad. Nauk SSSR 1916 (1968).

24 *Srivastara, R.M., Clapp, L.B.:* J. Heterocycl. Chem. **5**, 61 (1968).

Table XXXII. Cycloadditions of nitrile oxides to nitriles

$$R—C{\equiv}N{\rightarrow}O \quad + \quad R'—C{\equiv}N \quad \longrightarrow$$

Note: References up to No. 24 are to be found on pages 130 to 132, while Nos. 25—27 are at page 233.

Nitrile oxide		Nitrile	Ref.
R	method	R'	
CH_3	C	CCl_3	9
CH_3	C	CH_2Cl	9
CH_3	C	$5-NO_2-2$-furyl	9
CH_3	C	2-pyridyl	9
CH_3	C	3-pyridyl	9
CH_3	C	C_6H_5	9
CH_3	C	$p-NO_2C_6H_4$	9
CCl_3	D	C_6H_5	15
C_2H_5	C	CCl_3	9
$COOC_2H_5$	D	$NHC(=NH)NH_2$	15
$COOC_2H_5$	B	C_6H_5	5
C_6H_5	D	CCl_3	15
C_6H_5	B	CN	5
C_6H_5	B	CH_2Cl	5
C_6H_5	D, G	CH_3	16, 17
C_6H_5	A, B, D	$NHC(=NH)NH_2$	19, 20, 15
C_6H_5	A	CH_3CO	4, 5
C_6H_5	G	C_2H_5	17
C_6H_5	B	CH_3OCH_2	5
C_6H_5	B	$COOC_2H_5$	5
C_6H_5	G	$n-C_3H_7$	17
C_6H_5	B	$NHC(=NH)N(CH_3)_2$	22
C_6H_5	D	$5-NO_2-2$-furyl	15
C_6H_5	G	$CH_2COOC_2H_5$	17
C_6H_5	D	$(CH_3)_2CHCH_2$	15
C_6H_5	A	2-pyridyl	6, 7
C_6H_5	A, B	3-pyridyl	7, 5
C_6H_5	A, D	4-pyridyl	6, 7, 15
C_6H_5	B	$CH_3(CN)(CH_3COO)C$	5
C_6H_5	B	$NHC(=NH)N(C_2H_5)_2$	22
C_6H_5	A	$m-BrC_6H_4$	7
C_6H_5	A	$p-BrC_6H_4$	7
C_6H_5	A	$m-ClC_6H_4$	7
C_6H_5	A	$p-ClC_6H_4$	6, 7
C_6H_5	B	$p-ClC_6H_4O$	11
C_6H_5	A	$o-NO_2C_6H_4$	6, 7
C_6H_5	A	$m-NO_2C_6H_4$	6, 7
C_6H_5	A	$p-NO_2C_6H_4$	6, 7
C_6H_5	B	$p-NO_2C_6H_4O$	11
C_6H_5	A, B, C	C_6H_5	5, 6, 7, 15
	D, G		16, 25

Table XXXII. Cycloadditions of nitrile oxides to nitriles

Nitrile oxide	method	Nitrile R′	Ref.
C_6H_5	B, C	C_6H_5O	[10, 11]
C_6H_5	B	$NHC(=NH)N(CH_2)_5$	[22]
C_6H_5	B	C_6H_5CO	[5]
C_6H_5	B	$NHC(=NH)NH(2,4-Cl_2C_6H_3)$	[22]
C_6H_5	B	$NHC(=NH)NHC_6H_4Br-m$	[22]
C_6H_5	B	$NHC(=NH)NHC_6H_4Br-p$	[22]
C_6H_5	B	$NHC(=NH)NHC_6H_4Cl-o$	[22]
C_6H_5	B	$NHC(=NH)NHC_6H_4Cl-m$	[22]
C_6H_5	B	$NHC(=NH)NHC_6H_4Cl-p$	[22]
C_6H_5	B	$NHC(=NH)NHC_6H_4I-p$	[22]
C_6H_5	A	$o-CH_3C_6H_4$	[7]
C_6H_5	A	$m-CH_3C_6H_4$	[7]
C_6H_5	A	$p-CH_3C_6H_4$	[6, 7]
C_6H_5	B	$p-CH_3C_6H_4SO_2$	[26]
C_6H_5	A	$m-CH_3OC_6H_4$	[7]
C_6H_5	B	$C_6H_5OCH_2$	[5]
C_6H_5	B	$NHC(=NH)NHC_6H_4CH_3-m$	[22]
C_6H_5	B	$NHC(=NH)NHC_6H_4CH_3-p$	[22]
C_6H_5	B	$NHC(=NH)NHC_6H_4OCH_3-p$	[22]
C_6H_5	B	$NHC(=NH)NHC_6H_4NHCOCH_3-p$	[22]
C_6H_5	A	$m-C_6H_5SO_2C_6H_4$	[7]
C_6H_5	A	$p-C_6H_5SO_2C_6H_4$	[6, 7]
C_6H_5	B	3β-acetoxy-17-cyanomethyleneandrost-5-ene	[27]
C_6H_5	B	3β,17-diacetoxy-16-cyanoandrosta-5,16-diene	[27]
C_6H_5	D	3β,17β-diacetoxy-17α-cyanomethylandrost-5-ene	[27]
$o-ClC_6H_4$	B	$NHC(=NH)NH_2$	[20]
$p-ClC_6H_4$	B	$NHC(=NH)NH_2$	[20]
$p-ClC_6H_4$	B	C_6H_5	[5]
$m-NO_2C_6H_4$	A	1-CH_3-tetrazolyl	[8]
$m-NO_2C_6H_4$	A	2-CH_3-tetrazolyl	[8]
$m-NO_2C_6H_4$	D, G	C_6H_5	[16]
$p-NO_2C_6H_4$	B	$NHC(=NH)NH_2$	[20]
$p-NO_2C_6H_4$	A	1-CH_3-tetrazolyl	[8]
$p-NO_2C_6H_4$	A	2-CH_3-tetrazolyl	[8]
$p-NO_2C_6H_4$	D, G	C_6H_5	[16]
$p-NO_2C_6H_4$	B	C_6H_5O	[11]
$C_6H_5CH=CH$	B	$NHC(=NH)NH_2$	[21]
$2,4,6-(CH_3)_3C_6H_2$	A	$n-C_4H_9C(=NF)$	[18]
$2,4,6-(CH_3)_3C_6H_2$	A	$n-C_4H_9OC(=NF)$	[18]
$2,4,6-(CH_3)_3C_6H_2$	A	$p-ClC_6H_4C(=NF)$	[18]
$2,4,6-(CH_3)_3C_6H_2$	A	$C_6H_5C(=NF)$	[18]
$2,4,6-(CH_3)_3C_6H_2$	A	3β-acetoxy-17-cyanomethylene-androst-5-ene	[27]

Table XXXII (continued)

Nitrile oxide		Nitrile	Ref.
R	method	R'	
m-ONCC$_6$H$_4$	A	CN	[13]
m-ONCC$_6$H$_4$	A	NCCF$_2$CF$_2$	[13]
m-ONCC$_6$H$_4$	A	m-NCC$_6$H$_4$	[13]
m-ONCC$_6$H$_4$	A	p-NCC$_6$H$_4$	[13]
m-ONCC$_6$H$_4$	A	p-NCC$_6$H$_4$SO$_2$C$_6$H$_4$	[13]
p-ONCC$_6$H$_4$	A	NC	[13]
p-ONCC$_6$H$_4$	A	NCCF$_2$CF$_2$	[13]
p-ONCC$_6$H$_4$	A, D	C$_6$H$_5$	[12,15]
p-ONCC$_6$H$_4$	A	(CF$_2$)$_4$OCF(CF$_3$)CN	[13]
p-ONCC$_6$H$_4$	A	CF(CF$_3$)OCF$_2$CF$_2$OCF(CF$_3$)CN	[13]
p-ONCC$_6$H$_4$	A	m-NCC$_6$H$_4$	[13]
p-ONCC$_6$H$_4$	A	p-NCC$_6$H$_4$	[12,13]
p-ONCC$_6$H$_4$	A	p-NCC$_6$H$_4$SO$_2$C$_6$H$_4$	[13]
5-NO$_2$-2-furyl	D	NHC(=NH)NH$_2$	[15]
5-NO$_2$-2-furyl	D	5-NO$_2$-2-furyl	[15]
5-NO$_2$-2-furyl	D, G	C$_6$H$_5$	[16]

25 *Speroni, G., Bartoli, M.:* Sopra gli ossidi di nitrili, Nota VIII, Stab. Tip. Marzocco, Firenze, 1952.

26 *van Leusen, A. M., Jagt, J. C.:* Tetrahedron Letters 971 (1970).

27 *Ferrara, L.:* Private communication.

Subject Index

Note: If any entry is not further specified or subdivided, it usually refers to relation to nitrile oxides.
Individual compounds as well as reactants listed in the Tables of Chapter IX are not included in this index, since they are already arranged in a systematic manner.